Signals and Communication Technology

For further volumes:
http://www.springer.com/series/4748

Mikio Tohyama

Sound and Signals

Springer

Author

Mikio Tohyama
Graduate School of Fundamental Science
and Engineering
Waseda University
3-4-1, Okubo, Sinjuku-ku
Tokyo
169-8555 Japan
E-mail: m_tohyama@aoni.waseda.jp

ISBN 978-3-642-20121-9 e-ISBN 978-3-642-20122-6

DOI 10.1007/978-3-642-20122-6

Library of Congress Control Number: Applied for

Typeset: Scientific Publishing Services Pvt. Ltd., Chennai, India.

Printed on acid-free paper

9 8 7 6 5 4 3 2 1

springer.com

Preface

This book is a research monograph on acoustics from a signal theoretical point of view. Namely this book describes the elementary nature of sound waves and their signal theoretic signatures, which are very informative for communication. Essential issues about sound fields in closed space are dealt with according to classical wave physics and linear system theory, on which discrete signal analysis is based. Very introductory levels of university physics and mathematics, and fundamental formulations of digital signal analysis are assumed as preknowledge, but most of the issues are discussed in detail so as to be self-contained in this book as much as possible. Thus this book is for people who are interested in the scientific aspects of sound and sound waves, or motivated to research and/or develop acoustics. This might also be informative as an advance textbook for research oriented students. However, some other books are very much recommended to be used in parallel, for example,

- Jens Blauert and Ning Xiang: Acoustics for Engineers, Springer, 2008,
- William M. Hartmann: Signals, sound, and sensation, Springer, 1997,
- Thomas D. Rossing and Neville H. Fletcher: Principles of Vibration and Sound, Springer, 1995.

The author thanks Yoshinori Takahashi, Yoshihumi Hara and Tomomi Hasegawa for their research cooperation and assistance in preparing drawings. In particular, the author is very much grateful to Yoshihumi Hara for his great effort in setting the Tex manuscript with finalized drawings. The author also thanks Kurdyla and Associates Co.Ltd. for checking the author's written English.

The author acknowledges that this book is inspired and motivated by fruitful and informative discussions with Yoshimutsu Hirata and Mitsuo Matsumoto. Finally, the author appreciates all of the authors of the research articles referred to this book.

February 2011 Mikio Tohyama

Contents

Acronyms

Roman-Letter Symbols

A	Matrix
A, B	Magnitude or complex magnitude of oscillation
$A(\omega)$	Magnitude or complex magnitude of oscillation as function of angular frequency
$A_1(\omega), A_2(\omega)$	Magnitude or complex magnitude of oscillator 1 or 2 as function of angular frequency
A_b	Equivalent absorption area (m^2)
A_{b_3}	Equivalent absorption area in 3D space (m^2)
A_{b_2}	Equivalent absorption area in 2D space (m^2)
A_{b_1}	Equivalent absorption area in 1D space (m^2)
$A_{b_{ob}}$	Equivalent absorption area for oblique waves (m^2)
$A_{b_{tan}}$	Equivalent absorption area for tangential waves (m^2)
$A_{b_{ax}}$	Equivalent absorption area for axial waves (m^2)
B/A	Complex reflection coefficient
B	Region in space
A_n, B_n	n-th modal magnitude
B_p	Elastic property for bending plate
B_R	Region in space
B_M	Modal bandwidth or equivalent bandwidth (rad/s)
C	Euler's constant
$C_F(m)$	Correlation sequence (function)
$C_{F_3}(kr)$	Correlation coefficient between two receiving points in 3D field
$C_{F_2}(kr)$	Correlation coefficient between two receiving points in 2D field
$C_{F_1}(kr)$	Correlation coefficient between two receiving points in 1D field
C_P	Specific heat under constant pressure (J)

C_V	Specific heat under constant volume (J)
$C_f(X)$	Generating function for correlation sequence $c_f(n)$
$C_v(X)$	Generating function for convolved sequence $c_v(n)$
$D(\omega)$	Denominator of function
$D(k_x, k_y)$	Density of samples on disc in wavenumber space
D_{pm}	Directivity power spectrum
D_{R_0}	Ratio of direct and reverberant sound energy
$D_{R_{50}}$	Subjective energy ratio of direct and reverberant sound energy
$D_{R_{30}}$	Subjective energy ratio of direct and reverberant sound energy
D_0	Uniform density of samples on disc in wavenumber space
$D_{R_c} = K_c$	Ratio of direct and reverberant sound energy for circular array of sources
$D_{R_{dc}} = K_{dc}$	Ratio of direct and reverberant sound energy for double-circular array of sources
$D_{R_s} = K_s$	Ratio of direct and reverberant sound energy for spherical array of sources
$D_{R_{ds}} = K_{ds}$	Ratio of direct and reverberant sound energy for double-spherical array of sources
$D_1^2(\mathbf{r}'_1, \omega), D_2^2(\mathbf{r}'_2, \omega)$	Resonant response contributed from nearest resonance for point source at $\mathbf{r}'_1$ or $\mathbf{r}'_2$
E_0	Energy density (J/m^3)
$E_{0_{Av}}$	Average of energy density in single period (J/m^3)
E_{0_R}	Energy density for right-hand progressive wave (J/m^3)
E_P	Potential energy (J)
$E_1(\mathbf{r}'_1, \omega), E_2(\mathbf{r}'_2, \omega)$	Power response function for point source at $\mathbf{r}'_1$ or $\mathbf{r}'_2$
$< E(\omega) >$	Space average of $E(\mathbf{r}, \omega)$ with respect to $\mathbf{r}$
E_{P_0}	Potential energy density (J/m^3)
E_K	Kinetic energy (J)
E_{K_0}	Kinetic energy density (J/m^3)
$E_{Rev}(n)$	Reverberation energy decay curve
$E_{0_{st}}$	Energy density at steady state (J/m^3)
$E_{0_{50}}$	Energy density of subjective direct sound (J/m^3)
F	Magnitude of force (N)
F_T	Tension of pendulum or string (N)
F_X	Magnitude of external force (N)
$F_\psi^+(z)$	Primitive function of $z\psi_0(z)$
$G(\mathbf{r'}, \mathbf{r})$	Green function between source ($\mathbf{r'}$) and observation ($\mathbf{r}$) points in space
$H(z^{-1})$	Transfer function for discrete systems
$H(\mathrm{e}^{-\mathrm{i}\Omega})$	Frequency response function for discrete systems

$H(\omega)$	Frequency response function for continuous systems
$H(x', x, \omega)$	Frequency response function for continuous systems with respect to source and observation points, x' and x.
$H(x', x, k)$	Frequency response function of wavenumber for continuous systems with respect to source and observation points, x' and x.
$H_{A_{\mathrm{tf}}}(x', x, k)$	Transfer acoustic impedance between source and observation points x' and x. (Pa · s/m^3)
I	Density of sound energy flow or sound intensity (W/m^2)
I_3	Density of sound energy flow or sound intensity from three-dimensionally arranged source (W/m^2)
I_2	Density of sound energy flow or sound intensity from two-dimensionally arranged source (W/m^2)
I_1	Density of sound energy flow or sound intensity from one-dimensionally arranged source (W/m^2)
$I_{3_{st}}$	Density of sound energy flow at steady state in three-dimensional field (W/m^2)
I_D	Density of direct sound energy flow at steady state (W/m^2)
K	Spring constant (N/m)
L	Length of circumference or interval (m)
L_x, L_y, L_z	Lengths of sides for rectangular room (m)
L_{3D}	Length of circumfernece of 3D region (m)
L_{2D}	Length of circumfernece of 2D region (m)
L_{1D}	Length of 1D region (m)
L_p	Sound pressure level (dB)
M	Mass (kg)
MPR	Minimum power response for primary and secondary sources
M_{sf}	Surface density of plate (kg/m^2)
M_{FP_3}	Mean free path in 3D space (m)
M_{FP_2}	Mean free path in 2D space (m)
M_{FP_1}	Mean free path in 1D space (m)
$M_{FP_{xy}}$	Mean free path in xy-2D space (m)
$M(\omega)$	Modal overlap of sound field at angular frequecy ω
$N(\omega)$	Numerator of function
N_{c_x}	Number of collisions of sound with x-walls perpendicular to x-axis
N_{c_y}	Number of collisions of sound with y-walls perpendicular to y-axis
N_{c_z}	Number of collisions of sound with z-walls perpendicular to z-axis

N_{c3}	Average of number of collisions of sound with walls in 3D space
N_{c2}	Average of number of collisions of sound with walls in 2D space
$N_{ims_3}(t)$	Number of mirror image sources in sphere with radius of ct
N_e	Number of molecules
N_n	Number of nodes
N_{u_c}	Number of uncorrelated samples on circle
N_{u_s}	Number of uncorrelated samples on sphere
N_z	Number of zeros
N_z^+	Number of non-minimum-phase zeros
N_z^-	Number of minimum-phase zeros
$N_z^+(\eta_0, \omega)$	Number of non-minimum-phase zeros below angular frequency ω
N_p	Number of poles
$N_{\nu_{3D}}(k)$	Number of eigenfrequencies in 3D-space lower than k
$N_{\nu_{ob}}(k)$	Number of eigenfrequencies for oblique waves lower than k
$N_{\nu_{xy}}(k)$	Number of eigenfrequencies for xy-tangential waves lower than k
$N_{\nu_x}(k)$	Number of eigenfrequencies for x-axial waves lower than k
$N_{\nu_{2D}}(k)$	Number of eigenfrequencies in 2D-space lower than k
$N_{\nu_{tan}}(k)$	Number of eigenfrequencies for tangential waves lower than k
$N_a, N_b, N_c, N_d, N_T, N_\eta, N_\gamma$	Number of samples in time interval
P	Magnitude of pressure (Pa)
P_0	Pressure at initial state (Pa)
P_a	Atomic pressure of the air (Pa)
$P(x, y, z), P(\mathbf{r})$	Pressure wave as spatial function (Pa)
$P_{lmn}(x, y, z)$	Orthogonal function for sound field in rectangular room
$\hat{P}_{lmn}(x, y, z)$	Normalized orthogonal function for sound field in rectangular room
$P(x', x, \omega)$	Magnitude of sound pressure as function of source position x', obserbation point x, and angular frequency ω (Pa)
$P(x, \omega)$	Magnitude of sound pressure as function of x and ω (Pa)
P_i	Magnitude of incident pressure wave (Pa)
P_r	Magnitude of reflected pressure wave (Pa)
P_{sc}	Probability of residue sign change

P_{sc_1}	Probability of residue sign change for one-dimensional systems
P_{sc_2}	Probability of residue sign change for two-dimensional systems
P_{sc_3}	Probability of residue sign change for three-dimensional systems
P_t	Magnitude of transmitted pressure wave (Pa)
P_M	Magnitude of pressure of minimum audible sound (Pa)
Q	Magnitude of volume velocity (m^3/s)
$Q_d(\mathbf{r})$	Distribution density function of volume velocity source (1/s)
Q_1, Q_2	Magnitude of volume velocity of external source 1 or 2 (m^3/s)
Q_0	Magnitude of volume velocity of point source (Q_0 can be set unity for Green function) (m^3/s)
Q_{ss}	Magnitude of volume velocity of dipole source (m^3/s)
Q_{0d}	Distribution density function for point source (1/s)
$\hat{Q}_0$	Spectral density of impulsive point source, $\hat{Q}_0 \Delta\omega = Q_0$ (m^3)
R	Friction constant ($N \cdot s/m$)
$R_1^2(\mathbf{r}'_1, \omega), R_2^2(\mathbf{r}'_2, \omega)$	Response contributed by non-resonant modes for point source at $\mathbf{r}'_1$ or $\mathbf{r}'_2$
R_A	Real part of acoustic impedance ($Pa \cdot s/m^3$)
$R_{M_{\text{in}}}$	Driving point mechanical impedance ($Pa \cdot s/m^3$)
$R_{A_{\text{rad}}}$	Real part of radiation acoustic impedance ($Pa \cdot s/m^3$)
$R_{A_{\text{rad}_1}}$	Real part of radiation acoustic impedance for source 1 (to be simplified to R_1) ($Pa \cdot s/m^3$)
$R_{A_{\text{rad}_2}}$	Real part of radiation acoustic impedance for source 2 (to be simplified to R_2) ($Pa \cdot s/m^3$)
$R_{A_{\text{rad}_{12}}}$	Real part of mutual radiation acoustic impedance between source 1 and 2 (to be simplified to R_{12}) ($Pa \cdot s/m^3$)
$R_{A_{\text{rad}_a}}$	Real part of radiation acoustic impedance of spherical source with radius a ($Pa \cdot s/m^3$)
$R_{A_{\text{rad}_0}}$	Real part of radiation acoustic impedance of point source in free field ($Pa \cdot s/m^3$)
R_{c_3}	Range of coherent field in three-dimensional reverberant space (m)
R_{c_2}	Range of coherent field in two-dimensional reverberant space (m)
R_{c_1}	Range of coherent field in one-dimensional reverberant space (m)
R_M	Real part of mechanical impedance ($N \cdot s/m$)
R_{gas}	Gas constant ($J/(mol \cdot K)$)

$Nc_{z/xy}$	Ratio of average of number of collisions with z-walls to that with other side walls
$Nc_{z/all}$	Ratio of average of number of collisions with z-walls to that with all walls
Nc_{a2}	Ratio of average of number of collisions with z-walls to that with all walls in almost-2D-field
$\mathbf{R}_P$	Vector for $(x \pm x', y \pm y', z \pm z')$
S	Area of surface or cross section (m^2)
S_{3D}	Area of surface for 3D region (m^2)
S_{2D}	Area of 2D region (m^2)
T	Period (s)
T_R	Reverberation time (s)
T_{R_3}	Reverberation time in 3D space (s)
T_{R_2}	Reverberation time in 2D space (s)
$T_{R_{a2}}$	Reverberation time in almost-2D space (s)
T_{R_1}	Reverberation time in 1D space (s)
T_s	Sampling period (s)
T_1	Fundamental period (s)
T_{o1}	Fundamental period for open-open tube (s)
T_{c1}	Fundamental period for open-close tube (s)
T_{12}, T_{21}	Quotient of two magnitude responses
T_{emp}	Temperature (K)
TPR	Total power response for primary and secondary sources
$U(x)$	Displacement as function of spatial variable x (m)
U_z	Random variable for Z^2
$U_1(\omega), U_2(\omega)$	Magnitude of displacement for oscillator 1 or 2 as function of ω (m)
V	Volume (m^3)
V_0	Volume of initial state (m^3)
$V(x)$	Magnitude of velocity as function of spatial position x (m/s)
V_X	Magnitude of velocity of external source (m/s)
$\|\mathbf{v}_z\|$	Magnitude of z-component of velocity (m/s)
$\|\mathbf{v}_y\|$	Magnitude of y-component of velocity (m/s)
$\|\mathbf{v}_x\|$	Magnitude of x-component of velocity (m/s)
W	Spectral matrix for window function $w(n)$
W_R	Power loss by friction (W)
$W_{Av}(t)$	Ensemble average of squared impulse response
$W_{R_{Av}}$	Average of W_R (W)
W_X	Work done by external force in unit time interval (W)
$W_{X_{Av}}$	Average of W_X (W)
W_0	Sound power output of point source in free field (W)

W_{X_1}, W_{X_2}	Sound power output of point source 1 or 2 (W)
W_{X_a}	Sound power output of spherical source with radius a (W)
$W_{X_{inv}}$	Sound power output of anti-phase pair of point sources (W)
W_{ims_2}	Sound power output of two-dimensionally arranged sources (W)
R, S, U, X, Y, Z	Random variables, $S = \ln R$, $R = U/N$, N is integer.
$X(k)$	Fourier transform of signal $x(n)$
X_A	Imaginary part of acoustic impedance ($\mathrm{Pa \cdot s/m^3}$)
$X_{A_{\mathrm{rad}}}$	Imaginary part of radiation acoustic impedance ($\mathrm{Pa \cdot s/m^3}$)
$X_{A_{\mathrm{rad}_0}}$	Imaginary part of radiation acoustic impedance of point source in free field ($\mathrm{Pa \cdot s/m^3}$)
X_M	Imaginary part of mechanical impedance ($\mathrm{N \cdot s/m}$)
$X_{M_{\mathrm{rad}}}$	Imaginary part of radiation mechanical impedance ($\mathrm{N \cdot s/m}$)
$X_{M_{\mathrm{in}}}$	Imaginary part of driving point mechanical impedance ($\mathrm{N \cdot s/m}$)
Y_M	Young's modulus ($\mathrm{N/m^2}$)
z	Complex frequency for discrete systems
Z_0	Ratio of specific impedance and cross section ($\mathrm{Pa \cdot s/m^3}$)
Z_M	Mechanical impedance ($\mathrm{N \cdot s/m}$)
$Z_{MR_{in}}$	Driving point mechanical impedance for right-hand side ($\mathrm{N \cdot s/m}$)
$Z_{ML_{in}}$	Driving point mechanical impedance for left-hand side ($\mathrm{N \cdot s/m}$)
Z_A	Acoustic impedance ($\mathrm{Pa \cdot s/m^3}$)
$\hat{Z}_M$	Normalized mechanical impedance by tension
$Z_{M_{\mathrm{in}}}(x', \omega)$	Driving point mechanical impedance at source position x' as function of ω (variables are often abbreviated.) ($\mathrm{N \cdot s/m}$)
$Z_{A_{\mathrm{rad}}}$	Radiation acoustic impedance ($\mathrm{Pa \cdot s/m^3}$)
$Z_{A_{\mathrm{rad}_a}}$	Radiation acoustic impedance for spherical source with radius a ($\mathrm{Pa \cdot s/m^3}$)
$Z_{A_{\mathrm{rad}_0}}$	Radiation acoustic impedance of point source in free field ($\mathrm{Pa \cdot s/m^3}$)
$Z_{A_{\mathrm{in}}}(x', \omega)$	Driving point acoustic impedance at source position x' as function of ω (variables are often abbreviated.) ($\mathrm{Pa \cdot s/m^3}$)
$Z_{AR_{\mathrm{in}}}$	Driving point acoustic impedance for right-hand-side tube ($\mathrm{Pa \cdot s/m^3}$)

$Z_{AL_{\mathrm{in}}}$	Driving point acoustic impedance for left-hand-side tube ($\mathrm{Pa \cdot s/m^3}$)
$Z_{A_{\mathrm{tf}}}$	Transfer acoustic impedance between source and observation positions ($\mathrm{Pa \cdot s/m^3}$)
$Z_{M_{\mathrm{rad}_a}}$	Radiation mechanical impedance for spherical source with radius a ($\mathrm{Pa \cdot s/m^3}$)
$\overline{a(x)}$	average of $a(x)$ with respect to x
a, b, c	Ratios of lengths in rectangular room
a	Initial displacement (m)
a_A, a_B	Initial displacement for oscillator A or B (m)
b	Initial velocity (m/s)
$\mathbf{a}, \mathbf{b}$	Arbitrary vectors
$a(x)$	Initial displacement as function of x (m)
$b(x)$	Initial velocity as function of x (m/s)
$a(x,t)$	Function expressing propagation of initial displacement (m)
$b(x,t)$	Function expressing propagation of initial velocity (m/s)
$\mathbf{a} \cdot \mathbf{b}$	Inner product of vectors **a** and **b**
a_x	component of vector **a** to vector **x**
c	Speed of sound (m/s)
c_b	Speed of transversal or bending wave (m/s)
$c_{ep}(n)$	Cepstral sequence
c_i	Coefficient of i-th component vector for vector composition
$c_f(n,m)$	sample of correlation sequence $C_F(m)$
$c_{f_{ob}}(n)$	sample of correlation sequence to be observed
$c_{f_s}(n)$	sample of correlation sequence for source signal
$c_{f_{path}}(n)$	sample of correlation sequence for impulse response between source and observation points
$dcos_x, dcos_y, dcos_z$	Direction cosine
d_{th}	Thickness of plate (m)
f	Force (N)
f_K	Restoring force of spring (N)
f_{K_i}	Restoring force for i-th spring (N)
$f_K(x,t)$	Restoring force as function of position x and time t for distributed system (N)
f_R	Friction force (N)
f_X	External force (N)
f_{X_R}	External force for right-hand-side tube (N)
f_{X_L}	External force for left-hand-side tube (N)
$f_\kappa(r,t)$	Restoring force for spherical wave (N)
$f(x,t), g(x,t)$	Function expressing waves
$f(x \pm ct), g(x \pm ct), f(ct \pm x), g(ct \pm x)$	Functions expressing progressive waves

g	Acceleration of gravity (m/s^2)
$h(n)$	Impulse response sequence
$h(\mathbf{x}', \mathbf{x}, t)$	Impulse response between source and receiving positions $\mathbf{x}', \mathbf{x}$
i	Unit of imaginary number
$\mathbf{i}, \mathbf{j}, \mathbf{k}$	Unit vectors of (x, y, z) space
k	Wavenumber (1/m)
k_0	Wavenumber for wave of frequency ν_0 (1/m)
$\hat{k}$	Normalized wavenumber for exponential horn
$\mathbf{k}$	Wavenumber vector (1/m)
k_r	Real part of complex wavenumber (1/m)
k_x, k_y, k_z	Wavenumber components for $x-, y-$ and $z-$axis (1/m)
k_l, k_m, k_n	Wavenumber components for angular eigenfrequency ω_{lmn} (1/m)
k_{lmn}	Wavenumber for eigenmode with wavenumber components (k_l, k_m, k_n) (1/m)
k_N	Wavenumber of N-th eigenmode with (k_l, k_m, k_n) (1/m)
k_{n_c}	z component of wavenumber when $N_{c_z} = n_c$ (1/m)
$\mathbf{k}_N$	Vector for (k_l, k_m, k_n)
l	Integer
n	Integer or harmonics order
$n_{ims_3}(t)$	Number of reflection sound arriving at receiving position in unit-time interval at time t in 3D space (1/s)
$n_{ims_2}(t)$	Number of reflection sound arriving at receiving position in unit-time interval at time t in 2D space (1/s)
$n_{ims_1}(t)$	Number of reflection sound arriving at receiving position in unit-time interval at time t in 1D space (1/s)
m_e	Mass of single molecule (kg)
m_h	Indicator of extension of acoustic horn (1/m)
m_o	Modulation index
n_e	Number of molecules in unit volume
$n_{\lvert\mathbf{v}\rvert}$	Number of molecules with velocity of $\lvert\mathbf{v}\rvert$ in unit volume
n_{max}	Density of maximal amplitude on angular frequency (s/rad)
n_z	Number of zero crossings in unit frequency interval
$n_z^+(\eta_0, \omega)$	Density of non-minimum-phase zeros at angular frequency ω
$n_z^-(\eta_0, \omega)$	Density of minimum-phase zeros at angular frequency ω

$n_{\nu_{3\mathrm{D}}}(k)$	Density of eigenfrequencies in 3D-space at k (m)
$n_{\nu_{2\mathrm{D}}}(k)$	Density of eigenfrequencies in 2D-space at k (m)
$n_{\nu_{3\mathrm{D}}}(\omega)$	Density of eigenfrequencies in 3D-space at ω (s)
$n_{\nu_{2\mathrm{D}}}(\omega)$	Density of eigenfrequencies in 2D-space at ω (s)
p	sound pressure (Pa)
$p(x', x, t)$	Sound pressure as function of source position x', obserbation point x, and time t (Pa)
$p(t)$	Sound pressure as function of time t (Pa)
$\overline{p^2(t)}$	Mean square sound pressure for sound pressure $p(t)$ in real function form (Pa^2)
$\overline{\lvert p^2(t)\rvert}/2$	Mean square sound pressure for sound pressure $p(t)$ in complex function form (Pa^2)
p_i	Incident pressure wave (Pa)
p_r	Reflection pressure wave (Pa)
p_t	Transmitted pressure wave (Pa)
$p(x, t)$	Function expressing sound-pressure oscillation at position x (Pa)
$p_D(x, t)$	Sound pressure for direct sound from source (Pa)
$p_R(x, t)$	Sound pressure oscillation for right-hand-side tube (Pa)
q	Volume velocity (m^3/s)
q_R	Volume-velocity of source for right-hand-side tube (m^3/s)
q_L	Volume-velocity of source for left-hand-side tube (m^3/s)
$\mathbf{r}$	Position vector (m)
r	Spatial distance between two positions (m)
r', **r**	Spatial position vector for source and receiving points
r_c	Critical distance (m)
r_s	Radius of small sphere (m)
r_B	Radius of region (m)
rms	Square root of mean square
s	Condensation
s_0	Condensation at initial state of medium
$s(r, t)$	Condensation as function of position r and time t
$u(t)$	Displacement as function of time t (m)
$u(i, t)$	Displacement of i-th oscillator as function of time t (m)
$u_a(i, t)$	Displacement propagated from displacement (m)
$u_c(i, t)$	Displacement converted from velocity (m)
$u(x, t)$	Displacement as function of spatial position x and time t (m)

$u_i(t)$	Displacement of i-th oscillator as function of time t (m)
$u_n(x,t)$	n-th modal function for n-th eigenfrequency
$u_\omega(x,t)$	Modal function for eigenfrequency ω
$< u^2 >$	Time and space average for $\lvert u(x,t)\rvert^2$
$v(t)$	Velocity or speed as function of time t (m/s)
v([i,i+1],t)	Velocity of mass between i-th and $(i+1)$-th oscillator as function of time t (m/s)
v_a([i,i+1],t)	Velocity propagated from velocity (m/s)
v_c([i,i+1],t)	Velocity converted from displacement (m/s)
$v(x,t)$	Velocity as function of spatial position x and time t (m/s)
v_{in}	Velocity component in-phase with sound pressure (m/s)
v_{out}	Velocity component out-of-phase to sound pressure (m/s)
v_b	Vibrating velocity of bending wave (m/s)
$\mathbf{v}$	Velocity (m/s) or vector
$\mathbf{v}_z$	z-component of velocity vector (m/s)
$\mathbf{v}_y$	y-component of velocity vector (m/s)
$\mathbf{v}_x$	x-component of velocity vector (m/s)
$w(n)$	Window function for signal analysis
$w(t)$	Acceleration as function of time t (m/s^2)
$w_\Gamma(x,n)$	Probability density function for Γ distribution
$w_{Cau}(x)$	Probability density function for Cauchy distribution
$w_{Exp}(x)$	Probability density function for exponential distribution
$w_{Norm}(x)$	Probability density function for normal distribution
$w_{Ray}(x)$	Probability density function for Rayleigh distribution
$w_{Wig}(x)$	Probability density function for Wigner distribution
w_{ims_3}	Spatial density function for mirror image sources in 3D field
w_{ims_2}	Spatial density function for mirror image sources in 2D field
$w(x,y)$	Probability density function for two-dimensional space
$w(z)$	Probability density function for random variable Z
$\mathbf{x}$	Vector for signal
x	Displacement, spatial position, or spatial distance (m)

x', x	Spatial position for source and receiving points
$x(t), y(t)$	Signal, displacement, spatial position, or spatial distance as function of t (m)
$x(Q, n), y(P, n)$	Signal at position Q or P as function of time n
$x_0(t)$	Free oscillation of eigenfrequency
$x_d(t)$	Sinusoidal free oscillation
$x_A(t), x_B(t)$	Displacement, spatial position, or spatial distance for point A or B (m)
x'	Source position (m)
x	Observation position (m)
y	Displacement, spatial position, or spatial distance (m)
y_{Ec}	End correction (m)
z	Complex frequency for discrete systems
z_0	Specific impedance of medium (Pa · s/m)
$z_{0_{12}}$	z_{0_2}/z_{0_1}
z_{0_k}	k-th zero on z-plane of function
z_{wall}	Wall impedance (Pa · s/m)
z_{00}	Ratio of sound pressure and particle velocity of sound (Pa · s/m)

Greek-Letter Symbols

α	Sound absorption coefficient
α_3	Averaged sound absorption coefficient in 3D space
α_2	Averaged sound absorption coefficient in 2D space
α_1	Averaged sound absorption coefficient in 1D space
α_{ob}	Averaged sound absorption coefficient for oblique waves
α_{tan}	Averaged sound absorption coefficient for tangential waves
α_{ax}	Averaged sound absorption coefficient in axial waves
α_{a2}	Averaged sound absorption coefficient in almost-2D space
α_{xy}	Averaged sound absorption coefficient for side walls in xy-2D space
α_z	Averaged sound absorption coefficient for z-walls
$\hat{\alpha}$	$-\ln(1-\alpha)$
β	Imaginary part of complex wavenumber (1/m)
γ	C_P/C_V
γ	Euler's constant
$\Gamma(r)$	Γ function of r
δ	Damping factor (1/s)
δ	Distance from real-frequency axis on complex frequency plane

$\delta(x)$	Delta function of x
δ_0	Damping factor for eigenfrequency (1/s)
δ_N	Imaginary part of Nth angular eigenfrequnecy (1/s)
δ_{ob}	Damping factor for oblique waves (1/s)
δ_t	Distance from pole line on complex frequency plane
δ_{tan}	Damping factor for tangential waves (1/s)
δ_{ax}	Damping factor for axial waves (1/s)
ΔE_0	Small amount of change in acoustic energy density (J/m^3)
$\Delta k_{N_{Av}}$	Average distance between the adjacent eigenfreqe-uncies within spherical shell in wavenumber space
Δp	Small amount of change in sound pressure (Pa)
Δq	Small amount of change in volume velocity (m^3/s)
Δr	Small amount of change in spatial distance (m)
Δs	Small amount of change in condensation (Pa)
Δt	Small element of time interval (s)
Δv	Small amount of change in velocity (m/s)
ΔV	Volume of small element (m^3)
$\Delta\omega$	Difference of two angular eigenfrequencies (rad/s)
$\Delta\omega_{N_{Av}}$	Average distance of adjacent angular eigenfrequen-cies (rad/s)
$\Delta\omega_o$	Difference of adjacent angular eigenfrequencies for weak coupling
$\Delta\omega_p$	Coupling effect on spacing of adjacent angular eigenfrequencies
$\Delta\omega_{12}, \Delta\omega_{21}$	Coupling effect on eigenfrequencies
Δ_λ	Matrix for eigenfrequencies
Δ_{λ_o}	Matrix for eigenfrequencies without purterbation
Δ_{λ_p}	Matrix for eigenfrequencies under perturbation
η_{12}	Transmission coefficient between medium 1 and medium 2
η_{t}	Dilation ($M(\omega)/\pi$)
$\epsilon_1^2(\mathbf{r}_1,\omega), \epsilon_2^2(\mathbf{r}_2,\omega)$	Response contributed by non-resonant modes for point source at $\mathbf{r}'_1$ or $\mathbf{r}'_2$
θ_A, θ_B	Angle at A or B (rad)
θ	Angle (rad)
θ, ϕ	Angles in spherical coordinate system (rad)
θ_i, θ_r	Incident or reflection angle
κ	Bulk modulus (Pa)
λ	Wavelength (m)
λ_b	Wavelength of transversal or bending wave (m)
λ_T	Trace wavelength on boundary (m)
λ_u	Eigenvalue for eigenfunction u
$\lambda_{\mathbf{x}}$	Eigenvalue for eigenfunction or eigenvector $\mathbf{x}$
λ_i	i-th eigenvalue

Λ_{lmn}	Normalizing factor for modal function
ρ	Density (kg/m^3), (kg/m^2), (kg/m)
ρ_0	Density when no oscillation occurs (kg/m^3), (kg/m^2), or (kg/m)
σ	Poisson's ratio
σ	Standard deviation
Σ	Normalized standard deviation
Σ	Surface of small sphere (m^2)
μ	Reflection coefficient
μ_{12}	Reflection coefficient from medium 2 to medium 1
ν	Frequency of oscillation (Hz)
ν_o	Frequency of sound from object without moving (Hz)
ν_0	Eigenfrequency (Hz)
ν_1	Fundamental frequency (Hz)
ν_{o1}	Fundamental frequency for open-open tube (Hz)
ν_{c1}	Fundamental frequency for open-close tube (Hz)
ν_{cn}	n-th eigenfrequency for open-close tube (Hz)
ν_{on}	n-th eigenfrequency for open-open tube (Hz)
ν_{Pn}	n-th eigenfrequency for tube with open- and pressure-source ends (Hz)
ν_{Vn}	n-th eigenfrequency for tube with open- and velocity-source ends (Hz)
ν_n	n-th eigenfrequency (Hz)
$\tau_{S_R}, \tau_{S_L}, , \tau_{O_R}, \tau_{O_L}, \tau_T$	Time interval of arrival sound (s)
τ_{AB}	Time interval for traveling wave between A and B (s)
τ_{dr}	Time delay between direct and reflection waves (s)
κ	Bulk modulus (Pa)
ϕ	Phase or initial phase (rad)
$\hat{\phi}(\mathbf{r}), \hat{\psi}(\mathbf{r})$	Scalar functions
$\phi(\omega)$	Phase or initial phase as function of angular frequency (rad)
$\phi(t)$	Modulated component of instantaneous phase (rad)
$\phi_v(r,t)$	Velocity potential as function of spatial position and time (m^2/s)
$\Phi_v(r)$	Magnitude of velocity potential as function of spatial position (m^2/s)
$\Phi_{v_D}(r)$	Magnitude of velocity potential for direct wave (m^2/s)
$\Phi_{v_R}(r)$	Magnitude of velocity potential for reflection wave (m^2/s)
$\Phi_{v_0}(r)$	Magnitude of velocity potential for symmetric spherical wave as function of spatial position (m^2/s)
$\Phi^+_{v_0}(k)$	Spatial Fourier transform of $\Phi_{v_0}(r)$
$\Phi(t)$	Instantaneous phase (rad)

$\Phi(\omega)$	Accumulated phase up to ω (rad)
$\Phi_1(\omega)$	Accumulated phase up to ω for one-dimensional systems (rad)
$\Phi_2(\omega)$	Accumulated phase up to ω for two-dimensional systems (rad)
$\Phi_3(\omega)$	Accumulated phase up to ω for three-dimensional systems (rad)
$\psi(x, y, z)$	Arbitrary scalar function of x, y, z
$\psi_0(r)$	Function giving initial state ($t = 0$) of time derivative of velocity potential
$\Omega(t)$	Instantaneous angular frequency (rad/s)
Ω	Normalized angular frequency by sampling angular frequency (rad)
$d\Omega$	Solid angle for portion of small sphere (rad)
ω_0	Angular eigenfrequency (rad/s)
ω_B	Angular eigenfrequency where magnitude response becomes half of resonance response (rad/s)
ω_c	Angular eigenfrequency of coupling spring (rad/s)
ω_1, ω_2	Angular eigenfrequency of coupled pendulum (rad/s)
ω_{co}	Angular frequency of modulated sinusoidal component (rad/s)
ω_d	Angular frequency of damped free oscillation (rad/s)
ω_{lmn}	Angular frequency of mode with wavenumber (k_l, k_m, k_n) (rad/s)
ω_N	Angular frequency of N-th eigenmode with (k_l, k_m, k_n) (rad/s)
ω_{N_0}	Angular frequency of N-th free oscillation (rad/s)
ω_{s_n}	n-th complex frequency of damped free oscillation (rad/s)
ω_s	Complex-frequency plane for continuous systems (rad/s)
ω_p	Pole on complex-frequency plane (rad/s)
ω_{PL}	Complex frequency on pole line on complex frequency plane (rad/s)
ω_{p_N}	N-th pole in complex frequency plane for continuous systems (rad/s)
ω_M	Resonant angular frequency for displacement (rad/s)

Chapter 1
Introduction

Sound itself is vibration of a body or a medium. Chapters 2-8 explain the fundamental nature of vibration and sound-wave propagation to construct physical basis of waves with mathematical expressions of wave signals that are necessary for the discrete signal analysis of sound and sound waves. A system constructed of a mass and a spring that is called a **simple oscillator** is a good example to consider the fundamental issues of sound and vibration. Chapter 2 describes the fundamental notions of sound and vibration by taking the simple oscillator as an example. It is reconfirmed that a **sinusoidal function** with **frequency**, **magnitude**, and an **initial phase** plays a fundamental role in sound and vibration. In particular, beats exemplify the significance of the phase, which has received less attention. A sinusoidal wave or function is the fundamental basis of the signal analysis.

Sound propagating in an elastic medium is an **elastic wave**. Chapter 3 briefly explains the fundamental nature of a gas from the point of view of the elastic and chemical properties of the medium. The nature of a gas such the **specific heat** or the **elastic modulus** is necessary to specify the **speed of sound**.

A resonator, the so-called **Helmholtz resonator**, is introduced in Chapter 3. A Helmholtz resonator can be constructed of an air-mass and air-spring. It is shown that a loudspeaker that is installed into its own enclosure can be assumed as a Helmholtz resonator in the low frequency range.

A simple oscillator is a basic structure that describes sound and vibration, but it does not express the waves or sound propagation in a medium. An intuitive image for wave propagation can be obtained according to a model composed of a series of simple oscillators. Chapter 4 discusses propagation of sound waves from the viewpoint of energy exchange between the potential and kinetic energies. The exchange of the energies according to the energy preservation law specifies the finite speed of sound propagation in a medium. Consequently, a mathematical equation, which is called the **wave equation**, is introduced to represent the one-dimensional wave propagation for a simple form of a wave, namely a **plane wave**. The model of the series of simple oscillators and propagation of its initial disturbance provide the fundamental basis for understanding the wave propagation in a medium. In particular,

M. Tohyama: *Sound and Signals*, Signals and Communication Technology, pp. 1–8.
springerlink.com

effects of the difference in the initial conditions, namely, the initial displacement or velocity, on the wave propagation are surprising.

Chapter 5 considers the wave propagation on a vibrating string. The vibration of a string is most fundamental issue of acoustic, as is the sound traveling in a tube. The finite speed of wave propagation caused by the vibration of a string is given by the tension and the density of the string. This chapter expresses how the initial disturbance propagates along the string as time passes, assuming two types of initial disturbance, i.e., the initial displacement like for a harp or the velocity of the initial motion, such as for a piano.

There is an infinite number of eigenfrequencies in the vibration of a finite length of a string. Periodic vibrations are displayed for a finite length of a string that are composed of the **fundamental**, given by the length and speed of the wave, and its **harmonics**, the so-called **eigenfrequencies**. The harmonic vibration defines the **eigenfunctions** corresponding to the eigenfrequencies. The eigenfunctions visualize the patterns of vibrations of the eigenfrequencies and consequently show the patterns of **standing waves. Nodes** of the vibration patterns are introduced, which are positions that rest in the whole periods of vibration. The nodes equally divide the string, and the number of nodes denotes the order of the harmonics. The eigenfrequencies or eigenfunctions depend on the **boundary conditions** of the string. The effects of the boundary conditions on the eigenfrequencies can be interpreted by referring to the **coupled oscillator** mentioned in Chapter 2.

The external source is necessary to keep stationary vibration of a string. The energy transfer to the string from the external source is formulated in terms of the **driving point impedance**. No energy is transmitted to the stationary string when the source is applied to the node of the string vibration with the freguency of the external source. Consequently, the **energy preservation law** is formulated between the vibration of the string and the external source.

The standing wave is composed of two **progressive waves** that travel in opposite directions to each other. The periodic vibration of a string can be expressed as the superposition of the eigenfunctions. This is the fundamental basis of **Fourier analysis** that provides the fundamental concept of the signal theory. The basic notions of the signal analysis, such as the **impulse response**, **transfer functions**, and **frequency characteristics**, are introduced in this chapter.

Chapter 6 describes the **plane waves** propagating in a medium. The plane wave is the most fundamental mode of sound propagation in a medium such as air. First, the **speed of sound** is discussed by recalling that the **bulk modulus** depends on the **dilation** or **condensation** process of the medium mentioned in Chapter 3. Second, the **specific impedance** is defined as the ratio of the sound pressure and the particle velocity for the plane wave in a medium. It is shown that the specific impedance is defined uniquely to a medium. Following these issues, the **radiation impedance** is introduced to consider the sound radiated by vibrating objects. The sound pressure, rendered on the surface of the vibrating object, must contain in-phase components with the surface velocity of the vibration for radiating sound from the object. This is the fundamental nature of sound radiation from a source, and it can be applied to the radiated sound and even to **shock waves** from a moving object.

Chapter 7 describes the sound propagation in a tube. Sound waves in an open or closed tube are considered in terms of eigenfrequencies and eigenfunctions similar to the vibration of a string mentioned in Chapter 5. In addition, the radiation impedance and **boundary correction** for an open tube are explained. **Sound radiation from the open end** states that another wave model is necessary to represent the wave phenomena; namely, the **spherical wave** is the other wave mode for sound propagation in a medium. However, the propagation of plane waves in a tube is basically important to theoretically formulate that the eigenfrequencies and eigenfunctions depend on the boundary condition of the tube. Thus the difference between the fundamental frequencies and their harmonics are explained by taking examples from the flute and clarinet types of sound propagation in a tube.

Chapter 8 summarizes basic phenomena of sound-wave propagation in three-dimensional space. **Reflection** and **transmission** of waves are displayed according to geometrical interpretation and **Fermat's principle**. **Spherical waves** are described by referring to the **incompressibility** of the medium that makes the difference in the plane and spherical waves in terms of the **phase relationship between the sound pressure and the particle velocity**. **Interference** by reflection waves is briefly mentioned, and thus the notion of the frequency characteristics of the sound field in space is introduced. The **frequency characteristics** of sound field plays an important role in this book.

As stated above, very elementary and fundamental issues of sound and waves are presented with basic mathematical formulations in the chapters mentioned. The mathematical expressions with their physical or geometrical images, related to sinusoidal functions in the complex function forms, provide the fundamental basis for representation of signals such as in Fourier analysis.

Chapter 9 formulates sound propagation and radiation from a source into three-dimensional space. Namely, it provides the basis of room acoustics theory. In particular, propagation of the spherical waves caused by the initial disturbance might be interesting for seeing the example in which the **compression waves** followed by the **dilation waves** are propagated. Sound power output can be formulated using the radiation impedance including the effects of surroundings on the radiated sound from the source. The phase relationship between the sound pressure and the velocity on the surface is reconfirmed as mentioned in Chapter 6. The **radiation impedance of the open end of an acoustics tube** is formulated, and travelling of sound waves in an **exponential horn** is briefly mentioned from the viewpoint of the radiation impedance of the open end.

Chapter 10 considers the **wave equation** for sound waves in rooms. Eigenfrequencies and eigenfunctions are also the central issues for the sound waves in rooms. A big difference between the waves in rooms and on strings (or in tubes) is that the eigenfrequencies are not distributed uniformly on the frequency for the waves in rooms. Therefore, estimation of the **density of the eigenfrequencies** is a big issue in room acoustics. The **Green functions** are introduced according to the **orthogonality of the eigenfunctions** or **mirror-image theory**. The Green function can be interpreted as the "spatial" impulse response in wave-propagating space, while the impulse response is defined on the time region for a linear system.

Fourier analysis can be extended into the generalized orthogonal expansion. The sound field in rooms can be formulated based on the orthogonality of the eigenfunctions. However, another formulation is also possible based on the **integral representation** of the solutions. The integral representation might be intuitively understandable rather than the solutions by the **modal expansion using the eigenfunctions**.

The **reverberation time** might be the most well-known parameter of the sound field, but reverberation is never a simple issue. Chapter 11 explains the reverberation process in rooms by starting from the **energy balance equation** for the sound field in rooms. The random nature of sound propagation in rooms is a key issue to understand the reverberation process of the sound field. Such a random nature could be intuitively understood by introducing **wavenumber space**.

The transient response, such as the reverberation sound, can be theoretically formulated according to the **linear-system theory**, namely using the **impulse response** and the **convolution** scheme. Detailed discussions about the reverberation formulas are developed for the three-, two-, and one-dimensional fields. The **mean free path** is a key for the theoretical formulations of the reverberation process. The typical frequency characteristic appears in an **almost-two-dimensional sound field** in a rectangular room, due to the arrangement of sound-absorbing materials.

The mean free path of a sound field is a concept originally based on **geometrical acoustics** in terms of the mirror-image theory. However, it can be also interpreted following the **modal wave theory**. Thus, a **hybrid formula for the reverberation process** is possible, bridging modal and geometric acoustics. The frequency characteristics of the sound energy at the steady state can be developed using the **modal density** of the **oblique**, **tangential**, and **axial** waves and the mean free paths for the three-, two-, and one-dimensional fields with corresponding **sound absorption coefficients**, respectively. This hybrid formula for the reverberation response is applicable to many practical situations of room acoustics.

The sound field is never uniform in a room. As stated above, the sound energy response can be formulated by both the geometrical and wave-theoretical approaches. In the geometrical approach to the sound field in rooms, sound is represented as a **sound ray** or a particle of sound. In contrast, the sound field can be represented as the response of a linear system following the linear wave equation. The two approaches, however, exemplify the noticeable difference in sound propagation in a closed space.

Chapter 12 describes the **spatial distribution of sound** in space. The chaotic properties must be a noticeable phenomenon observed by the geometric approach based on the **ray theory** for sound propagation. A boundary that is composed of plane and curved surfaces makes sound propagation complicated in space surrounded by the boundary. The sound propagation, for example, in a so-called **stadium field**, manifests **chaotic properties** as the reflection process goes on, if the reflection process follows the traveling of the sound rays.

In contrast, the sound field under any boundaries follows the wave equation without non-linearity from a wave theoretic viewpoint. Namely, chaotic properties cannot be expected as long as the wave theoretic approach is taken to the sound field.

A bridge over this gap between the sound rays and sound waves seems to be hidden in the distribution of the eigenfrequencies of the sound field. That is, the so-called **scar of the chaotic properties**, namely the leftovers of the non-linearity in the linear field, can be seen in the eigenfrequency distribution. If a family of the **Gamma distributions** including the freedom of non-integers is introduced into the distribution of the eigenfrequencies, the sound field is possibly characterized from the regular (non-chaotic trajectories for travelling of the sound ray) to chaotic (chaotic trajectories) fields by the freedom of the distribution.

The sound field can also be expressed as superposition of plane waves with random magnitudes and phases. According to this model, the distribution of sound energy in a closed space can be estimated including its variance. The standard deviation of sound pressure levels, 5.5 dB, is a key number representing the random sound field with a single frequency. Note here that the standard deviation of the sound pressure records assumes random sampling of the sound pressure records from the sound field. By randomly sampling the sound pressure from the sound field, the outcome, namely the obtained sound-pressure records, can be assumed random samples even if the sound field is created by a pure tone with a single frequency in a room.

Another key number of the random sound field in a room is the **spatial correlation coefficients** of sound pressure records. The cross-correlation coefficient between two sound pressure records sampled at two points in a three-dimensional **random sound field** with spacing of r (m) is given by the sinc function of kr, where k denotes the wavenumber. The correlation coefficients, however, depend on the dimensionality of the space (two- or one-dimensional space). Note that the spatial correlation coefficients in a rectangular room can be noticeably different from the sinc function because of the symmetric geometry even in the three-dimensional field.

Sound propagation from a source position to a receiving point in a room can be represented by the **transfer function** from a linear-system theoretic viewpoint. Chapter 13 describes the transfer function in the sound field in rooms in terms of the **poles and zeros** or the **magnitude and phase responses**.

The **driving-point impedance** is defined as a limit of the transfer function when the receiving position approaches the source location. The **sound power response** radiated from the source in the space can be characterized by the driving-point impedance. The poles and zeros are interlaced with each other in the driving point impedance, even after the spatial average is taken with respect to the source positions. Namely, the power response can be basically represented by the poles that are almost equal to the eigenfrequencies under the small damping condition.

The **energy balance equation** can be derived following the driving point impedance. The energy balance equation states that the power response can be estimated by the spatial average of the squared sound-pressure records. Estimation of the spatial average for the squared sound-pressure records is difficult in general, when the frequency band of a source becomes narrow. However, the **spatial variance of the records** decreases on the room boundaries, namely on the walls, edges, or at the corners, in the sound field excited by a narrow-band noise source.

By recalling the driving-point impedance at the source position, so-called **active power minimization** is possible to some extent in a closed space by setting the secondary sources close to the primary source. However, the achieved power reduction decreases as the **modal overlap** increases in the space.

The frequency characteristics of the magnitude and phase are governed by the poles and zeros of the transfer function. The **zeros** are complicated, and they depend on the **signs of the residues** for adjacent pairs of the poles that correspond to the eigenfrequencies. The **probabilities for the sign changes** are analyzed in one-, two-, and three-dimensional spaces. In one-dimensional space, the probability increases in proportion to the distance from the source. However, this is not the case for the two- or three-dimensional space.

The range in which the probability increases in proportion to the distance defines the **coherent field of the sound space**. The **coherent length of a sound path** is a fundamental parameter that specifies to what extent the sound travels as a spherical wave, as if it travels in a free field, from the source in the space. The phase progresses in proportion to the distance from the source for a spherical wave even in a closed space as long as the distance is within the coherent length. This regular phase increase is called the **propagation phase**. However, the regular phase increase stops, but jumps and remains stationary within the random fluctuations out of the coherent field. This random phase fluctuation is called the **reverberation phase**. The residue sign changes, and the propagation and reverberation phases are discussed in detail according to the poles and zeros for the transfer functions.

Sound signals are characterized in both the time and frequency domains. Chapter 14 considers the signal signatures according to the **correspondence between the temporal and frequency characteristics**. In general, sound, like speech signals, is represented by the **magnitude and phase spectral components**. However, the magnitude spectrum is considered important in almost all types of applications of speech processing, while the phase has received less attention. This chapter, first, demonstrates the **phase dominance** rather than the magnitude spectrum for shorter or longer analysis/synthesis windows. The dominance for the longer time windows is intuitively understandable; however, for the shorter frames it is surprising. Following this outcome, it is reconfirmed that the preservation of the **narrow-band envelopes** constitutes an important factor for **speech intelligibility**.

Speech waveforms are reconstructed by **magnitude-spectral peak selection** on a frame-by-frame basis subject to the frame length being chosen appropriately. For example, an intelligible speech waveform can be reconstructed for every short frame, if more than five dominant-peak spectral components are selected in every frame.

Fundamental frequencies are important signatures. The fundamental frequencies and their harmonics are estimated by **auto-correlation** analysis for the sequences of the frequencies by peak spectral selection. The harmonic structure of sound is displayed on the frame-by-frame basis even when the fundamental is missing.

Time-envelopes of speech or musical sound are generally characterized on the time domain. However, the effect of the envelope can also be observed on the frequency domain. **Beats** are good examples that the envelopes are constructed of clustered sinusoidal components. Clustered line spectral modeling (**CLSM**) of sound is formulated in this chapter. CLSM represents a signal modulated by the envelope as a superposition of the clustered sinusoidal components according to the **least-square-error** (LSE) solutions on the frequency domain. Those clustered components cannot be separately estimated by the conventional frame-wise discrete Fourier transformation. Decaying musical sound is represented by CLSM. The fundamental and its harmonics can be decomposed into clustered sinusoidal components, respectively.

The CLSM approach can be converted into the time domain according to the correspondence between the time and frequency regions. By recalling the correspondence, the **instantaneous magnitude and phase on the time domain** can be interpreted as the **spectral magnitude and phase on the frequency plane**. Consequently, the sign change of the instantaneous frequency can be understood according to the **minimum- and non-minimum-phase** properties on the spectral domain.

Clustered time series modeling (CTSM) can be formulated by converting the CLSM approach on the frequency domain to the temporal plane. A brief signal in a short time interval, like a percussion sound, or reflection waves from a boundary for the traveling waves on a string, can be appropriately represented by CTSM.

Chapter 15 develops fundamental issues of estimation and creation of transfer functions that can be interpreted as the **path information** between the sound source and receivers. The spectral peak selection stated in the previous chapter can be applied to estimate the path information. The power spectral properties can be estimated on the frame-by-frame basis, even if the short frame length is taken. If the spectral peaks are closely located to each other, the frame length must be longer than the inverse of the frequency interval of the clustered peaks.

Source waveform recovery is developed as an example of **inverse filtering** that is only possible for the **minimum-phase path information**. Inverse filtering can be interpreted using pairs of the poles and zeros on the frequency plane. **Sound image projection** in a three-dimensional space is a good example of the inverse filtering for the so-called **head-related transfer function**. Inverse filtering for the minimum-phase component of the path information provides stable sound image control from a theoretical viewpoint.

Stabilization of a sound path including the **feedback loop**, namely a closed loop, is another example that requires **equalization of the path information** for avoiding an unstable closed loop. **Instability of the closed loop** can be understood in terms of the poles and zeros for the transfer function.

The **ratio of the direct and reverberant sound energy** of the reproduced sound field is a key to achieving reproduction of intelligible speech. Interestingly, there is a **minimum ratio** in every reverberant space that is given by the averaged sound absorption coefficient of the field. According to this fact, **multi-channel reproduction of speech** is a possible way for improving speech intelligibility in a reverberant space.

As a whole, a sound field and sound observed in the field are described in detail, according to the modal and the linear-system theory represented by the poles and zeros for the transfer functions, or the magnitude and phase in the time and frequency planes. In particular, the phase receives great attention, in contrast to conventional articles. For that purpose, it is described in detail how the envelopes are created by the phase or clustered sinusoidal components by recalling the correspondence between the temporal and frequency planes. In addition, how the phases are accumulated is described following the distributions of the poles and zeros.

Consideration of the poles and zeros might also be a key to understand the sound and waves in a closed sound space that can be represented by the impulse responses and Green functions. The contents of this book might show a possible way to bridge the modal theory, geometrical acoustics, and signal analysis.

Chapter 2
Oscillation and Resonance

When sound travels in a medium, it is called a sound wave in physics terms. A wave in a medium is excited by vibration or oscillation of a portion of the medium. An example of vibration or oscillation that illustrates basic physical properties of sound is motion of a mass attached to a spring. A study of simple oscillation of a mass with a spring reminds us of the historical investigation about motion and force by Galileo and Newton. This chapter describes fundamental properties of a simple oscillator composed of a mass with a spring. The period of free oscillation, eigenfrequencies, decaying oscillation, response to external force, and resonant frequency of the external force will be described in terms of the fundamental law of preservation of dynamical energy. Further examples are presented, which illustrate so-called beats and modulation from the view point of energy transfer in a coupled oscillator.

2.1 Harmonic Oscillator

A portion of a medium oscillates when sound travels in the medium. A wave is a physical concept that defines such motion of the medium. A visible image of oscillation can be obtained by taking an example of a simple oscillator.

2.1.1 Mass-Spring System

Suppose that a mass is attached to a spring, which is called a mass-spring system or a **simple oscillator** or **harmonic oscillator**, as shown in Fig. 2.1. If the spring is stretched (or strained) by the amount of x (m) from the original length, the **stress** (or **restoring force**) (N) is proportional to the amount of stretch such that

$$f_K = -Kx. \qquad \text{(N)} \tag{2.1}$$

This is called **Hooke's law**, where K (N/m) denotes a **spring constant**. If the spring is hard (soft), the spring constant is large (small). Note here N denotes the unit that represents force, which can be specified by $\mathrm{kg} \cdot \mathrm{m/s^2}$, following the Newtonian law

M. Tohyama: *Sound and Signals*, Signals and Communication Technology, pp. 9–30.
springerlink.com

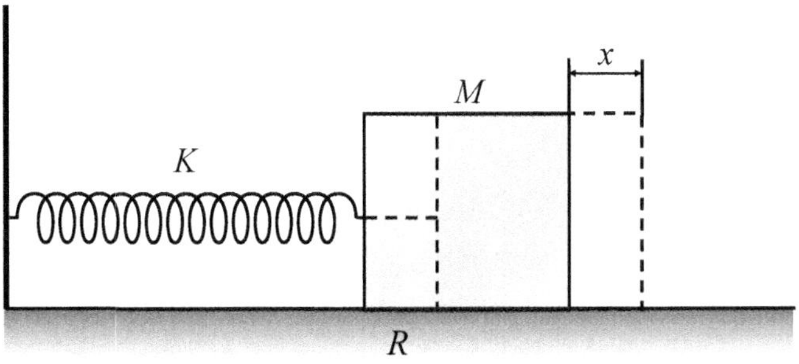

Fig. 2.1 Mass-spring system

with respect to a mass and its acceleration. If the external force that was necessary to stretch the spring is released, the mass continues its motion of oscillation. This type of oscillation is called **free oscillation**.

2.1.2 *Free Oscillation*

Motion of a mass can be observed by variation of the **position** of the mass with respect to time, if a sign ($\pm$) is applied to an amount of stretch so that a position of the mass might be specified. Express a time-variant position of the mass as a function of time, $x(t)$. Motion of the mass, however, is characterized by three types of variables in general: position (m), **speed** (or **velocity**) (m/s), and **acceleration** ($\mathrm{m/s^2}$). Speed represents a rate of temporal change in the position, which can be calculated by $v = \Delta x/\Delta t$, and similarly acceleration shows the rate of change of speed, such as $w = \Delta v/\Delta t$. By introducing here mathematical notations in terms of differential calculus[1], which is based on a mathematical sense of a limit as the Δt approaches 0, to describe the motion $x(t)$, speed $v(t)$ and acceleration $w(t)$ can be expressed as

$$v(t) = \frac{dx(t)}{dt} \qquad (\mathrm{m/s}) \tag{2.2}$$

$$w(t) = \frac{dv(t)}{dt} = \frac{d^2x(t)}{dt^2}. \qquad (\mathrm{m/s^2}) \tag{2.3}$$

In general, speed including a sign (plus or minus) is called **velocity**, which indicates a direction of motion. Speed and acceleration represent local properties of dynamical motion in time, and are the first and second derivatives respectively of a function that might represent motion.

Free oscillation follows the **Newtonian law of motion**, which states that acceleration $d^2x(t)/dt^2$ of a mass M(kg) is proportional to force working on the mass, and is formulated in the mathematical equation

$$M\frac{d^2x(t)}{dt^2} = -Kx(t) \qquad \text{(N)} \tag{2.4}$$

or equivalently

$$\frac{d^2x(t)}{dt^2} = -\frac{K}{M}x(t). \qquad (\text{m/s}^2) \tag{2.5}$$

The equation above indicates that the second derivative of a function that represents the motion of free oscillation must be the same as that for motion itself except dilation including a sign.

If a function remains similar after a mathematical manipulation (linear operator, if it is said strictly) such as taking the derivative is applied to the function, then the function is called the **eigenfunction** for the manipulation (linear operator). Here the scalar, such as $-\frac{K}{M}$ in the equation above, is called the **eigenvalue** for the eigenfunction[2][3]. This is the theoretical background for $\frac{1}{2\pi}\sqrt{\frac{K}{M}}$ (Hz) being called the **eigenfrequency**.

Motion of the mass can be expressed as a **sinusoidal function** that displays periodic motion of a definite period. Here a sinusoidal function is defined as

$$x(t) = x_0(t) = A\sin(\omega_0 t + \phi) = A\sin\Phi(t). \qquad \text{(m)} \tag{2.6}$$

It is confirmed that the sinusoidal function defined above satisfies the Newtonian law expressed as Eq. 2.4 according to the properties of a sinusoidal function. A cyclic property for iterative differentiation is a typical characteristic of a sinusoidal function.

It is quite interesting that free oscillation could be expressed as a sinusoidal function. A sinusoidal function is **periodic**, such as

$$A\sin(\omega_0 t + \phi) = A\sin(\omega_0(t + lT) + \phi), \tag{2.7}$$

and is the simplest periodic function. Here l is an integer, $T = 2\pi/\omega_0 = 1/\nu_0$ is the **period**, and ν_0 denotes the **frequency**. It will be shown in Chapters 5 and 7 that a periodic function can be represented by superposition of sinusoidal functions whose frequencies ν_n are specified as $\nu_n = n\nu_0 = n/T$. A sinusoidal function is a periodic function that can be expressed by a single frequency. This explains why the mass-spring system is called a **simple oscillator**.

Free oscillation for a simple oscillator is periodic with a single frequency. In the sinusoidal function, A is the **magnitude** or **amplitude** and ϕ denotes the **initial phase**. Both magnitude and the initial phase are determined according to initial conditions that specify the initial state of motion of the mass, such as

$$x(t)|_{t=0} = x(0) = a \qquad \text{(m)} \tag{2.8}$$

$$v(t)|_{t=0} = v(0) = b. \qquad \text{(m/s)} \tag{2.9}$$

If the mass is released quietly (without velocity) after being stretched, then $b = 0$ can be set. The period of motion is defined independent of the initial conditions as described in the following section. Therefore, the frequency of free oscillation is called **eigenfrequency**.

2.2 Frequency of Free Oscillation

The frequency of free oscillation can be determined following the **energy preservation law of a dynamical system**. The energy preservation law is a fundamental law as well as the Newtonian law in physics.

2.2.1 *Potential and Kinetic Energy*

The motion of a simple oscillator has two types of dynamical energy: potential and kinetic energy. Potential energy is due to stretch (or strain) of the spring. External force is needed to stretch the spring against the stress (restoring force) given by Kx. Thus it can be interpreted that there is static energy preserved in a stretched (or contracted) spring. Such energy is called **potential energy**. The potential energy E_P (N·m=J) preserved in a spring can be expressed by integration (intuitively accumulation) of the work done due to the external force, such as

$$E_P = \int_0^x Kxdx = \frac{1}{2}Kx^2. \qquad \text{(J)} \qquad (2.10)$$

The integrand $Kxdx$ stated above shows the **work done** by external force that is needed to stretch (or contract) the spring from x to $x + dx$, where dx represents a very small increment. The integration can be intuitively understood as accumulation of such a small amount of work done.

The other type of energy, **kinetic energy**, specifies the work done due to motion itself instead of static deformation such as stretch of a spring. Express the velocity of motion by $v(t) = dx(t)/dt$ (m/s). The kinetic energy (J) can be defined by

$$E_K = \int_0^x Mvdv = \frac{1}{2}Mv^2, \qquad \text{(J)} \qquad (2.11)$$

where M denotes mass (kg) and $vdt = dx$, which shows a small displacement due to the motion. The integrand above corresponds to the work done, such that

$$Mvdv = M\frac{d^2x(t)}{dt^2} \cdot dx, \qquad \text{(J)} \qquad (2.12)$$

where $M\frac{d^2x(t)}{dt^2}$ represents the force working on the mass.

2.2.2 *Energy Preservation Law and Eigenfrequency*

The **energy preservation law** states that total energy, which is the sum of the potential and kinetic energy, must be constant independent of time. Again suppose that motion of a mass in a simple oscillator is expressed as a sinusoidal function. Potential and kinetic energy are expressed as

$$E_P = \frac{1}{2}KA^2\sin^2(\omega_0 t + \phi) \qquad \text{(J)} \tag{2.13}$$

$$E_K = \frac{1}{2}M\omega_0^2A^2\cos^2(\omega_0 t + \phi) \qquad \text{(J)}, \tag{2.14}$$

respectively. The periods of two energy functions are the same, and the sum of the two types of energy must be constant independent of time so that the energy might be preserved.

The eigenfrequency is determined following the energy preservation law. The **eigen-(angular)-frequency** can be given by

$$\omega_0 = \sqrt{\frac{K}{M}}, \qquad \text{(rad/s)} \tag{2.15}$$

subject to the total energy being preserved:

$$E_P + E_K = \frac{1}{2}KA^2 = constant, \qquad \text{(J)} \tag{2.16}$$

which corresponds to the initial potential energy when the motion starts.

The potential energy takes its maxima when the displacement of oscillation is maximum, and it takes its minima at null displacement. In contrast, the kinetic energy takes its maxima at the point of maximum oscillation speed corresponding to null displacement, and it takes its minima when the oscillation stops at positions for displacement of the maximum. Thus, the total energy has alternate maxima of the potential and kinetic energy so that the total energy might be constant. In Chapter 4 it will be shown that this type of energy exchange is a key issue for understanding sound traveling in a medium.

2.3 Damped Oscillation

2.3.1 *Mathematical Expression for Damped Oscillation*

Free oscillation, as stated in the previous section, decays as time passes and will eventually stop. This is because energy exchange between potential and kinetic energy does not last for long due to a loss of dynamical energy. The loss of dynamical energy results from interaction between an oscillation system and its surroundings, and consequently the dynamical energy that an oscillation system loses changes to thermal energy, such as friction heat. In general, a loss of dynamical energy is proportional to oscillation speed. Thus **friction force** (N), which is denoted by f_R here and causes energy conversion (loss), can be assumed as

$$f_R = Rv, \qquad \text{(N)} \tag{2.17}$$

where v denotes oscillation velocity (m/s) and R is called a **friction constant** $(\mathrm{N} \cdot \mathrm{s/m})$.

The Newtonian law that specifies the relationship between acceleration of a mass and force working on the mass can be expressed by including friction:

$$M\frac{d^2x(t)}{dt^2} + R\frac{dx(t)}{dt} + Kx(t) = 0. \qquad \text{(N)} \tag{2.18}$$

Oscillation that follows the equation above can be obtained by

$$x(t) = A\mathrm{e}^{-\delta_0 t}\sin(\omega_d t + \phi). \qquad \text{(m)} \tag{2.19}$$

Set the expression for a damped free oscillation as[4]

$$x(t) = A\mathrm{e}^{-\delta_0 t}x_d(t), \tag{2.20}$$

where $\delta_0 = R/2M$. By substitution of the expression above into the equation

$$\frac{d^2x(t)}{dt^2} + 2\delta_0\frac{dx(t)}{dt} + \omega_0^2 x(t) = 0, \tag{2.21}$$

the equation

$$\frac{d^2x_d(t)}{dt^2} + \left(\omega_0^2 - \delta_0^2\right)x_d(t) = 0 \tag{2.22}$$

is obtained. The function $x_d(t)$ can be written as

$$x_d(t) = A\sin(\omega_d t + \phi) \qquad \text{(m)} \tag{2.23}$$

$$\omega_d = \sqrt{\omega_0^2 - \delta_0^2}. \qquad \text{(rad/s)} \tag{2.24}$$

The mathematical expression above, however, does not indicate that the oscillation eventually stops. Instead the magnitude decays as time goes on, and it approaches the limit zero. This is a possible way to mathematically represent the damped oscillation using a smooth function. The quotient δ_0 indicates **speed of decay** of damped oscillation. As it increases, the life of the oscillation decreases.

2.3.2 *Frequency of Damped Oscillation*

Damped oscillation is no longer periodic in the mathematical sense. However, from a perceptual view point, it makes sense to define the angular **frequency of damped oscillation** denoted by ω_d. A listener perceives pitch even for slowly decaying sound. Note here the frequency or pitch becomes lower than the eigenfrequency without loss of energy, as the speed of decay increases.

Figure 2.2 shows examples of decaying oscillations. Speed of decay increases as the damping increases. Moreover, if the damping becomes too strong, not even a single cycle of oscillation is observed. This faded cycle can also be interpreted following the frequency given by ω_d. If damping (a quotient such as $R/2M$) increases over the eigenfrequency, then the frequency of damped oscillation is lost as indicated by a square root of a negative number in mathematical terms. An increase of damping creates a limit in the frequency of oscillation.

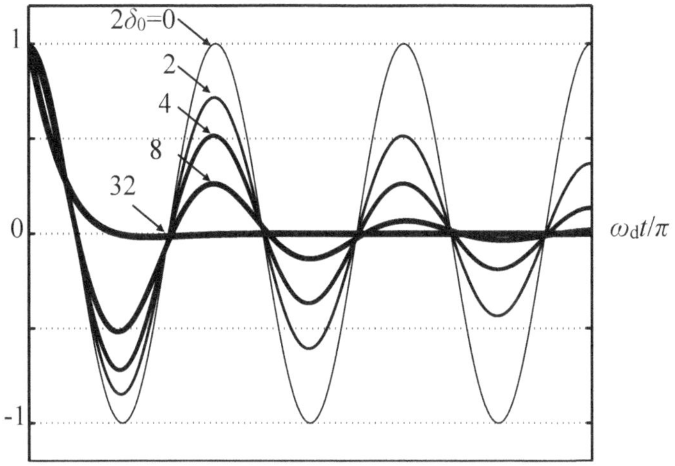

Fig. 2.2 Samples of damped oscillation

2.4 Forced Oscillation and Resonance

2.4.1 *Newtonian Law for Oscillation by External Force*

Damped free-oscillation eventually stops because of energy loss (or conversion from dynamical to thermal energy) such as friction heat. This means that oscillation lasts as long as the loss of dynamical energy is supplied by an external energy source. Stationary oscillation is called **forced oscillation**, when it is excited by an external force and remains stationary. The external force, which is needed to excite steady oscillation, is decomposed into three types of force: inertial force due to acceleration of a mass, stretching force against restoration (stress) of a spring, and compensating force against friction. Therefore, motion of a simple oscillator, which is represented by a time-variant position of the mass $x(t)$, follows

$$M\frac{d^2x(t)}{dt^2}+R\frac{dx(t)}{dt}+Kx(t)=f_X(t), \qquad \text{(N)} \tag{2.25}$$

where $f_X(t)$ denotes the external force. The expression shown above is the same as that for free oscillation, if external force is not available.

There are many possibilities for external force. Periodic, non-periodic, or random, and transient or pulse-like forces are examples of possible external force. A common phenomena of responses to such a variety of external forces is **resonance**. The usual approach to resonance is to analyze responses to periodic external force. A simple example of periodic force is sinusoidally alternating force.

2.4.2 Oscillation Excited by Periodic Force of Single Frequency

Suppose that the external force is represented by a sinusoidal function with an angular frequency of ω, such that

$$f_X(t)=F_X\sin\omega t. \qquad \text{(N)} \tag{2.26}$$

An expression of oscillation such as

$$M\frac{d^2x(t)}{dt^2}+R\frac{dx(t)}{dt}+Kx(t)=F_X\sin\omega t \qquad \text{(N)} \tag{2.27}$$

is obtained for the external force. Motion of the mass, which can be represented by the position of the mass $x(t)$ as well as free oscillation, is expressed using a sinusoidal function such that [5]

$$x(t)=A(\omega)\sin(\omega t+\phi(\omega)), \qquad \text{(m)} \tag{2.28}$$

subject to

$$A(\omega)=\frac{|F_X|/M}{\sqrt{(\omega_0^2-\omega^2)^2+4\delta_0^2\omega^2}} \tag{2.29}$$

$$\tan\phi(\omega)=\frac{2\delta_0\omega}{\omega_0^2-\omega^2}. \tag{2.30}$$

It should be noted that, different from free oscillation, the frequency of stationary oscillation is the same as that for external force. This explains why the oscillation above is called **forced oscillation**. The variable $A(\omega)$ defined above is called **magnitude response** of the oscillator of interest, and similarly $\phi(\omega)$ indicates **phase response**.

2.4.3 Magnitude and Power Response to Sinusoidal Force

Magnitude $A(\omega)$ is a variable of the frequency of the external force as shown in Fig. 2.3a. **Resonance**, in which magnitude response takes its maximum as a frequency of external force approaches the eigenfrequency, is not noticeable under damped conditions. In general, a loss of energy is proportional to a friction constant R. Approximating the equations of motion for the external source of a frequency that is close to the eigenfrequency such that

$$\frac{d^2x(t)}{dt^2} + \frac{K}{M}x(t) \cong 0 \tag{2.31}$$

$$\frac{R}{M}\frac{dx(t)}{dt} \cong \frac{f_X(t)}{M} = \frac{F_X}{M}\sin\omega t, \tag{2.32}$$

it can be seen that the velocity of oscillation denoted by $dx(t)/dt$ is in-phase with the external force at the **resonance frequency**(see Appendix)[5][6]. In other words, the external force must be synchronized with the velocity of motion in order to build up the resonant oscillation. Recall that motion of oscillation (displacement) can be understood as in-phase with the velocity in the quarter period after its maxima. This indicates the maxima of motion occur at good times for applying the external force in order to excite the resonant motion.

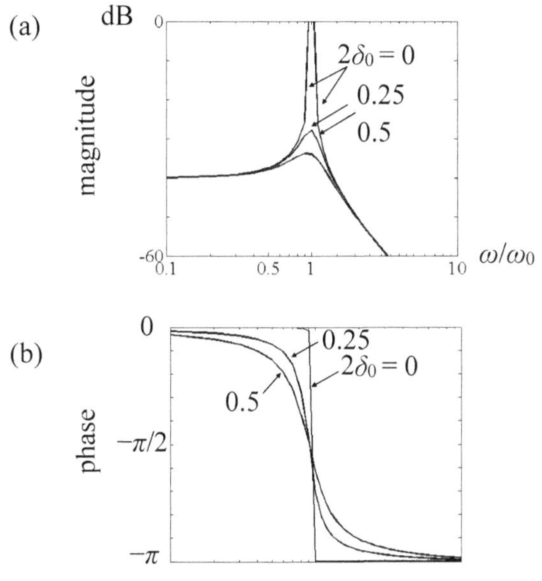

Fig. 2.3 Magnitude (a) and phase (b) responses to sinusoidal external function

The velocity is inversely proportional to the friction constant at the resonance frequency. Then the work done for the motion of oscillation in a unit time interval by the external force can be written as

$$W_X(t) = f_X(t)\frac{dx(t)}{dt} = \frac{F_X^2}{R}\sin^2 \omega t. \qquad (\mathrm{W} = \mathrm{J/s}) \tag{2.33}$$

By taking an average over a single period of oscillation,

$$W_{X_{Av}} = \overline{W_X(t)} = \frac{F_X^2}{2R} \qquad (\mathrm{W}) \tag{2.34}$$

is obtained. The work done increases (decreases) as the friction constant decreases (increases). On the other hand, power consumed by the friction force is

$$W_R(t) = R\left(\frac{dx(t)}{dt}\right)^2 = \frac{F_X^2}{R}\sin^2 \omega t \qquad (\mathrm{W}), \tag{2.35}$$

and thus, by taking an average of a single period,

$$W_{R_{Av}} = \overline{W_X(t)} = \frac{F_X^2}{2R} = W_{X_{Av}}. \qquad (\mathrm{W}) \tag{2.36}$$

That is, an energy loss, which denotes the conversion from dynamical to thermal energy, is equal to the work done by an external force and consequently can be compensated by an external source at the resonance frequency.

It seems somewhat puzzling that the energy loss by friction heat decreases in inverse proportion to the friction constant. This is a consequence of the balance between friction and external force at the resonance frequency. In fact velocity of motion increases as the friction constant decreases under the balance of force. An increase in the energy loss despite a small friction constant is an effect of this increase of velocity.

The energy compensation based on a balance between imported and consuming power actually holds well independent of the frequency of an external source. Power that can be supplied by an external source is always equal to the loss of energy. This is called the **energy balance principle of a dynamical system**. If a loss of energy is very small and thus resonance is very noticeable, then oscillation excited under an out-of-resonance frequency is very weak. This is because only a little energy may be supplied to the system of interest from an external source. In contrast when the loss becomes significant, the resonance is not noticeable and consequently out-of-resonance oscillation can be excited instead.

2.4.4 *Phase and Power Response to Sinusoidal Force*

Phase response is dependent on frequency as well as the magnitude response, as shown in Fig. 2.3(b). The phase response to the external force can be intuitively interpreted in two frequency regions: lower or higher frequencies than the resonance frequency. Motion of oscillation can be approximately expressed for lower frequencies such that

$$Kx(t) \cong f_X(t) \qquad \text{(N)} \tag{2.37}$$

where the resonance frequency is sufficiently high, implying that K is large but M is small. Here displacement of the mass whose magnitude is normally very small because of large K is in-phase with the external force.

On the other hand, motion of oscillation can be approximated for higher frequencies such that

$$M\frac{d^2x(t)}{dt^2} \cong f_X(t), \qquad \text{(N)} \tag{2.38}$$

where the resonance frequency is low assuming that K is small but M is large. Acceleration of the mass instead of displacement, which might be very small because of large M, is in-phase with the external force. Thus, displacement is anti-phase with a sinusoidal force.

Consequently, the phase of displacement from a sinusoidal force varies from in-phase to anti-phase, as the frequency of the external source increases. The phase lag is only $\pi/2$ at the resonance frequency as shown in Fig. 2.3(b). Velocity of the mass is in-phase with external force, so power might be imported from the external source at the resonance frequency.

The power, which corresponds to the work done by external force in a unit time interval, becomes

$$W_X(t) = f_X(t)\frac{dx(t)}{dt} \cong \frac{F_X^2\omega}{2K}\sin 2\omega t \qquad \text{(W)} \tag{2.39}$$

in a lower frequency than the resonance frequency. The expression above shows that the sign of power is periodically alternate. This outcome, alternative change in the sign of power, indicates that power goes back and forth between the external source and the oscillator of interest; this means the external source supplies energy to the oscillator in a quarter cycle, and it receives energy from the oscillator in the next quarter cycle. Therefore, by taking an average over a single cycle,

$$W_{X_{Av}} = \overline{W_X(t)} = 0 \tag{2.40}$$

can be obtained showing that no power is continuously supplied to the oscillator.

Similarly, the work done by external force in a unit time interval becomes

$$W_X(t) \cong \frac{F_X^2}{2M\omega}\sin 2\omega t \qquad \text{(W)} \tag{2.41}$$

in a higher frequency than the resonance frequency. Again the alternative sign of power indicates that power goes back and forth between an external source and the oscillator of interest. By taking an average over a single period,

$$W_{X_{Av}} = \overline{W_X(t)} = 0 \qquad (2.42)$$

indicates that no power is continuously supplied at higher frequencies or at lower ones.

Recall generalized expressions of the magnitude and phase responses given by Eqs. 2.29 and 2.30. The average of work done is expressed as[4][7]

$$W_{X_{Av}} = \frac{\omega F_X^2}{2MD(\omega)} \sin\phi(\omega) = \frac{F_X^2}{2R} \sin^2\phi(\omega) \qquad \text{(W)} \qquad (2.43)$$

$$D(\omega) = \sqrt{(\omega_0^2 - \omega^2)^2 + 4\delta_0^2\omega^2} = \frac{2\delta_0\omega}{\sin\phi(\omega)}. \qquad (2.44)$$

The work done, corresponding to the consuming power due to friction heat, depends on the phase difference between oscillation and the external force. When the phase difference is $\pi/2$, corresponding to the resonance, the power is maximum. As a friction constant increases, however, a small amount of oscillation is possible in a wider range of out-of-resonance frequencies.

2.5 Coupled Oscillation

An external source is needed to excite continuous oscillation as described in the previous section. However, the energy of dynamical oscillation can be transferred between dynamical systems as implied by the alternative sign of power. A vibrating source in a surrounding medium is another example that demonstrates such energy interaction. This section describes energy interaction in oscillation of the coupled pendulum shown in Fig. 2.4. Here two pendulums are connected so that a

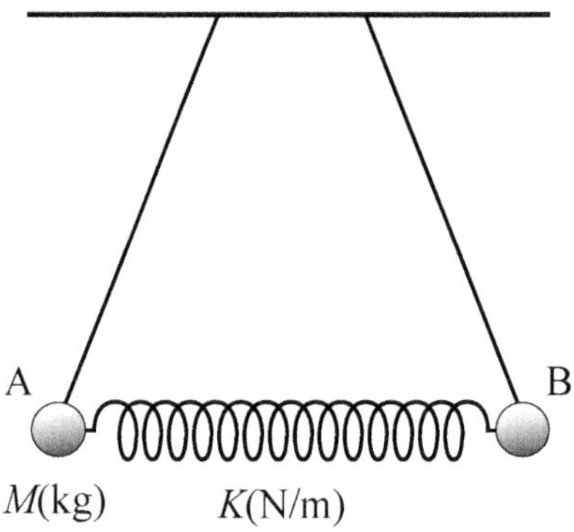

Fig. 2.4 Coupled pendulum

coupled oscillator is created[8]. The pendulums alternate the roles of energy sender and receiver as described below. Consequently, it is impossible to define which is the source. However, the system has two eigenfrequencies, which are different from the original eigenfrequency of each pendulum, and thus oscillation of those eigenfrequencies takes typical patterns of motion called **eigenmotion** corresponding to eigenfunctions.

2.5.1 Oscillation of Pendulum

Figure 2.5 shows a single pendulum. The restoring force of a pendulum is a cause of oscillation as is a simple oscillator composed of a mass and a spring. Here the restoring force (N) is due to tension of the thread of the pendulum.

As shown in the figure, a mass of M (kg) is attached to the pendulum, and the length of the thread is L (m). The other point of the thread marked by O is fixed. Suppose that the pendulum oscillates with small magnitude around the central point A in the vertical plane including the pendulum. By taking a small displacement $x(t)$ of the pendulum as illustrated in the figure, the restoring force that returns the mass to the equilibrium position accelerates the mass such that

$$M\frac{d^2x(t)}{dt^2} = -F_T \sin\theta = -F_T\frac{x(t)}{L} \cong -Mg\frac{x(t)}{L}. \quad \text{(N)} \qquad (2.45)$$

Here, F_T (N) denotes the tension of the thread, g(m/s^2) means acceleration due to gravity, and

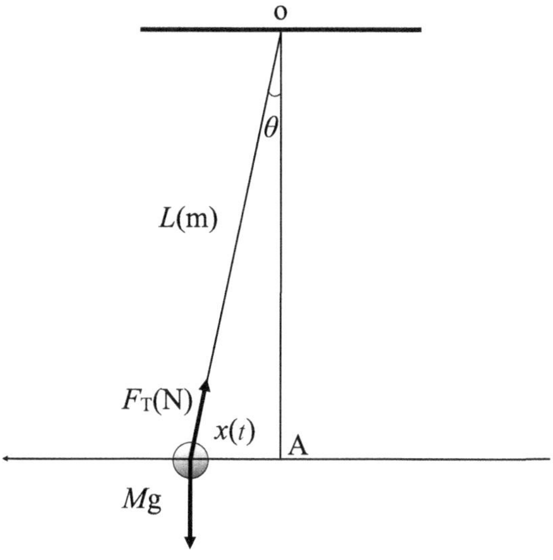

Fig. 2.5 Oscillation of pendulum

$$Mg = F_T \cos\theta \cong F_T. \qquad \text{(N)} \tag{2.46}$$

Consequently, the **eigen(angular)frequency of free oscillation of a pendulum** is given by

$$\omega_0 = \sqrt{\frac{g}{L}} \qquad \text{(rad/s)} \tag{2.47}$$

as well as that for a simple oscillator. The eigenfrequency of a pendulum does not depend on the mass attached to the thread but the length of the thread. A period is long (short) for a long (short) thread. Consequently, the length of the thread of a pendulum can be estimated by the period of free oscillation.

2.5.2 Eigenfrequencies of Coupled Pendulum

Recall the coupled pendulum as shown in Fig. 2.4. Each pendulum has the same eigenfrequency; however, it should be noted here that a coupled pendulum has a pair of eigenfrequencies different from the eigenfrequency for each individual pendulum. This is because the restoring force working on each mass is the sum of the force due to tension of the pendulum and stress of the spring. A pair of equations of motion is obtained for the two masses respectively:

$$M\frac{d^2x_A(t)}{dt^2} = -\frac{Mg}{L}x_A(t) - K(x_A(t) - x_B(t)) \qquad \text{(N)} \tag{2.48}$$

$$M\frac{d^2x_B(t)}{dt^2} = -\frac{Mg}{L}x_B(t) - K(x_B(t) - x_A(t)) \tag{2.49}$$

where $x_A(t)$ and $x_B(t)$ denote displacement of each mass representing motion respectively.

By adding or subtracting the two equations above, the same type of expression as that for a simple oscillator is obtained for $x_A(t) + x_B(t)$ or $x_A(t) - x_B(t)$, such as

$$\frac{d^2}{dt^2}(x_A(t) + x_B(t)) + \omega_0^2(x_A(t) + x_B(t)) = 0 \tag{2.50}$$

$$\frac{d^2}{dt^2}(x_A(t) - x_B(t)) + (\omega_0^2 + 2\omega_c^2)(x_A(t) - x_B(t)) = 0. \tag{2.51}$$

Here, $\omega_0^2 = g/L$ and $\omega_c^2 = K/M$ correspond to the eigenfrequencies of the pendulum and the spring with a single mass. The first equation states that $x_A(t) + x_B(t)$ follows free oscillation with the eigenfrequency of $\omega_0 = \omega_1$, while the second one indicates free oscillation with the eigenfrequency of $\sqrt{\omega_0^2 + 2\omega_c^2} = \omega_2$.

A pair of expressions for the coupled pendulum is obtained as

$$x_A(t) = A_1 \cos\omega_1 t + A_2 \cos\omega_2 t \tag{2.52}$$

$$x_B(t) = A_1 \cos\omega_1 t - A_2 \cos\omega_2 t, \tag{2.53}$$

where coefficients A_1 and A_2 are determined following initial conditions. The expressions above indicate that free oscillation of a coupled pendulum is composed of two sinusoidal oscillations, which are called **eigenmotion** of vibration, corresponding to the eigenfrequencies. It should be noted that the first eigenmotions of the masses with the first eigenfrequency of ω_1 are in-phase with each other, while the second eigenmotions of the second eigenfrequency ω_2 are in anti-phase to each other. Eigenmotion of vibration is also called **modal vibration**.

2.5.3 *Effects of Coupling on Oscillation*

Suppose that an initial condition for starting the coupled pendulum can be written as

$$x_A(0) = a_A, \qquad x_B(0) = 0 \qquad \text{(m)} \tag{2.54}$$

under the condition that mass A is stretched to the amount a_A and is quietly released at $t = 0$. According to the initial condition above,

$$x_A(t) = a_A \cos\left(\frac{\omega_2 - \omega_1}{2}t\right)\cos\left(\frac{\omega_2 + \omega_1}{2}t\right) \tag{2.55}$$

$$x_B(t) = a_A \sin\left(\frac{\omega_2 - \omega_1}{2}t\right)\sin\left(\frac{\omega_2 + \omega_1}{2}t\right) \tag{2.56}$$

are derived. As shown above, free oscillation of a coupled pendulum is written as a product of two sinusoidal components. One has a frequency of an average of two eigenfrequencies, while the other has a frequency of one half the difference of the two eigenfrequencies.

Suppose that the effects of a coupling spring are weak. That is, $\omega_0 >> \omega_c$ and thus $\omega_1 \cong \omega_2$. Oscillation can be expressed as

$$x_A(t) \cong a_A \cos\left(\frac{\Delta\omega}{2}t\right)\cos\omega_0 t \tag{2.57}$$

$$x_B(t) \cong a_A \sin\left(\frac{\Delta\omega}{2}t\right)\sin\omega_0 t \tag{2.58}$$

$$\Delta\omega = \omega_2 - \omega_1. \tag{2.59}$$

Figure 2.6 shows an example of weak coupling.

Oscillation of A has a frequency that is almost the same as the eigenfrequency of a pendulum alone, but the effects of coupling appear in slow variation of the magnitude. That is, the oscillation energy goes back and forth between the two pendulums A and B. Actually, it can be seen that the phases of motion of the two pendulums are out-of-phase with each other, and thus it is impossible to identify which is an external source for oscillation, even if pendulum A was given initial displacement as shown in the initial condition.

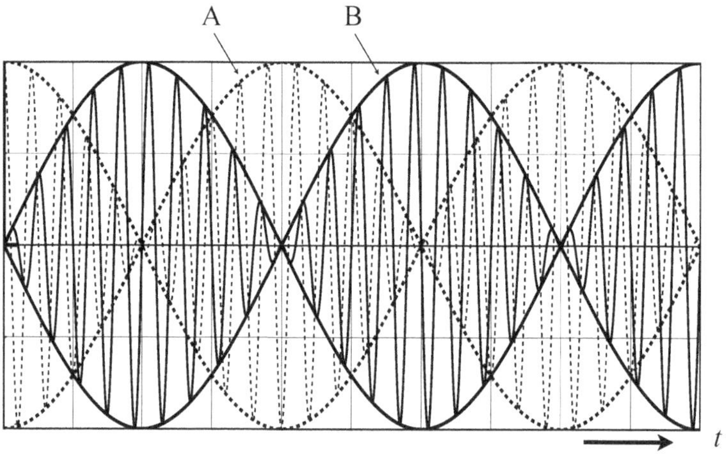

Fig. 2.6 Oscillation of coupled pendulum under weak coupling from [9] Fig. 2.9

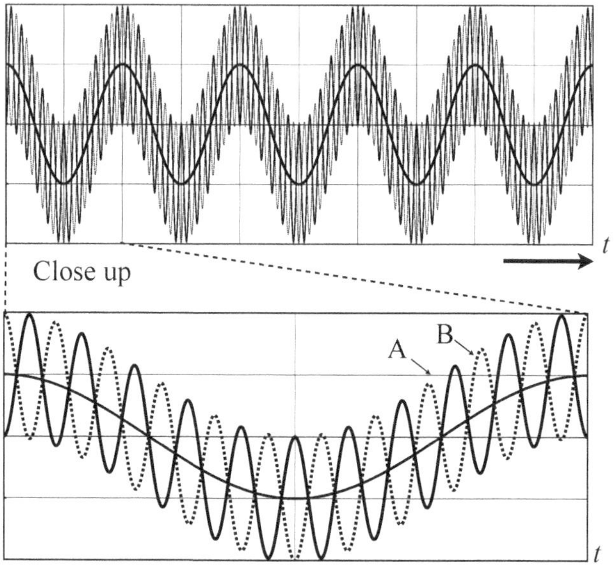

Fig. 2.7 Oscillation of coupled pendulum under tight coupling from [9] Fig.2.10

The frequency of the magnitude variation is given by half of the difference between the two eigenfrequencies. Speed of magnitude variation becomes slow as the difference becoems small. As coupling becomes very weak, pendulum A oscillates almost independently with its eigenfrequency without relation to pendulum B, while pendulum B almost stops.

Figure 2.7 shows another example under tight coupling. When coupling becomes tight, the eigenfrequencies are regarded as $\omega_0 << \omega_c$. As shown in the figure, speed of magnitude variation, which corresponds to energy transfer rate, becomes fast, which is very different from weak coupling. The frequency of oscillation is almost ω_2, and the phases of the motion of two pendulums are in anti-phase. This means pendulum B starts its anti-phase motion against pendulum A, even if pendulum B was stopped at the initial state.

2.5.4 *Beats*

Set the two eigenfrequencies as

$$\omega_1 = \omega_c - \Delta\omega/2, \qquad \omega_2 = \omega_c + \Delta\omega/2. \tag{2.60}$$

The coupled motion discussed in the previous section can be rewritten as

$$\begin{aligned} x_A(t) &= a_A \cos\left(\frac{\Delta\omega}{2}t\right)\cos\omega_c t \\ &= \frac{a_A}{2}\left(\cos(\omega_c - \Delta\omega/2)t + \cos(\omega_c + \Delta\omega/2)t\right) \end{aligned} \tag{2.61}$$

$$\begin{aligned} x_B(t) &= a_A \sin\left(\frac{\Delta\omega}{2}t\right)\sin\omega_c t \\ &= \frac{a_A}{2}\left(\cos(\omega_c - \Delta\omega/2)t - \cos(\omega_c + \Delta\omega/2)t\right). \end{aligned} \tag{2.62}$$

The expression above indicate that a compound signal composed of a pair of sinusoidal waves whose frequencies are very close must have slowly varied magnitude. If a listener perceives sound with slowly varied magnitude, that sound is usually called **beats**. It should be noted that a period of beats is just only the difference of frequencies between two sinusoidal components. Beats provide quite useful information in tuning sound instruments such as a piano because the difference from in the fundamental frequencies of simultaneously vibrating strings under test can be tuned by counting beats of the sound.

2.5.5 *Modulation*

Temporal variation of magnitude is also called **amplitude modulation**. In general, amplitude modulation is expressed using three sinusoidal components such as

$$x(t) = \sin \omega_{co} t + \frac{m_o}{2}(\sin(\omega_c - \Delta\omega/2)t + \sin(\omega_c + \Delta\omega/2)t)$$
$$= (1 + \frac{m_o}{2}\cos\frac{\Delta\omega}{2}t)\sin\omega_c t. \tag{2.63}$$

Here m_o is called a **modulation index**. It can be intuitively understood that by looking at Fig. 2.8. However, it would be interesting to see the effects of the phase relationship between sinusoidal components on the modulation[10].

Suppose that a signal is composed of three sinusoidal waves such that

$$x(t) = \sin \omega_{co} t + \frac{m_o}{2}(\cos(\omega_{co} - \Delta\omega/2)t + \cos(\omega_{co} + \Delta\omega/2)t). \tag{2.64}$$

Here the side-band components are of the same magnitude but are out-of-phase with the central component. By rewriting the equation above so that

$$x(t) = A\sin(\omega_{co} t + \phi(t)) \tag{2.65}$$
$$A = \sqrt{1 + \frac{m_o^2}{4}\cos^2\frac{\Delta\omega}{2}t} \tag{2.66}$$
$$\phi(t) = \tan^{-1}\left(\frac{m_o}{2}\cos\frac{\Delta\omega}{2}t\right), \tag{2.67}$$

it can be seen that the phase relationship surprisingly changes a signal waveform from the amplitude-modulated one as shown in Fig. 2.9. In fact, the amplitude seems essentially flat as the modulation index increases. Instead, the frequency of the central sinusoidal component seems modulated. Thus, this type of signal property is called **quasi frequency modulation (QFM)**[10].

Normally, angle $\Phi(t)$ of a generalized sinusoidal function such as $x(t) = \sin\Phi(t)$ is called **instantaneous phase**. Phase change during a small interval of time is denoted as **instantaneous angular frequency**, which can be defined by

$$\Omega(t) = \frac{d\Phi(t)}{dt} \tag{2.68}$$

using a mathematical formulation. The instantaneous angular frequency for the signal shown in Fig. 2.9 can be derived as

$$\Omega(t) = \frac{d\Phi(t)}{dt} = \omega_{co} + \frac{d\phi(t)}{dt} \tag{2.69}$$
$$\frac{d\phi(t)}{dt} = -\frac{1}{1 + \frac{m_o^2}{4}\cos^2\frac{\Delta\omega}{2}t}\frac{m_o\Delta\omega}{4}\sin\frac{\Delta\omega}{2}t \cong -\frac{m_o\Delta\omega}{4}\sin\frac{\Delta\omega}{2}t \tag{2.70}$$

for a QFM case. The frequency is sinusoidally modulated by the frequency of the difference between side-band components instead of amplitude in Fig. 2.10. This is a typical example of the effects of phase on a signal waveform.

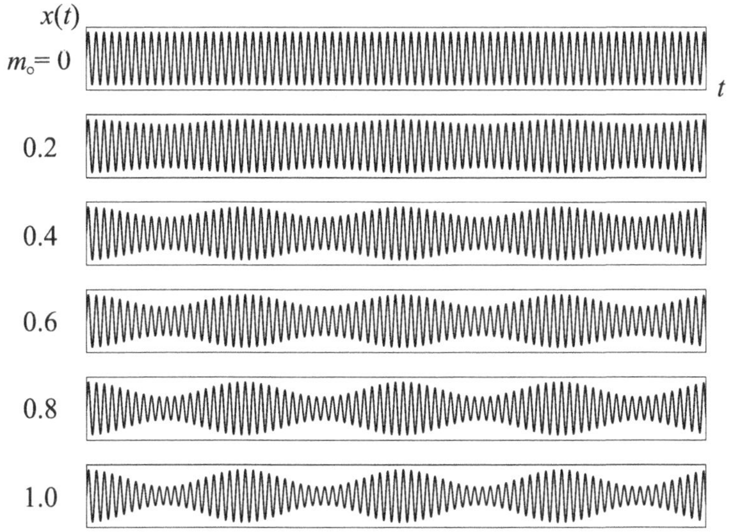

Fig. 2.8 Amplitude modulation and its modulation index

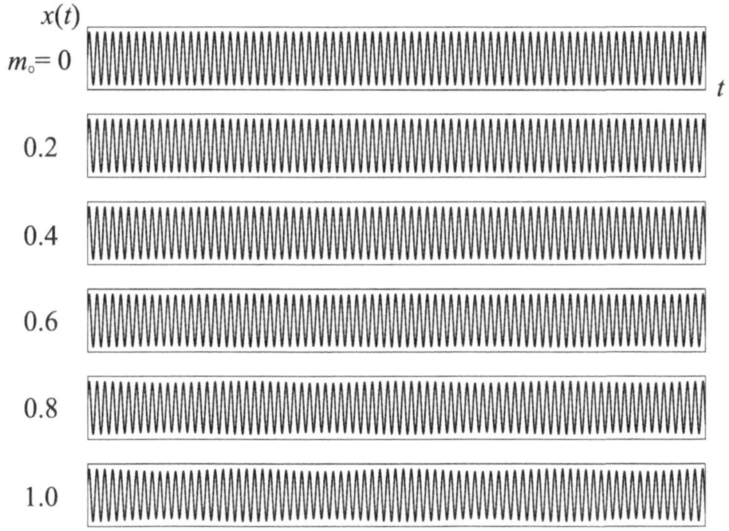

Fig. 2.9 Quasi frequency modulation and its modulation index

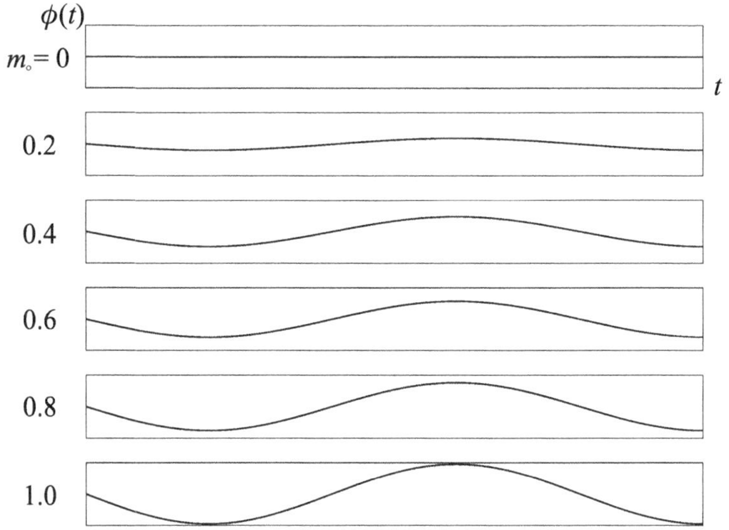

Fig. 2.10 Instantaneous angular frequency of QFM and its modulation index

Modulation in general as well as beats is a very important and significant issue for perceiving and understanding sound or speech signals[11][12][13]. Even if a signal is simply described from a physical view point, such as the sum of three sinusoidal components with different phases stated above, it often seems greatly different from a perceptual view point. Sometimes such a difference might be understood by mathematical analysis of the signal waveform, as if our hearing organ performed mathematical analysis of signals.

2.5.6 Vibration Transmission between a Coupled Oscillator

Vibration of a coupled spring, which is called a **2-degree-of-freedom (2DOF) system** in general as shown in Fig. 2.11, is a fundamental model of vibration transmission control such as machinery noise reduction. As described by oscillation of a

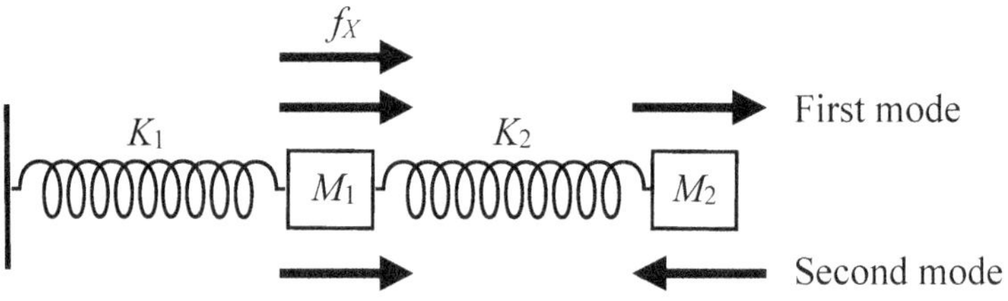

Fig. 2.11 Schematic of two-degree-of-freedom vibration system

coupled pendulum in section 2.5.2, the first modal vibration is in-phase, while the second one is in anti-phase between the two mass components.

Suppose that the external force is applied to mass component 1 as shown in Fig. 2.11. The motion of the two mass components follows a pair of equations in a complex function form such that[8]

$$M_1 \frac{d^2x_1(t)}{dt^2} = -K_1x_1 - K_2(x_1 - x_2) + F_X e^{i\omega t} \quad \text{(N)} \tag{2.71}$$

$$M_2 \frac{d^2x_2(t)}{dt^2} = -K_2(x_2 - x_1). \tag{2.72}$$

The transmission ratio of vibration magnitude might be quite informative for intuitively understanding the dynamical properties of a 2DOF system stated above[14]. Set

$$x_1(t) = U_1(\omega)e^{i\omega t} \tag{2.73}$$

$$x_2(t) = U_2(\omega)e^{i\omega t}, \tag{2.74}$$

then a quotient of the two magnitude responses,

$$T_{12} = U_2(\omega)/U_1(\omega) = \frac{K_2}{K_2 - \omega^2 M_2} = \frac{\omega_2^2}{\omega_2^2 - \omega^2} \tag{2.75}$$

$$\omega_2^2 = \frac{K_2}{M_2}, \tag{2.76}$$

is obtained.

Ratio T_{12} indicates that resonant vibration can be almost perfectly transmitted into the second component, in spite of the first component being excited by external force. In other words, mass component 1 remains stationary without motion at the resonance frequency for the second vibrator. In contrast, mass component 2 is nearly at rest at any out-of-resonance frequency, indicating that vibration of the first component does not propagate to the second one.

The sign of T_{12} changes at the resonance frequency of the second component. This can be interpreted in terms of change of the sign of motion. Motion of the first component decreases as the frequency of vibration approaches the resonance frequency, and comes to a stop at the resonance frequency. Such a state or frequency when vibration stops is called a **zero** of the vibrator of interest. Thus, when the frequency is over the resonance of the second component, the sign of motion of the first component changes because the frequency increases across the zero. Consequently, the vibration patterns of two components are in anti-phase for a frequency over the resonance of the second component.

Ratio T_{12} could be a good indicator for vibration control. Suppose that a vibrating machine whose mass is M_1 (kg) is located on a floor of mass M_2 (kg). If a heavy machine is directly located on a floor, vibration from the machine is propagated to the floor. However, if a spring is put between the machine and the floor, it is possible to reduce vibration transmission from the machine to the floor. Assuming that a floor

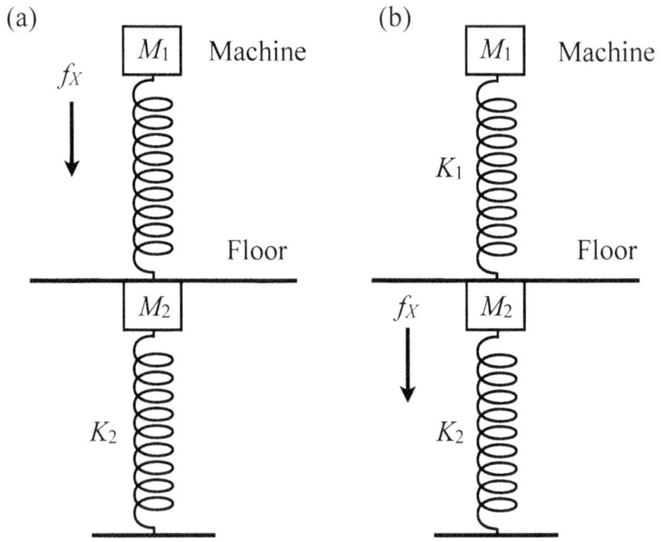

Fig. 2.12 Machine located on floor that is represented by simple vibrator

on which a machine is located is also represented by a simple vibrator as shown in Fig. 2.12 (a), then T_{12} should be small for vibration reduction.

If the position on which external force works is changed as shown in Fig. 2.11(b), then ratio T_{21} given by

$$T_{21} = U_1(\omega)/U_2(\omega) = \frac{K_1}{K_1 - \omega^2 M_1} = \frac{\omega_1^2}{\omega_1^2 - \omega^2} \tag{2.77}$$

$$\omega_1^2 = \frac{K_1}{M_1} \tag{2.78}$$

could be an indicator for the robustness of a machine to surrounding vibrations. A machine can be proofed against external vibration unless the frequency of an external noise is not close to the resonance of the machine with a spring, as shown in Fig. 2.12(b). Again T_{21} must be small for a machine to be vibration proof.

Chapter 3
Simple Resonators for Sound

Sound that a listener perceives is generally a wave propagating in the air. A sound wave is a pattern of vibration transmissioin in an elastic medium. However, vibration of a local portion in an elastic medium such as air is invisible, different from vibration of a spring. Therefore, intuitively understanding a process for wave propagation is difficult by listening to sound. This chapter describes elastic properties of a medium in which a sound travels as a wave. Dilation, condensation, and bulk modulus are the characteristics that represent the elastic properties of a gas such as air. Bulk modulus of a gas, which corresponds to a spring constant of a simple oscillator, is important for representing a sound wave in air. A Helmholtz resonator, composed of air, is introduced; its response is similar to a harmonic oscillator composed of a spring with a mass. An enclosure of a loudspeaker unit is an example of a Helmholtz resonator, although listeners usually may not be aware of this.

3.1 Elastic Properties of Gas

As described in the previous chapter, elastic properties of a spring are represented by its spring constant. Hooke's law states restoring force of a spring, which causes oscillation, is in proportion to the stretch of the spring. A gas such as air is also an elastic medium.

3.1.1 Dilation and Condensation of Gas

Macroscopic properties of a gas can be specified by **volume** $V_0(\mathrm{m}^3)$, **density** $\rho_0(\mathrm{kg/m}^3)$, **mass** $M = \rho_0 V_0(\mathrm{kg})$, **pressure** $P(\mathrm{Pa})$, and **temperature** $T_{emp}(\mathrm{K})$. Elastic properties of a gas are described using these macroscopic parameters. Let the volume of a gas be dilated to V from the initial state of V_0. **Dilation** ε is defined by a quotient as

$$\varepsilon = \frac{V - V_0}{V_0}, \tag{3.1}$$

M. Tohyama: *Sound and Signals*, Signals and Communication Technology, pp. 31–41.
springerlink.com

which is normalized by the initial volume. Similarly **condensation** of a gas is defined as

$$s = \frac{\rho - \rho_0}{\rho_0} \tag{3.2}$$

when the volume-density is changed to ρ from the initial state of ρ_0. If the differences in the volume or density are small, then it holds that

$$s \cong -\varepsilon. \tag{3.3}$$

A similar formulation to Hooke's law for a spring also holds well for the elastic properties of a gas. Pressure, such as the stress of a stretched spring, increases in proportion to the condensation such that

$$P = \kappa s. \qquad \text{(Pa)} \tag{3.4}$$

Here κ(Pa), which corresponds to a spring constant, is called the **bulk modulus** of a gas. Note that the unit of bulk modulus is different from the spring constant. When condensation of a local area in a medium travels as a wave, the pressure caused by the condensation is called **sound pressure**.

The bulk modulus mentioned above is not a simple parameter, however. It depends on the thermal conditions of a gas during the dilation or condensation process. The relationship between the pressure and volume of a gas is considered in the next section[15]. It is closely related to sound speed in the air, which has been discussed for over 100 years since Newton first tried to theoretically estimate sound speed.

3.1.2 State Equation of Gas

The volume of a gas is inversely proportional to its pressure, as stated by **Boyle's law**, such that

$$PV = \text{constant}. \qquad \text{(J)} \tag{3.5}$$

The above equation implies that the product of the pressure and volume of a gas is related to kinetic energy of the gas molecules. Thus Boyle's law can be interpreted as a kind of energy preservation law for gases. Note however that Boyle's law holds well only under the constant temperature of a gas, which is called an **isothermal process**.

The state of an **ideal gas** can be described using three variables, pressure P(Pa), volume $V(\mathrm{m}^3)$, and temperature T_{emp}(K), such that[15]

$$PV = R_{gas} T_{emp}, \qquad \text{(J)} \tag{3.6}$$

where $R_{gas} = 8.314(\mathrm{J}/(\mathrm{mol} \cdot \mathrm{K}))$ is called the **gas constant**, and the temperature is defined as 273 K at 0 C and 0 K at -273 C. A unit **mol** of a gas occupies a volume of 22.414 liters $(10^{-3}\mathrm{m}^3)$ at 273 K under an atomic pressure of $1(1.0133 \times 10^5$ Pa).

Thus, a gas dilates by 1/273 of the volume for 0 C, when the temperature increases by 1 C.

The equation above is called **Boyle and Charles' law**. The product of the pressure and volume of a gas is no longer constant under Boyle's law when the temperature changes. This suggests that the temperature of a gas might not be constant when the pressure or volume varies. This is closely related to the state of a gas in which a sound wave travels because such a variation of the pressure or volume of a local portion of a medium results in sound propagation as a wave.

The pressure of a gas is a result of random motion of the molecules of which the gas is composed. Consider a gas in a bottle. The effects of motion of the molecules on the pressure can be visualized as bouncing billiard balls (Fig. 3.1). The molecules, under irregular motion, randomly hits the inner surface of the bottle. The velocity of a molecule, which can be expressed as a vector **v**, is decomposed into

$$\mathbf{v} = \mathbf{v}_x + \mathbf{v}_y + \mathbf{v}_z \qquad (\mathrm{m/s}) \tag{3.7}$$

in three-dimensional space. Here by setting the perpendicular component to $\mathbf{v}_z$, only the perpendicular component changes to $-\mathbf{v}_z$ at every collision with the inner surface in Fig. 3.1. Therefore, $2m_e|\mathbf{v}_z|(\mathrm{N \cdot s})$ is conveyed to the inner surface where m_e is the mass of the molecule.

Assume that the number of molecules with velocity of $|\mathbf{v}|$ is $n_{|\mathbf{v}|}$ in a unit volume of the gas. The number of collisions with a unit area of the inner surface is then given by $n_{|\mathbf{v}|}|\mathbf{v}_z|$ every unit time interval, as shown in Fig. 3.2. Thus if by summing $2m_e n_{|\mathbf{v}|}|\mathbf{v}_z|^2$ over all the incident directions on the hemisphere S of the radius $|\mathbf{v}|$, and again integrating the result over all the speeds of the molecules,

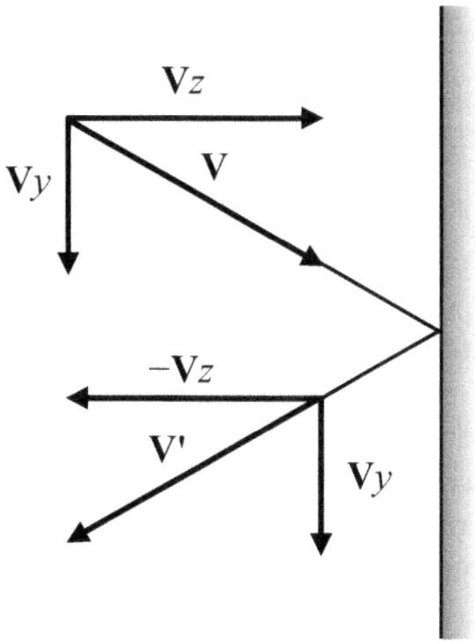

Fig. 3.1 Collision of molecule with inner surface of bottle from Fig. 1.2[15]

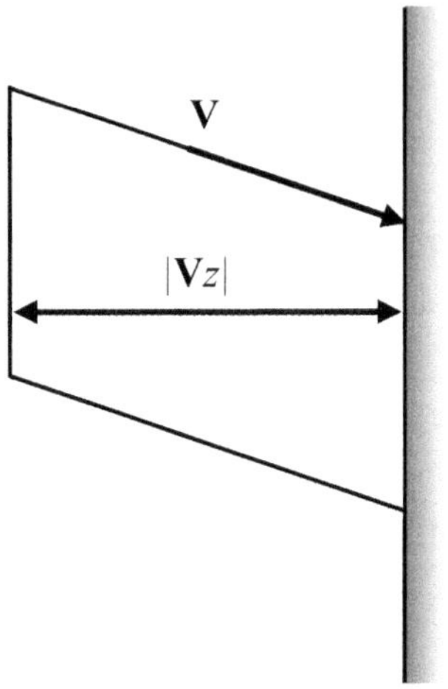

Fig. 3.2 Volume occupied by molecules with velocity **v** that collides with unit area in unit time interval from Fig. 1.1 [15]

$$PV = \frac{M}{3}\overline{|\mathbf{v}|^2} = \frac{2}{3}E_K = R_{gas}T_{emp} \qquad \text{(J)} \tag{3.8}$$

can be derived[15], according to

$$2m_e \int_{|\mathbf{v}|} n_{|\mathbf{v}|}|\mathbf{v}_z|^2 dS = 2m_e \cdot \frac{1}{2} n_{|\mathbf{v}|}\overline{|\mathbf{v}_z|^2} = m_e n_{|\mathbf{v}|}\frac{\overline{|\mathbf{v}|^2}}{3} \tag{3.9}$$

$$\frac{m_e}{3}\int_{\text{unit-volume}} n_{|\mathbf{v}|}|\mathbf{v}|^2 dv = \frac{m_e}{3} n_e \overline{|\mathbf{v}|^2} = P. \tag{3.10}$$

Here, E_K denotes the average of the kinetic energy of molecules,

$$E_K = \frac{3}{2}R_{gas}T_{emp} = \frac{1}{2}M\overline{|\mathbf{v}|^2}, \qquad \text{(J)} \tag{3.11}$$

$$M = m_e N_e \qquad \text{(kg)} \tag{3.12}$$

$$\overline{|\mathbf{v}_x|^2} = \overline{|\mathbf{v}_y|^2} = \overline{|\mathbf{v}_z|^2} = \frac{\overline{|\mathbf{v}|^2}}{3} \qquad (\text{m/s})^2, \tag{3.13}$$

n_e is the number of molecules in a unit volume of the gas, N_e denotes the number of molecules in the gas, and $\overline{(*)}$ indicates that a long-term or ensemble average is taken. Equality of the average velocity in the three dimensions shows a symmetric distribution of random motion of the molecules.

It is customary to take a long-term (or an ensemble) average of the squared quantity for representing a random variable. Therefore by taking the square root of the average, the moving velocity can be expressed as

$$\sqrt{\overline{|\mathbf{v}|^2}} = \sqrt{\frac{3R_{gas}T_{emp}}{M}}. \qquad (\text{m/s}) \tag{3.14}$$

The square root of the average of a squared quantity is called a **root mean square (RMS)**. An RMS corresponds to the standard deviation in terms of statistics. As shown above, the speed of the molecules of a gas is proportional to the temperature. If the temperature is 0 K, all the molecules stop moving. This explains the origin of the unit of **Kelvin** for measuring temperature.

3.1.3 Specific Heat of Gas

Specific heat of a gas means the energy needed to increase the temperature of the gas by 1 C. However, the specific heat depends on the heating process used on the gas, namely which of the volume or the pressure is constant in the process. More energy is needed to increase the temperature of a gas under constant pressure than a constant volume condition. This is because a heated gas dilates under the constant pressure condition, so the energy of the gas is possibly spent on dilation of the volume instead of increase of the temperature. Consequently, the specific heat C_P under the constant pressure condition is generally greater than C_V under a constant volume.

Take a simple example of kinetic energy for a gas composed of single atoms. The specific heat C_V needed under constant volume without dilation or condensation of the volume is given by[15]

$$C_V = \frac{\Delta E_K}{\Delta T_{emp}} = \frac{3}{2}R_{gas}. \qquad (\text{J}) \tag{3.15}$$

In contrast, more energy is needed to increase the temperature of the gas under a constant pressure condition. Suppose a change of the volume to be ΔV, and an increase in temperature to be $\Delta T_{emp} = 1(\text{C})$. The state of the gas can then be expressed as

$$P\Delta V = R_{gas}\Delta T_{emp} = R_{gas}. \qquad (\text{J}) \tag{3.16}$$

Here, R_{gas} is a part of the specific heat, which shows the energy spent increasing the temperature and dilating the gas. Therefore, the total specific heat is given by

$$C_P = C_V + R_{gas} = \frac{5}{2}R_{gas}. \qquad (\text{J}) \tag{3.17}$$

Consequently, the **ratio between the two specific heats** for an ideal single atomic gas is given by

$$\gamma = \frac{C_P}{C_V} = \frac{5}{3} \cong 1.667. \tag{3.18}$$

However, this ratio generally depends on the gas.

3.1.4 Volume and Temperature of Gas under Adiabatic Process

Following the state equation of an ideal gas, a different relationship holds well under an adiabatic condition from Boyle's law for the relationship between the pressure and volume of a gas. Here the **adiabatic condition** means that there is no communication of thermal energy between the gas and surroundings, i.e., no energy is imported from the environment into the gas and vice versa. Therefore, the equation

$$C_P \Delta T_{emp} = C_V \Delta T_{emp} + R_{gas} \Delta T_{emp} = C_V \Delta T_{emp} + P \Delta V = 0 \qquad \text{(J)} \quad (3.19)$$

holds well. Consequently, variation of the temperature of the gas ΔT_{emp} is written as

$$\Delta T_{emp} = -\frac{R_{gas} T_{emp} \Delta V}{V C_V} \qquad \text{(K)} \quad (3.20)$$

where $PV = R_{gas} T_{emp}$. This equation indicates temperature decreases (increases) when the volume of a gas dilates (condenses) under an adiabatic condition.

By introducing the ratio of specific heat γ, the relation above can be rewritten as

$$\frac{\Delta T_{emp}}{T_{emp}} + (\gamma - 1)\frac{\Delta V}{V} = 0. \quad (3.21)$$

Consequently we have

$$PV^{\gamma} = constant \qquad \text{(J)} \quad (3.22)$$

is derived, recalling that $T_{emp} = PV/R_{gas}$ and

$$\frac{d \log T_{emp}}{d T_{emp}} \Delta T_{emp} + (\gamma - 1)\frac{d \log V}{dV} \Delta V = 0 \quad (3.23)$$

$$\log T_{emp} + (\gamma - 1) \log V = constant. \quad (3.24)$$

The relationship above is used for estimating the bulk modulus of a gas under an adiabatic condition.

3.1.5 Bulk Modulus of a Gas

Bulk modulus of a gas κ can be written as

$$\kappa = \frac{\Delta P}{s} \cong -V_0 \frac{\Delta P}{\Delta V}. \qquad \text{(Pa)} \quad (3.25)$$

This can be interpreted as a coefficient between the pressure and condensation of a gas and also depends on the process of condensation. Consider an isothermal condition following $PV = constant$, i.e., the temperature of the gas is constant. The pressure and volume satisfy the relationship such that

$$(P_0 + \Delta P)(V_0 + \Delta V) = P_0 V_0 \qquad \text{(J)} \tag{3.26}$$

under the isothermal condition, where P_0(Pa) and V_0(m^3) denote the pressure and volume in the initial stage, respectively. Then a relationship between the pressure and volume of a gas such that

$$\frac{\Delta P}{\Delta V} \cong -\frac{P_0}{V_0} \qquad (\text{Pa/m}^3) \tag{3.27}$$

holds well by assuming $\Delta P \Delta V << PV$ so that the second order term $\Delta P \Delta V$ might be neglected. Thus, the **bulk modulus under an isothermal condition** can be expressed as

$$\kappa \cong P_0. \qquad \text{(Pa)} \tag{3.28}$$

On the other hand, if an adiabatic case following $PV^\gamma = constant$ is assumed, it holds that

$$(P_0 + \Delta P)(V_0 + \Delta V)^\gamma = P_0 V_0^\gamma. \qquad \text{(J)} \tag{3.29}$$

By neglecting the second order quantity and introducing approximation such as

$$(1 + \frac{\Delta V}{V_0})^\gamma \cong 1 + \gamma \frac{\Delta V}{V_0}, \tag{3.30}$$

another relationship is derived between the pressure and volume such that

$$\frac{\Delta P}{\Delta V} = -\gamma \frac{P_0}{V_0} \qquad (\text{Pa/m}^3). \tag{3.31}$$

Therefore, the **bulk modulus under an adiabatic condition** can be estimated as

$$\kappa \cong \gamma P_0. \qquad \text{(Pa)} \tag{3.32}$$

This bulk modulus is suitable for estimating sound speed in the air.

3.2 Resonators

3.2.1 Helmholtz Resonators

Pressure of the air is in proportion to the condensation. The elastic properties of the air that can be represented by the bulk modulus also cause resonant vibration such as in a harmonic oscillator. Figure 3.3 is an example of a Helmholtz resonator. The mass of the air in the neck vibrates like a piston of mass M(kg), and the large volume of the air in the cavity V_0(m^3) works like a spring in a harmonic oscillator. When a piston-like motion of the mass is excited in the neck by sound that comes to the neck by traveling in the air surrounding the resonator, the pressure of the air

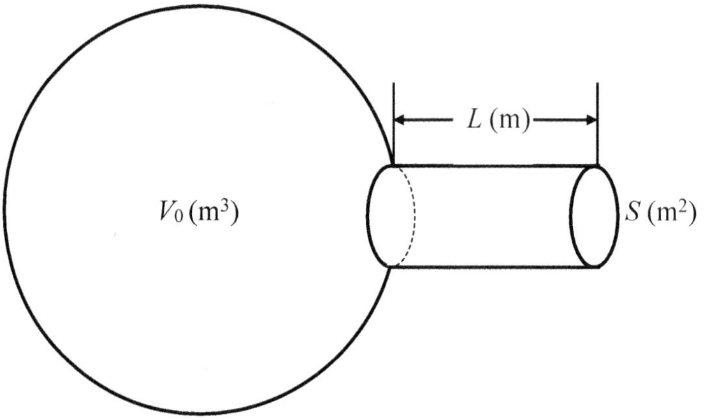

Fig. 3.3 Example of Helmholtz resonator

uniformly changes in the cavity. Uniform distribution of pressure in the cavity is likely for reasonably slow oscillation of the air.

Suppose that a length and a cross section of the neck are $L(\text{m})$ and $S(\text{m}^2)$, respectively. If the mass of the air in the neck moves by $x(\text{m})$, then the pressure in the cavity is uniformly varied because of condensation or dilation of the air. This corresponds to a simple harmonic resonator composed of a spring and a mass. Therefore, the **eigenfrequency of a Helmholtz resonator** can be given by

$$\omega_0 = \sqrt{\frac{\kappa S}{\rho_0 L V_0}} = \sqrt{\frac{\kappa S^2/V_0}{M}} = \sqrt{\frac{K}{M}} = 2\pi\nu_0, \qquad (\text{rad/s}) \qquad (3.33)$$

where $M = \rho_0 SL(\text{kg})$ denotes the mass of air in the neck with $\rho_0(\text{kg/m}^3)$ as air density, and $K = \kappa S^2/V_0$ (N/m) corresponds to a spring constant of a harmonic oscillator. Thus the restoring force of air in the cavity decreases (increases) as the volume of the cavity increases (decreases), and consequently the eigenfrequency becomes lower (higher).

3.2.2 Enclosure of Loudspeaker

Sound from a loudspeaker is a sound wave that is radiated from vibration of the diaphragm of the loudspeaker unit and travels to a listener in the air. A loudspeaker unit is generally put into an enclosure as shown in Fig. 3.4. Such an enclosure can also be considered a Helmholtz resonator. The diaphragm of a loudspeaker unit works as a piston corresponding to the air in the neck for a Helmholtz resonator, and the enclosure works as the cavity. Consequently, sound of lower frequencies than the resonance of a loudspeaker unit itself can be somewhat strengthened by the resonance of the enclosure.

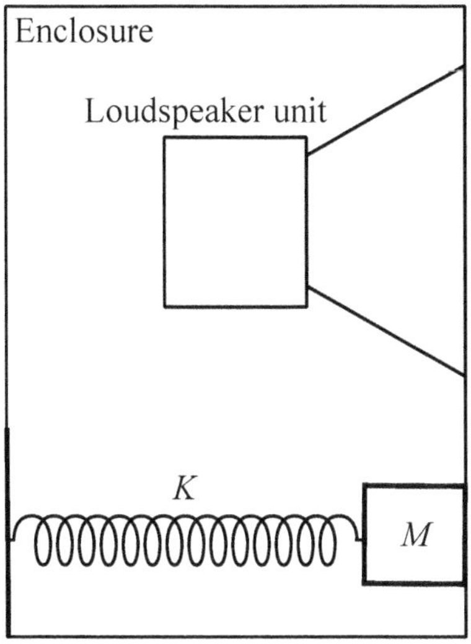

Fig. 3.4 Loudspeaker unit in enclosure

A loudspeaker system can be deconstructed as shown in Fig. 3.4 into a simple oscillator made of a spring and a mass. Suppose that external force works on the mass so that the mass moves $x(\mathrm{m})$, which is interpreted as displacement of the diaphragm in the loudspeaker unit. As the force increases and the diaphragm moves into the enclosure, the air inside the enclosure is condensed and consequently sound pressure increases. On the other hand, if the diaphragm moves outside the enclosure, the air dilates and sound pressure decreases inside the enclosure. Consequently, variation of sound pressure inside the enclosure and movement of the diaphragm are in opposite directions; in other words, the sound pressure is anti-phase to displacement of the diaphragm.

It is interesting to see increment of the sound pressure just in front of the diaphragm outside the enclosure. Sound pressure in front of the diaphragm, which is caused by diaphragm oscillation, is understood differently from the pressure inside the enclosure. This is because the volume is finite and space is closed inside the enclosure, while space in which a sound wave travels from the diaphragm is essentially open outside the enclosure. When there is a piston-like movement of a plate in such an open space, condensation of the air is not always in proportion to the motion displacement of the plate. This is due to **incompressibility** of the medium in which a sound wave travels. Inside a Helmholtz oscillator there are essentially no sound "waves" and thus the pressure is always uniform throughout the cavity.

Sound pressure in front of the loudspeaker unit is not in proportion to displacement of the diaphragm, different from the pressure inside the enclosure. Air particles that contact the surface of the diaphragm move with the same velocity as that of the diaphragm. From the Newtonian law of motion, force working on the vibrating

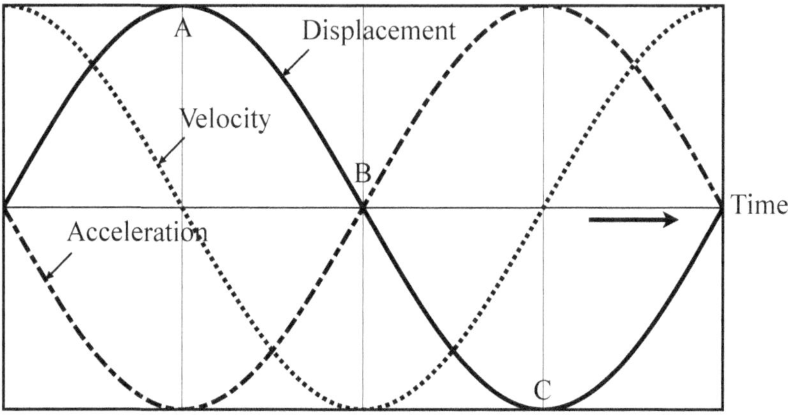

Fig. 3.5 Displacement, velocity, and acceleration of oscillation

diaphragm is in proportion to acceleration of the diaphragm. This indicates that sound pressure on the surface of the diaphragm results from its acceleration.

Figure 3.5 shows an example of displacement, velocity, and acceleration of oscillation. Particles close to the diaphragm vibrate with the same velocity as the diaphragm, as the diaphragm oscillates. Note that displacement and velocity are out-of-phase, and thus acceleration is in anti-phase to displacement, as seen in Fig.3.5. As the diaphragm moves outside the enclosure, air particles close to the diaphragm also move toward the outside. In such a movement of particles, acceleration decreases as shown in the figure, and consequently force working on the body of particles decreases. Such a decrease of force causes condensation to be lower. When the diaphragm vibration comes to point A in the figure, condensation as well as sound pressure become the minimum. This indicates that sound pressure in front of the loudspeaker unit becomes the minimum when displacement of the diaphragm reaches the maximum outside the enclosure.

In contrast, when the diaphragm moves into the enclosure, air particles also enter. In such a movement of particles, acceleration changes direction and increase, and consequently sound pressure changes to also increase. When the diaphragm vibration comes to point B in the figure, condensation returns to the normal (neutral) state. Furthermore, the diaphragm reaches the maximum displacement inside the enclosure, as shown by point C in the figure, and acceleration and increase of the sound pressure are also the maximum. Consequently, increase of sound pressure is the maximum in front of the loudspeaker unit, when the diaphragm maximally moves into the enclosure. Consideration of sound pressure and the motion of the diaphragm described above suggests that sound pressure is in anti-phase to displacement of the diaphragm. This means variations of the sound pressure are in-phase inside and outside the enclosure.

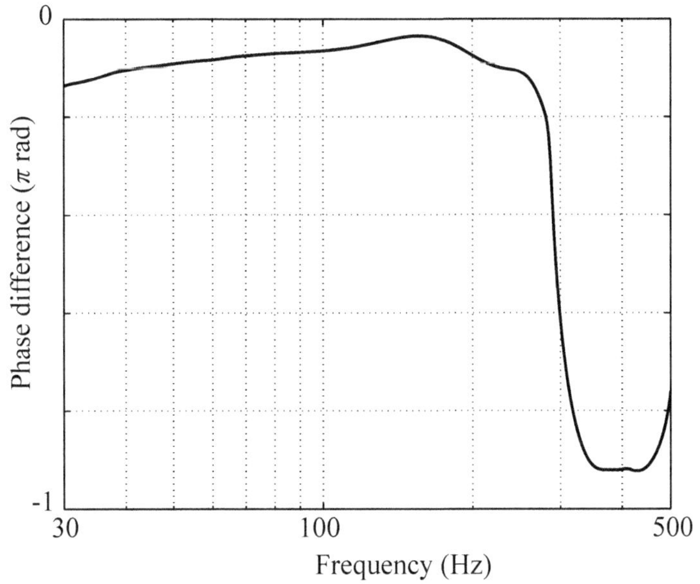

Fig. 3.6 Phase difference in sound pressure responses between inside and outside of enclosure from [9] (Fig.3.4)

Figure 3.6 confirms the phase relationship of the sound pressure inside and outside the enclosure[16]. In Fig. 3.6, certainly there is no significant phase-difference between the inside and outside in the frequencies lower than 300 Hz, while a phase change is observed over 300 Hz[16]. This frequency of 300 Hz seems to be the upper limit of the frequency range in which the Helmholtz resonator works well in the enclosure.

Chapter 4
Propagation of Oscillation

A series of simple harmonic oscillators, which represents masses connected by springs, is a good conceptual medium in which sound travels as a wave. This chapter focusses on oscillation propagating via connected oscillators as a wave that originated in initial disturbance in a local portion of a medium. A sound wave travels by a process in which the kinetic energy alternates with the potential energy of oscillation. Such an energy exchange system following the energy preservation law defines the speed of a wave traveling in a medium. Displacement or velocity of a medium in which a wave travels are converted into each other in the wave. This type of propagation scheme is a key for sound wave propagation. Finally a mathematical expression that governs wave motion travelling with a finite speed is derived by taking a limit case where harmonic oscillators are "embedded" into continuous distributions of masses with the elastic restoring force.

4.1 Propagation of Initial Disturbance

A wave such as sound is excited by initial disturbance that occurs in a local portion of a medium and travels through the medium without motion of the whole body of the medium. Such initial disturbance, e.g., stretch of a spring, arrives at an observation point in a finite time interval and passes with a finite speed. That is, local disturbance or deformation of an elastic medium manifests itself in a traveling wave through the medium. This is an important notion different from that of a Helmholtz resonator in which no local disturbance is observed as described in the previous chapter; consequently, no traveling waves are in the resonators. This section describes the travel of initial disturbance in a series of connected harmonic oscillators[17].

4.1.1 Propagation of Oscillation Energy

Suppose that harmonic oscillators such as springs are connected in a series as shown in Fig. 4.1. The two halves of the initial disturbance (displacement) at $t = 0$, which is given by

M. Tohyama: *Sound and Signals*, Signals and Communication Technology, pp. 43–53.
springerlink.com

$$u(i,t)|_{i=0,t=0} = u(0,0) = a \qquad \text{(m)} \tag{4.1}$$

as shown in Fig. 4.2, are equally propagated to the right and left, respectively. Namely, if the time at the second stage is labelled $t = 1$, then displacements of two oscillators 1 and -1 can be written as

$$u(-1,1) = \frac{a}{2} = u(1,1). \qquad \text{(m)} \tag{4.2}$$

Now consider the potential energy accumulated in the springs due to the disturbance. The stretch of every spring is caused by a difference in the displacement between a pair of adjacent springs. The difference is a between -1 and 0 and is $-a$ between 0 and 1 at the initial condition $t = 0$. Therefore, the potential energy can be expressed as

$$E_P(0) = \frac{1}{2}Ka^2 + \frac{1}{2}K(-a)^2 = Ka^2 \qquad \text{(J)} \tag{4.3}$$

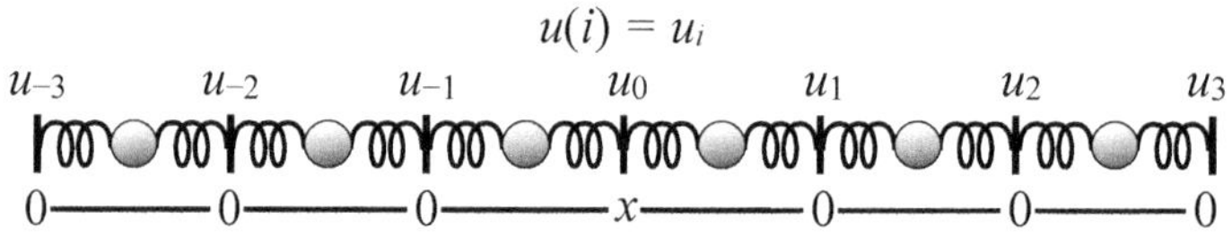

Fig. 4.1 Series of simple oscillators

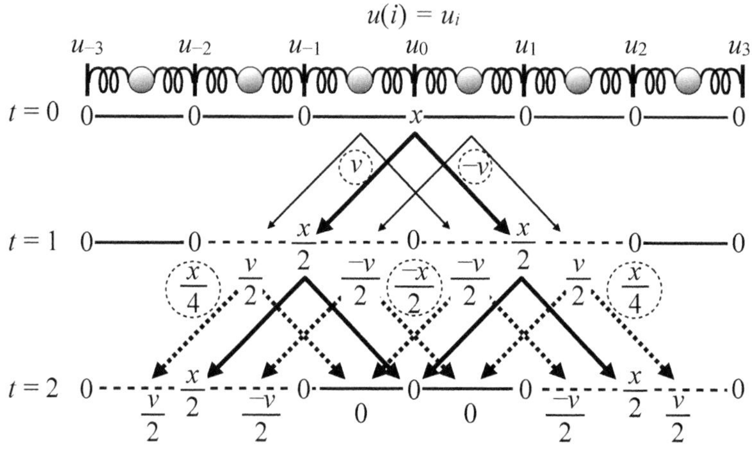

Fig. 4.2 Initial displacement ($x = a$) at $t = 0$ and its propagation ($v = b$) from [17] Fig. 22.3

due to the initial disturbance. Here $K(\mathrm{N/m})$ denotes spring constants of the springs connected as shown in Fig. 4.1.

Consider similarly the state at $t = 1$ after propagation of the disturbance. The difference between -2 and -1 is $a/2$, between -1 and 0 is $-a/2$, between 0 and 1 is $a/2$, and between 1 and 2 is $-a/2$. Consequently the potential energy traveled as

$$E_P(1) = \frac{1}{2}K\left(\left(\frac{a}{2}\right)^2 + \left(\frac{-a}{2}\right)^2 + \left(\frac{a}{2}\right)^2 + \left(\frac{-a}{2}\right)^2\right) = \frac{1}{2}Ka^2. \qquad \text{(J) (4.4)}$$

By comparison of these two results of potential energy at $t = 0$ and $t = 1$, the energy at $t = 1$ is found to be half as much as the initial energy at $t = 0$. This indicates that the other half of the energy changed into kinetic energy during propagation of the initial disturbance. In other words, when the initial disturbance is propagated, the initial displacement causes motion of a mass attached to a spring by restoring force.

Suppose that the initial difference a between -1 and 0 is converted to the velocity of motion of a mass denoted by b. One half of the velocity, $b/2$, is propagated to the left interval between -2 and -1, while $-b/2$ travels to the right between 0 and 1. Similarly, one half of the velocity, $-b/2$, is propagated to the left interval between -1 and 0, while $b/2$ travels to the right between 1 and 2. Therefore, the kinetic energy propagated from the initial disturbance is expressed as

$$E_K(1) = \frac{1}{2}M\left(\left(\frac{b}{2}\right)^2 + \left(\frac{-b}{2}\right)^2 + \left(\frac{-b}{2}\right)^2 + \left(\frac{b}{2}\right)^2\right) = \frac{1}{2}Mb^2. \qquad \text{(J) (4.5)}$$

Here $M(\mathrm{kg})$ denotes mass connected as shown in Fig. 4.1.

Consequently, an equation can be derived between the potential and kinetic energy such that

$$E_P = \frac{1}{2}Ka^2 = \frac{1}{2}Mb^2 = E_K \qquad \text{(J)} \qquad (4.6)$$

following the **energy preservation law**. This equation indicates that potential energy alternates with kinetic energy while the initial disturbance is propagated.

4.1.2 Propagation of Initial Displacement through Series of Connected Oscillators

As described above, the initial disturbance is propagated by alternating the potential energy with kinetic energy. A further step must be taken to confirm the exchange process between the displacement of a spring and velocity of motion of a mass following Fig. 4.2. Just as the initial displacement is equally propagated to oscillator $i = -1$ and $i = 1$ by $a/2$, velocity b is also converted from the initial displacement a and is equally propagated to the left and right with positive and negative signs, respectively. As shown in the figure denoting the velocity between i and $i+1$ at t as $v([i, i+1], t)$, a distribution of the velocity at $t = 1$ can be written as

$$v([-2,-1],1) = \frac{b}{2} \tag{4.7}$$

$$v([0,1],1) = \frac{-b}{2} \tag{4.8}$$

$$v([-1,0],1) = \frac{-b}{2} \tag{4.9}$$

$$v([1,2],1) = \frac{b}{2}. \qquad \text{(m/s)} \tag{4.10}$$

It seems as if a virtual "source" of velocity were hidden at $t = 0$ such that

$$v([-1,0],0) = b \tag{4.11}$$

$$v([0,1],0) = -b. \qquad \text{(m/s)} \tag{4.12}$$

The displacement of 0-th oscillator becomes zero at $t = 1$ because the initial displacement $u(0,0) = a$ already moved to the left and right.

However, the questions of whether the wave returns to the initial place and whether 0-th oscillator is quiet after the initial displacement occurs remain. Wave propagation is thus an interesting topic. The velocity is propagated at $t = 2$ from the velocity that was actually caused at $t = 1$ as shown in Fig. 4.2. From the figure, velocity propagation is seen to be

$$v_a([-3,-2],2) = \frac{b}{4} \quad \text{from} \quad [-2,-1] \tag{4.13}$$

$$v_a([-2,-1],2) = -\frac{b}{4} \quad \text{from} \quad [-1,0] \tag{4.14}$$

$$v_a([-1,0],2) = \frac{b}{4} - \frac{b}{4} = 0 \quad \text{from} \quad [-2,-1] \quad \text{and} \quad [0,1] \tag{4.15}$$

$$v_a([0,1],2) = -\frac{b}{4} + \frac{b}{4} = 0 \quad \text{from} \quad [-1,0] \quad \text{and} \quad [1,2] \tag{4.16}$$

$$v_a([1,2],2) = -\frac{b}{4} \quad \text{from} \quad [0,1] \tag{4.17}$$

$$v_a([2,3],2) = \frac{b}{4} \quad \text{from} \quad [1,2]. \qquad \text{(m/s)} \tag{4.18}$$

In addition to the velocity propagation stated above, the velocity is also converted as if the virtual velocity source were contained at $t = 1$ such that

$$v_c([-2,-1],1) = \frac{b}{2} \tag{4.19}$$

$$v_c([-1,0],1) = \frac{-b}{2} \tag{4.20}$$

$$v_c([0,1],1) = \frac{b}{2} \tag{4.21}$$

$$v_c([1,2],1) = \frac{-b}{2}. \qquad \text{(m/s)} \tag{4.22}$$

Again the velocity stated above is equally propagated to the left and right with positive and negative signs, respectively, such that

$$v_c([-3,-2],2) = \frac{b}{4} \quad \text{from} \quad [-2,-1] \tag{4.23}$$

$$v_c([-2,-1],2) = -\frac{b}{4} \quad \text{from} \quad [-1,0] \tag{4.24}$$

$$v_c([-1,0],2) = -\frac{b}{4} + \frac{b}{4} = 0 \quad \text{from} \quad [-2,-1] \quad \text{and} \quad [0,1] \tag{4.25}$$

$$v_c([0,1],2) = \frac{b}{4} - \frac{b}{4} = 0 \quad \text{from} \quad [-1,0] \quad \text{and} \quad [1,2] \tag{4.26}$$

$$v_c([1,2],2) = -\frac{b}{4} \quad \text{from} \quad [0,1] \tag{4.27}$$

$$v_c([2,3],2) = \frac{b}{4} \quad \text{from} \quad [1,2]. \qquad (\text{m/s}) \tag{4.28}$$

After summation of the velocity components v_a and v_c at $t = 2$, the distribution of velocity at $t = 2$ is observed as shown in Fig. 4.2.

Similarly, right after the initial displacement is equally propagated to $i = -1$ and $i = 1$, propagation is repeated at $t = 2$ to $i = -2$, $i = 0$, and $i = 2$, where the displacements are

$$u_a(-2,2) = \frac{a}{4} \quad \text{from} \quad i = -1 \tag{4.29}$$

$$u_a(0,2) = \frac{a}{2} \quad \text{from} \quad i = -1 \quad \text{and} \quad i = 1 \tag{4.30}$$

$$u_a(2,2) = \frac{a}{4} \quad \text{from} \quad i = 1. \qquad (\text{m}) \tag{4.31}$$

It appears that the initial displacement returns to $i = 0$; however, this propagation of displacement is cancelled by displacement converted from the velocity. Displacement is also caused at $t = 2$ by conversion from the velocity actually propagated at $t = 1$.

Recall the distribution of velocity v at $t = 1$. Conversion from velocity to displacement occurs after an average is taken of both sides of a spring of interest. After the average for every spring is taken and the velocity of b is replaced by displacement of a, a distribution of the displacement is obtained such that

$$u_c(-2,2) = \frac{a}{4} \quad \leftarrow (0 + b/2)/2 \tag{4.32}$$

$$u_c(-1,2) = 0 \quad \leftarrow (b/2 - b/2)/2 \tag{4.33}$$

$$u_c(0,2) = -\frac{a}{2} \quad \leftarrow (-b/2 - b/2)/2 \tag{4.34}$$

$$u_c(1,2) = 0 \quad \leftarrow (-b/2 + b/2)/2 \tag{4.35}$$

$$u_c(2,2) = \frac{a}{4} \leftarrow \quad (-b/2 + 0)/2 \qquad (\text{m}) \tag{4.36}$$

as seen in the figure. After summation of the components of displacements u_a and u_c at $t = 2$ as well as for velocity, finally the distribution of displacement is obtained as shown in Fig. 4.2.

The propagation process described above is repeated at every propagation stage, and it can be confirmed that a passing wave does not return. Propagation of a wave is a remarkable outcome of conversion between the displacement and velocity of motion.

4.1.3 Propagation of Initial Velocity

Propagation of the initial velocity can also be illustrated in addition to propagation of the initial displacement. Suppose that the initial velocity between $[0,1]$ at $t = 0$ is given as shown in Fig. 4.3.

The initial disturbance of velocity is equally propagated to the left and right. Just as the initial velocity is propagated to $[-1,0]$ and $[1,2]$ by $b/2$, displacement is caused by conversion from the average of the velocity between the left and right sides of a spring of interest. As shown in the figure, a distribution of the displacement occurs at $t = 1$ such that

$$u(0,1) = \frac{a}{2} \leftarrow (0+b)/2 = b/2 \tag{4.37}$$

$$u(1,1) = \frac{a}{2} \leftarrow (b+0)/2 = b/2 \qquad \text{(m)} \tag{4.38}$$

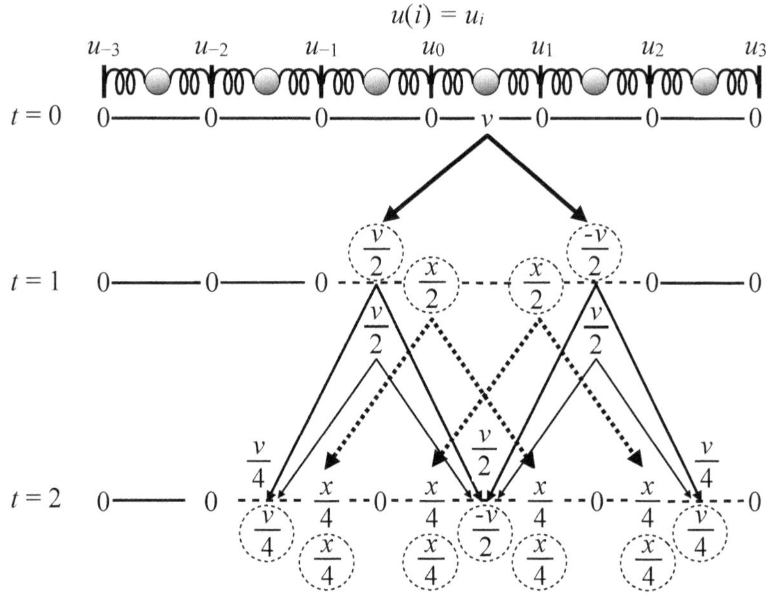

Fig. 4.3 Initial velocity ($v = b$) at $t = 0$ and its propagation ($x = a$) from[17] Fig. 22.4

Right after the initial velocity is propagated, propagation is repeated at $t = 2$ to $[-2,-1]$, $[0,1]$, and $[2,3]$ such that

$$v_a([-2,-1],2) = \frac{b}{4} \quad \text{from} \quad [-1,0] \tag{4.39}$$

$$v_a([0,1],2) = \frac{b}{4} + \frac{b}{4} = \frac{b}{2} \quad \text{from} \quad [-1,0] \quad \text{and} \quad [1,2] \tag{4.40}$$

$$v_a([2,3],2) = \frac{b}{4}. \quad \text{from} \quad [1,2]. \qquad \text{(m/s)} \tag{4.41}$$

In addition to the velocity propagation stated above, velocity is converted from the displacement as if the virtual velocity source were hidden at $t = 1$ such that

$$v_c([-1,0],1) = \frac{b}{2} \tag{4.42}$$

$$v_c([1,2],1) = -\frac{b}{2}. \qquad \text{(m/s)} \tag{4.43}$$

The velocity above is equally propagated to the left and right at $t = 2$ with positive and negative signs, respectively, such that

$$v_c([-2,-1],2) = \frac{b}{4} \quad \text{from} \quad [-1,0] \tag{4.44}$$

$$v_c([0,1],2) = -\frac{b}{4} - \frac{b}{4} = -\frac{b}{2} \quad \text{from} \quad [-1,0] \quad \text{and} \quad [1,2] \tag{4.45}$$

$$v_c([2,3],2) = \frac{b}{4} \quad \text{from} \quad [1,2]. \qquad \text{(m/s)} \tag{4.46}$$

After summation of v_a and v_c at $t = 2$, a velocity distribution is obtained as shown in Fig. 4.3.

Here consider propagation of displacement u_a from the displacement actually caused at $t = 1$ as shown in Fig. 4.3. From the figure, a distribution at $t = 2$ such that

$$u_a(-1,2) = \frac{a}{4} \quad \text{from} \quad i = 0 \tag{4.47}$$

$$u_a(0,2) = \frac{a}{4} \quad \text{from} \quad i = 1 \tag{4.48}$$

$$u_a(1,2) = \frac{a}{4} \quad \text{from} \quad i = 0 \tag{4.49}$$

$$u_a(2,2) = \frac{a}{4} \quad \text{from} \quad i = 1 \qquad \text{(m)} \tag{4.50}$$

can be seen. In addition to the distribution stated above, conversion from velocity to displacement u_c occurs following an averaging of both sides of a spring of interest. After the average for every spring is taken and the velocity of b is replaced by displacement of a, a distribution of the displacement is obtained such that

$$u_c(-1,2) = \frac{a}{4} \quad \leftarrow (0+b/2)/2 \tag{4.51}$$

$$u_c(0,2) = \frac{a}{4} \quad \leftarrow (b/2+0)/2 \tag{4.52}$$

$$u_c(1,2) = \frac{a}{4} \quad \leftarrow (0+b/2)/2 \tag{4.53}$$

$$u_c(2,2) = \frac{a}{4} \quad \leftarrow (b/2+0)/2. \qquad \text{(m)} \tag{4.54}$$

By addition of the components u_a and u_c, finally a distribution of displacement at $t = 2$ is obtained as shown in Fig. 4.3.

In comparing the two cases, namely wave propagation due to initial disturbance given by displacement or velocity, there is a remarkable difference. A wave caused by the initial velocity leaves a "scar". In other words, a displacement, which does not exist before a wave passes, is moved to $a/2$ in the example above when a wave arrives at a point of interest, and it does not disappear even after the wave passes. This phenomenon is a typical characteristic of wave propagation in a one-dimensional medium. One-dimensional wave propagation is important in particular for basic study of the physics of musical instruments. Differences in the initial conditions are closely related to sound sources exciting waves in instruments.

4.2 Equation of Wave Propagation

4.2.1 Speed of Wave Propagation

In the previous section, propagation of initial disturbance was described step by step. However, waves continuously travel in a medium with a finite speed, if taken from a physical point of view. Now consider a limit case where harmonic oscillators are very densely connected so that a distribution of displacement or velocity might be expressed as a "continuous" function of spatial position x in the medium such as $u(x)$ or $v(x)$, and consequently mass in a small interval of the medium can be written as $M = \rho_0 \Delta x$ by using the density of $\rho_0(\text{kg/m})$. Recall that the energy preservation relationship in wave propagation is

$$\frac{1}{2}K(\Delta u)^2 = \frac{1}{2}M(\Delta v)^2. \qquad \text{(J)} \tag{4.55}$$

The energy preservation law stated above can be written as

$$\frac{1}{2}K(\frac{\Delta u}{\Delta x}\cdot \Delta x)^2 = \frac{1}{2}\rho_0 \Delta x \cdot (\frac{\Delta u}{\Delta t})^2 \qquad \text{(J)} \tag{4.56}$$

in the limit (continuous) case. From this expression, the equation below is derived:

$$\frac{(\Delta u/\Delta t)^2}{(\Delta u/\Delta x)^2} = \frac{K\Delta x}{\rho_0} = c^2, \qquad (\text{m/s})^2 \tag{4.57}$$

which indicates **speed of wave propagation** in a medium by $c(\text{m/s})$.

Wave speed is given by a quotient of vibration velocity and slope of deformation at a very local portion of a medium in which a wave travels. Vibration speed is actually in proportion to wave propagation speed or the slope of deformation. It is intuitively accepted that vibration speed increases as the deformation slope becomes steeper or as propagation speed increases, while vibration slows down as the deformation slope becomes gentler or wave speed decreases. Note also that wave propagation speed is given by a quotient of the elastic property, such as a spring constant and mass of a medium of interest. Therefore, high propagation speed is possible, even if the density of a medium is high. For example, sound speed is faster in water than in air because of the difference of the elastic properties.

4.2.2 Wave Equation

Wave propagation speed can be formulated in terms of mathematics. In actually describing wave propagation from a mathematical view point, an equation called the **wave equation** is derived. The wave equation is a source from which acoustic phenomena are theoretically predicted or estimated.

Recall Figs. 4.1 and 4.2 and the Newtonian equation of motion, which says that mass times acceleration creates:

$$M\frac{d^2u_i(t)}{dt^2} = f_{K_i}, \qquad \text{(N)} \tag{4.58}$$

where u_i denotes displacement of the i-th spring and f_{K_i} is force working on the mass attached to the i-th spring. The restoring force f_{K_i} against stretch of the i-th spring can be written as

$$f_{K_i} = K(u_{i+1} - u_i) - K(u_i - u_{i-1}). \qquad \text{(N)} \tag{4.59}$$

Assume that restoring force f_K and displacement u are smooth functions of time t and place x such that

$$f_{K_i} = f_K(x,t) \qquad \text{(N)} \tag{4.60}$$

$$u_i = u(x,t) \qquad \text{(m)} \tag{4.61}$$

for a limit case. Consequently, the restoring force $f_K(x,t)$ can be rewritten further as

$$\begin{aligned} f_K(x,t) &= K(u(x+\Delta x,t) - u(x,t)) - K(u(x,t) - u(x-\Delta x,t)) \\ &= K\Delta x \left(\left.\frac{\partial u(z,t)}{\partial z}\right|_{z=x} - \left.\frac{\partial u(z,t)}{\partial z}\right|_{z=x-\Delta x} \right) \\ &\cong K\frac{\partial^2 u(x,t)}{\partial x^2}(\Delta x)^2. \qquad \text{(N)} \end{aligned} \tag{4.62}$$

Stretch, which occurs at the side of a spring, is the difference from the normal position. Thus the difference between the stretch at both sides, which is expressed as the difference of the differences, expresses the whole stretch for a spring of interest. Such deformation effect of a local area in a medium, which causes waves, can generally be expressed by the second derivatives with respect to spatial variables such as x above.

Using smooth functions, the equation of motion can be rewritten as

$$\frac{\partial^2 u(x,t)}{\partial t^2} = c^2 \frac{\partial^2 u(x,t)}{\partial x^2} \tag{4.63}$$

where c denotes speed of a wave, which is expressed as

$$c^2 = \frac{K\Delta x}{\rho_0} \qquad (\text{m/s})^2 \tag{4.64}$$

and $M = \rho_0 \Delta x(\text{kg})$. Although the equation above seems to be nothing but an outcome of the Newtonian law of motion, it is an important equation called the **wave equation** that governs wave propagation in an elastic medium. It is interesting to see that wave motion, which is observed in a wide area of a medium, is expressed following the Newtonian low and elastic properties in a very local area of a medium. Speed of wave propagation from the local area to the whole medium is expressed using a ratio of the elastic properties of the medium representing restoring force against deformation and density related to acceleration caused by the restoring force.

4.2.3 *Propagation of Wave*

Waves are formulated in terms of mathematics following the wave equation where c represents propagation speed. A solution of the equation can be represented by

$$u(x,t) = f(x+ct) + g(x-ct), \qquad (\text{m}) \tag{4.65}$$

where $f(x+ct)$ and $g(x-ct)$ represent waves travelling to the left and right with speed c, respectively. A function $f(z)$ always takes the same value when the variable z is equal. This means the variable $x-ct$ really indicates the position x where $x-ct$, taking the same value, moves to the right at speed c as time passes. In contrast, the variable $x+ct$ indicates the position x taking the same value as it moves to the left at the same speed c. This type of movement of the spatial position corresponding to the time variable indicates that there is a traveling wave in a medium.

Suppose that the initial disturbance is given by

$$u(x,0) = f(x) + g(x) = a(x) \qquad \text{for displacement} \qquad (\text{m}) \tag{4.66}$$

$$v(x,0) = \left.\frac{\partial u(x,t)}{\partial t}\right|_{t=0} = c(f'(z) - g'(z))|_{z=x} = b(x) \text{ for velocity.} \quad (\text{m/s}) \tag{4.67}$$

Here $f'(z)|_{z=x} = f'(x)$ and $g'(z)$ denote the derivatives of the functions with respect to z, respectively. If the functions $f(x)$ and $g(x)$ can be given by $a(x)$ and $b(x)$, which follow the initial disturbance, then propagation of the wave can be determined by the initial disturbance.

Such functions can be found such that

$$f(x) = \frac{1}{2}\left(a(x) + \frac{1}{c}\int_{\alpha}^{x} b(z)dz\right) \tag{4.68}$$

$$g(x) = \frac{1}{2}\left(a(x) - \frac{1}{c}\int_{\alpha}^{x} b(z)dz\right) \tag{4.69}$$

according to

$$f(x) + g(x) = a(x) \tag{4.70}$$

$$f(x) - g(x) = \frac{1}{c}\int_{\alpha}^{x} b(z)dz. \tag{4.71}$$

Here $\int b(z)dz$ denotes integration of the function $b(z)$ with respect to z, and α is a constant. Consequently, the solutions for displacement $u(x,t)$ and velocity $v(x,t)$ are given by

$$\begin{aligned} u(x,t) &= f(x+ct) + g(x-ct) \\ &= \frac{1}{2}(a(x+ct) + a(x-ct)) + \left(\frac{1}{2ct}\int_{x-ct}^{x+ct} b(z)dz\right)t \qquad \text{(m)} \end{aligned} \tag{4.72}$$

$$\begin{aligned} v(x,t) &= cf'(x+ct) - cg'(x-ct) \\ &= \frac{c}{2}(a'(x+ct) - a'(x-ct)) + \frac{1}{2}(b(x+ct) + b(x-ct)), \qquad \text{(m/s)} \end{aligned} \tag{4.73}$$

respectively.

The solution analytically describes the process of wave propagation that was considered schematically using connected oscillators in the previous sections. Displacement of a position x observed at time t, namely denoted by $u(x,t)$, indicates that the initial displacement $a(x)$ is equally propagated to the left and, right keeping its initial shape without distortion, at speed c. In addition, displacement converted from the initial velocity is superposed. Such converted displacement can be expressed based on the average of initial velocity $b(x)$ between $x-ct$ and $x+ct$. This expanding speed of the area over which the average is taken is c, which explains why c denotes the wave propagation speed. Similarly velocity at a position x observed at time t denoted by $v(x,t)$ shows that the initial velocity is equally propagated to the left and right. Furthermore, velocity converted from the initial displacement $a(x)$ is added. Such velocity is expressed as the slope of the initial displacement $a'(x)$, which is equally propagated to both sides with positive and negative signs, respectively. The examples taken above using a series of oscillators are good examples for interpreting wave propagation processes.

Chapter 5
Vibration of String and Wave Propagation

Strings are naturally musical sound instruments as are acoustic pipes. This is because free vibration of one-dimensionally extending media such as a string can be composed of a fundamental and its harmonics, which are important for musical sound and with scales. Scientific studies on the resonance of string vibrations date from Pythagoras, and Galileo mentioned eigenfrequencies of free vibration of a string. This chapter starts by focussing on progressive waves and their propagation speed along a long string, similar to the waves on a linear array of harmonic oscillators considered in the previous chapter. However, considering the waves along a finite length of string, vibration is made of countless "round trips" between both ends of a string. Therefore, resonance can be found to occur at not only a single frequency but many ones. Finally, this chapter discusses the expression of free vibration as superposition of eigenmodes (or modal functions) with harmonic frequencies as well as graphical representation of wave propagation. The results are physical basis for mathematical representation of a periodic function by Fourier series expansion.

5.1 Wave Propagation of Infinitely Long String

Suppose a long string has density $\rho_0(\mathrm{kg/m})$ for a unit length and its tension is $F_T(\mathrm{N})$. Assign coordinate axis of x along the string, and also assign y for the perpendicular direction to x. A wave along the string $u(x,t)$ is a **transversal wave** in which vibration is perpendicular to the wave propagation. The initial disturbance is propagated as transversal waves on the string with definite speed.

5.1.1 Speed of Transversal Wave

Restoring force due to elasticity of a medium is necessary to excite wave propagation in a medium. The y-component for tension of a string as shown in Fig. 5.1 yields restoring force. Potential energy can be estimated as work done that is necessary to stretch a string against the restoring force. According to the figure, stretch of a string is estimated as[4]

M. Tohyama: *Sound and Signals*, Signals and Communication Technology, pp. 55–87.
springerlink.com

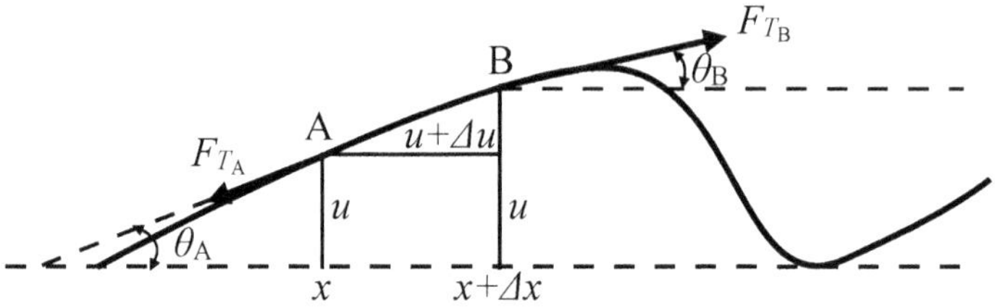

Fig. 5.1 Deformation of string and its tension from [4] Fig. 2.1

$$\sqrt{(\Delta x)^2 + (\Delta u)^2} - \Delta x \cong \frac{1}{2}\left(\frac{\Delta u}{\Delta x}\right)^2 \Delta x \qquad \text{(m)} \tag{5.1}$$

for a small interval of the string shown by Δx. Therefore, the work done can be written as

$$E_P = \frac{1}{2} F_T \left(\frac{\Delta u}{\Delta x}\right)^2 \Delta x, \qquad \text{(J)} \tag{5.2}$$

where F_T denotes the tension (N). Similarly, kinetic energy can be expressed as

$$E_K = \frac{1}{2} \rho_0 \Delta x v^2, \qquad \text{(J)} \tag{5.3}$$

where v denotes vibration velocity (m/s) along the y-axis such that $v = \Delta u/\Delta t$.

If both energies are set to be equal, following the energy preservation law,

$$\frac{(\Delta u/\Delta t)^2}{(\Delta u/\Delta x)^2} = \frac{F_T}{\rho_0} = c^2 \qquad (\text{m/s})^2 \tag{5.4}$$

is obtained showing the square of transversal wave speed along a string. Transversal wave speed on a string can be written as the ratio of vibration velocity $(\Delta u/\Delta t)$ and slope of displacement with respect to x $(\Delta u/\Delta x)$. Here y shows displacement perpendicular to the x-direction along a string of interest. If the slope is the same, the vibration velocity increases with the wave speed, given by the ratio of the tension and density of a string. Wave speed decreases as the density increases under a constant tension, while it increases when the tension is strong under a constant density.

5.1.2 Equation of Wave Propagation on String

A wave traveling on a string is governed by the wave equation in one-dimensional space. Suppose that transverse vibration with small displacement (deflection) travels a long string with the uniform density ρ_0(kg/m). Recall Fig. 5.1, and assume that the angle θ is small so that $\sin\theta \cong \theta$. The tension that acts on a small portion Δx of the string can then be written as

$$F_{T_B}\sin\theta_B - F_{T_A}\sin\theta_A \cong F_T(\theta_B-\theta_A) \tag{5.5}$$

$$\theta_B-\theta_A \cong \frac{\partial\theta(x)}{\partial x}\Delta x \tag{5.6}$$

$$\tan\theta(x) \cong \theta(x) \cong \frac{\partial u(x,t)}{\partial x} \tag{5.7}$$

where $u(x,t)$ denotes the deflection of the string at the small portion. Then the equation of motion for the small portion of the string becomes

$$\rho_0\Delta x\frac{\partial^2 u(x,t)}{\partial t^2} = F_T\Delta x\frac{\partial^2 u(x,t)}{\partial x^2}; \tag{5.8}$$

namely,

$$\frac{\partial^2 u(x,t)}{\partial t^2} = c^2\frac{\partial^2 u(x,t)}{\partial x^2} \tag{5.9}$$

is obtained as the **wave equation for transversal waves on a string** where

$$c^2 = \frac{F_T}{\rho_0} \tag{5.10}$$

and $c(\mathrm{m/s})$ is the speed of the waves on a string. Vibration of a string is an important example of wave propagation along a one-dimensional axis.

5.1.3 Initial Displacement and Its Propagation

A progressive wave can be interpreted as free vibration of an oscillator, but its frequency is arbitrary. External force to create initial disturbance, from which progressive waves are generated, can be characterized in terms of initial conditions. Suppose that a portion of a string is picked up and released quietly as shown in Fig. 5.2. This is a simplified example for an initial condition of string vibration such as in a harp. The displacement is equally separated into two parts, both of which are in the shape of the initial disturbance. As the peak of the central part of the initial displacement decreases, two equally divided progressive waves are propagated to the right and left, respectively.

Given the initial displacement and velocity (initial condition) such that

$$u(x,0) = a(x) \qquad (\mathrm{m}) \tag{5.11}$$

$$v(x,0) = 0, \qquad (\mathrm{m/s}) \tag{5.12}$$

the initial displacement is propagated to the left and right separately such that

$$u(x,t) = \frac{1}{2}\left(a(x+ct)+a(x-ct)\right) \qquad (\mathrm{m}) \tag{5.13}$$

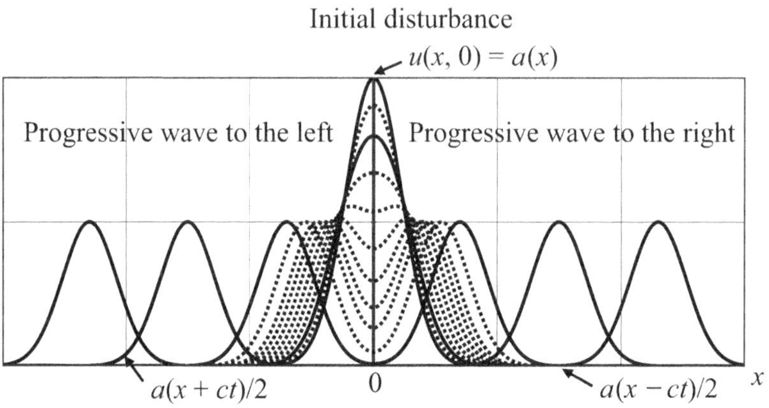

Fig. 5.2 Initial displacement of string and propagation waves from [9] (Fig.5.1)

after the string is quietly released. Here the first term represents the progressive wave to the left (negative side of x-direction), and the second one represents a wave in the opposite direction. The waves are propagated without change in their forms as time passes.

5.1.4 Propagation of Initial Velocity

Suppose that initial velocity is given to a portion of a string as shown in Fig. 5.3 instead of the example above. It represents a simplified initial condition for string vibration such as in a piano. The condition is expressed as

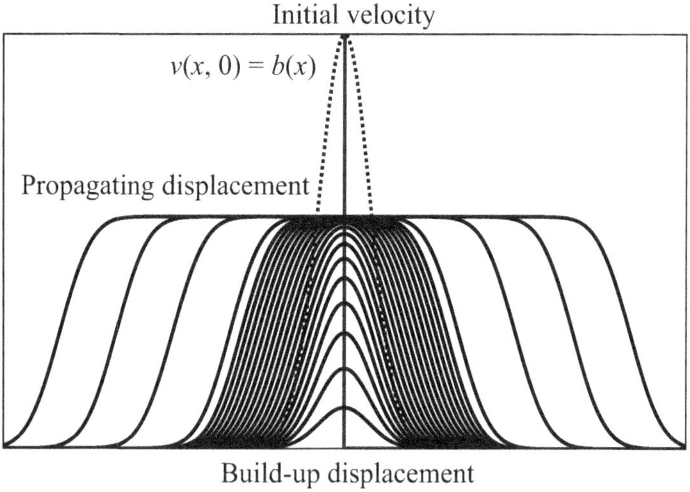

Fig. 5.3 Initial velocity of string and propagation waves from [9] (Fig.5.2)

$$u(x,0) = 0 \qquad \text{(m)} \tag{5.14}$$
$$v(x,0) = b(x). \qquad \text{(m/s)} \tag{5.15}$$

The initial velocity is also equally separated into two parts; however, displacement excited by the initial velocity is different. It is expressed as

$$u(x,t) = \left(\frac{1}{2ct}\int_{x-ct}^{x+ct} b(z)dz\right)t, \qquad \text{(m)} \tag{5.16}$$

which shows that displacement is propagated according to the average of the initial velocity between $x-ct$ and $x+ct$.

Displacement at the position of $x=0$, the center to which initial velocity is given as shown by the broken line in the figure, increases up as time passes. Similarly, the build-up process at position x is expressed as a product of the time and average of the initial velocity between $x-ct$ and $x+ct$. The interval over which the average is taken increases as time passes; this indicates that the area in which a progressive wave travels spreads as time passes. The spreading speed of the area indicates the wave speed. The displacement converted from the initial velocity does not disappear as pointed out in the previous chapter, even after the progressive wave passes.

5.1.5 Generalized Initial Conditions and Propagation of Waves

Initial conditions can be expressed by both displacement and velocity. This means propagating waves on a string can be predicted by the initial displacement and velocity. If the initial conditions are given such that

$$u(x,0) = a(x) \quad \text{for displacement} \qquad \text{(m)} \tag{5.17}$$
$$v(x,0) = b(x) \quad \text{for velocity,} \qquad \text{(m/s)} \tag{5.18}$$

then formal expressions of progressive waves of displacement and velocity

$$u(x,t) = \frac{1}{2}(a(x+ct)+a(x-ct)) + \left(\frac{1}{2ct}\int_{x-ct}^{x+ct} b(z)dz\right)t \qquad \text{(m)} \tag{5.19}$$
$$v(x,t) = \frac{c}{2}\left(a'(x+ct)-a'(x-ct)\right) + \frac{1}{2}(b(x+ct)+b(x-ct)), \qquad \text{(m/s)} \tag{5.20}$$

can be obtained, respectively. Displacement (or velocity) is a superposition of the propagating initial displacement (velocity) and converted from the initial velocity (displacement). Therefore, it is possible that superpositions of waves might cancel each other out, and consequently a one-way traveling wave is possible. Figure 5.4 is an example of such a one-way wave propagation, where the initial conditions are given as[18]

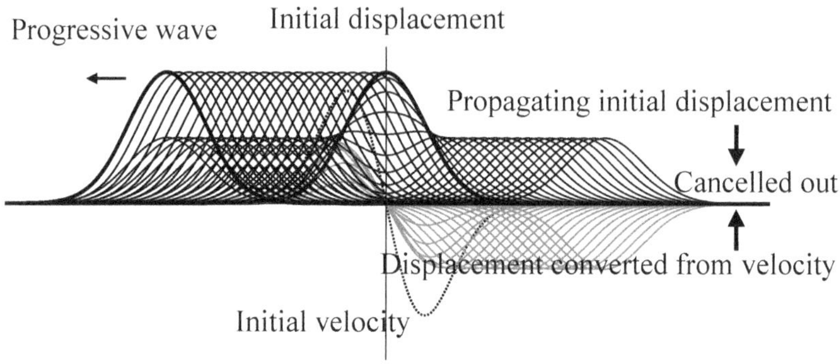

Fig. 5.4 Example of one-way propagating wave from [9] (Fig.5.3)

$$u(x,t) = a(x) \quad \text{for displacement} \qquad \text{(m)} \tag{5.21}$$

$$v(x,t) = c \cdot a'(x) \quad \text{for velocity.} \qquad \text{(m/s)} \tag{5.22}$$

A one-way progressive wave such as

$$u(x,t) = \frac{1}{2}(a(x+ct)+a(x-ct)) + \left(\frac{1}{2ct}\int_{x-ct}^{x+ct} b(z)dz\right)t = a(x+ct) \qquad \text{(m)} \tag{5.23}$$

has been confirmed possible due to cancellation between the initial and converted displacements.

5.2 Boundary Conditions and Harmonic Vibration

The previous section described progressive waves along an infinitely long string caused by initial disturbance to the string. However, in a practical situation, a string normally has a finite length. Therefore, a progressive wave will eventually come to the ends of a string of interest. At such a boundary, a progressive (incidental) wave returns as a reflection wave. Consequently, propagation of the initial disturbance without change of the waveform is no longer expected. Thus, this section describes the normal modes of vibration that are composed of a fundamental and its harmonics.

5.2.1 Wave Reflection at Boundary

Consider a string with one end clamped as shown in Fig. 5.5, where the endpoint (boundary) is set at $x = 0$. A clamped boundary at $x = 0$ specifies a condition at a boundary (or a boundary condition) such that the string does not move at the boundary. That is, displacement of a string can be set at the boundary such that

$$u(0,t) = 0. \qquad \text{(m)} \tag{5.24}$$

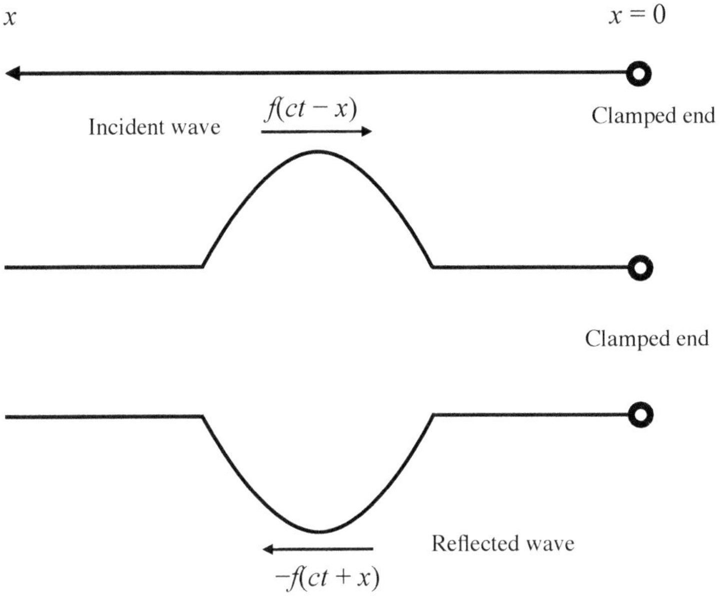

Fig. 5.5 Progressive wave reflected at clamped end ($x = 0$) of string

The boundary condition above regulates a progressive wave that approaches boundary from the negative side of the string as shown in Fig. 5.5, so the displacement might be cancelled at the boundary. The boundary condition or the regulation needs another wave that cancels the coming wave at the boundary. The other wave is called a **reflection wave**, which returns to the left in the figure.

Take an approaching progressive wave as

$$u_1(x,t) = g(ct - x), \tag{5.25}$$

and set an opposite (returning) progressive wave to be

$$u_2(x,t) = f(ct + x). \tag{5.26}$$

By superposition of these two waves such that

$$u(x,t) = u_1(x,t) + u_2(x,t) = g(ct - x) + f(ct + x), \tag{5.27}$$

the reflective wave is obtained such that

$$f(ct + x) = -g(ct - x) \tag{5.28}$$

following the boundary condition.

Note here that there are two ways of representing a progressive wave: $g(ct - x)$ or $g(x - ct)$. There is no significant difference between these two. The difference exists in taking a "reference" for representation of time or spatial distance. When a progressive wave is expressed as $g(ct - x)$, the wave is represented based on a "sense of time", such as how long the wave will take to arrive at a position x. That is, the origin $x = 0$ is always taken as a reference. In contrast, a progressive wave in the form $g(x - ct)$ follows the sense of how long ago the wave left the origin $x = 0$. An observation point at x is taken as a reference in this case. Both ways are used to represent progressive waves in this book.

5.2.2 Vibration Patterns of Finite Length of String

Suppose that a finite length of string is clamped at both ends, denoted by $x = 0$ or $x = L$. Here L(m) is the length of the string. It seems natural to suppose a pattern of vibration as shown in Fig. 5.6($n = 1$). This pattern meets the boundary conditions at both ends and is called a **fundamental mode** of free vibration of a finite length of string. With the assumption that progressive waves simultaneously leave the center of the string at $t = 0$ and respectively head to the right and left, both waves meet again at the center after traveling back and forth between both ends of the string.

The travel of waves on a string is **periodic**. A **period** for the back and forth travel of a wave is given by $2L/c$ (s), where c denotes the wave speed (m/s). Therefore, a period $T_1 = 2L/c$ (s) is called the **fundamental period**, and its inverse, i.e., $\nu_1 = c/2L$ (Hz), is called the **fundamental**. The fundamental is interpreted as the

Order	Vibration pattern	Wavelengh: λ_n	Resonant freq: ν_n
$n = 1$	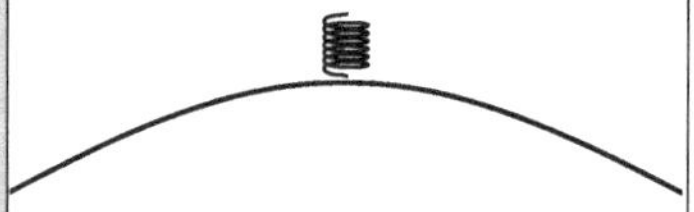	$\lambda_1 = 2L$	$\nu_1 = c/2L$
$n = 2$	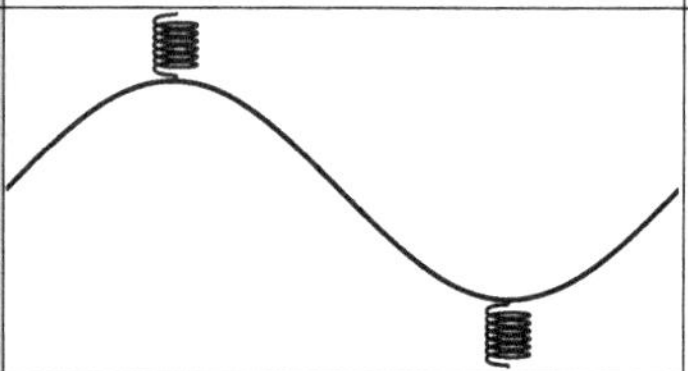	$\lambda_2 = L$	$\nu_2 = 2\nu_1$
$n = 3$	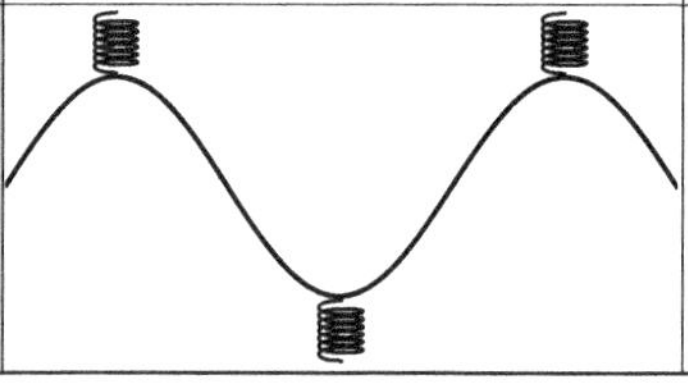	$\lambda_3 = 2L/3$	$\nu_3 = 3\nu_1$

Fig. 5.6 Vibration patterns for finite length of string with clamped ends from [19] (Fig.2.34)

lowest eigenfrequency of a finite length of string, and thus there are countless eigenfrequency for vibration of a finite length of string as described below.

The pattern of vibration shown in Fig. 5.6 ($n = 1$) is periodic with a period of $2L/c$. A vibration pattern does not necessarily have to be the pattern stated above, however. Figure 5.6 ($n = 2$) is another pattern of vibration. By equal division of a string into two parts, the "fundamental for each part" is given by c/L. Thus if the divided segments are smoothly reconnected, a vibration pattern could be obtained as shown in Fig. 5.6 ($n = 2$). Here the fundamental for each of the divided parts, i.e., $\nu_2 = 2\nu_1 = c/L$ (Hz), is called the **2nd harmonics**, i.e., the second eigenfrequency. There is a dividing point, which is called a **node**, where the displacement is null as if that point were a virtual boundary (end). On the other hand, the position on a string where displacement takes its maximum is called a **loop**.

Note however that another condition is needed for connecting the divided parts; namely, vibration patterns are in anti-phase with respect to the connecting point (virtual boundary) as implied by smooth connection. By repetition of this dividing and connecting process, the n-th harmonics (eigenfrequency) whose frequency is given by $\nu_n = n\nu_1$ can be obtained. Figure 5.6 ($n = 3$) shows the 3rd harmonics. Vibration patterns like n-th harmonics are called **normal modes** of vibration.

There are $n-1$ nodes in the n-th mode in general. Any pair of adjacent parts with respect to the node makes vibrations that are in anti-phase to each other. In other words, vibration reverses its phase at every node. As a result, the vibration patterns suggest that free vibration of a finite length of string might be represented by odd and periodic functions. Here the term "periodic" has a double meaning; namely, the functions are periodic with respect to time with a period of $2L/c$(s), and in addition, that the functions are also spatially periodic with an interval of $2L$.

Now consider free vibration of a finite string whose length is L (m) such that

$$u(x,t) = f(x+ct) + g(x-ct) \qquad \text{(m)} \tag{5.29}$$

under the initial conditions for the displacement and velocity such as

$$u(x,0) = u_0(x) = a(x) \qquad 0 \le x << L \quad \text{(m)} \tag{5.30}$$

$$v(x,0) = 0. \qquad \text{(m/s)} \tag{5.31}$$

By following the general solution for an infinitely long string derived in the previous section, the displacement can be written as

$$u(x,t) = \frac{1}{2}(u_0(x+ct) + u_0(x-ct)). \qquad \text{(m)} \tag{5.32}$$

Here, recalling the boundary conditions for a clamped string at $x = 0$ and $x = L$ such that

$$u(0,t) = \frac{1}{2}(u_0(ct) + u_0(-ct)) = 0 \tag{5.33}$$

$$u(L,t) = \frac{1}{2}(u_0(L+ct) + u_0(L-ct)) = 0, \qquad \text{(m)} \tag{5.34}$$

the relation

$$u_0(-ct) = -u_0(ct) \qquad \text{(m)} \tag{5.35}$$
$$u_0(L-ct) = -u_0(L+ct) \qquad \text{(m)} \tag{5.36}$$

can be obtained, and thus

$$u_0(L+ct) = -u_0(L-ct) = u_0(ct-L) = u_0(L+ct-2L) \qquad \text{(m)} \tag{5.37}$$

holds well. Consequently, it can be confirmed that functions which represent the propagating waves are odd and are spatially periodic with an interval of $2L$.

5.2.3 Generalized Boundary Conditions: End Correction

Boundary conditions for vibration of a string can be generalized to represent a more realistic situation. This type of modification of the boundary conditions is well known as a historical issue about end correction for an acoustic tube[7]. If the end of a tube is "open" where there is no acoustic pressure, then no sound is radiated from the tube. In addition, the fundamental of an open tube is different from that estimated by $c/2L$. Namely, the "effective length" of the tube must be modified from the geometrical length of the tube, in general, to appropriately estimate the fundamental. This is due to the fact that the end of a tube seems to be open but that from an acoustical point of view it is not. This is a very important issue in designing musical instruments like organs.

It is also an important issue with regard to the waves travelling on a string in a musical instrument such as a piano[20]. If the string of a piano is perfectly fixed at the end points, then no vibration is transmitted to the sound board of the piano, and consequently no sound is radiated. The boundary conditions can be formulated in terms of the **impedance** of vibration[21].

Suppose that a string is terminated at $x = 0$ by **mechanical impedance** Z_M as shown in Fig. 5.7. Here the mechanical impedance Z_M denotes a ratio between the

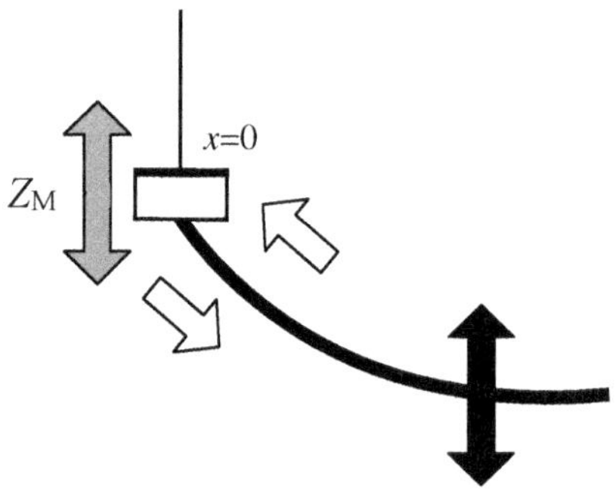

Fig. 5.7 End of string at $x = 0$ that is terminated by mechanical impedance Z_M

force and velocity at the point of interest. Recall that deflection travels on a string as sinusoidal waves by discarding the time components such that

$$U(x) = A\mathrm{e}^{-\mathrm{i}kx} + B\mathrm{e}^{\mathrm{i}kx}, \tag{5.38}$$

where k denotes the **wavenumber** (1/m) defined by $k = \omega/c$. Here the first term represents the **incident wave** and the second one shows the **reflection wave** from the boundary. Thus, the ratio B/A expresses the **reflection coefficient from the boundary**, which depends on the boundary condition.

The deflection waves described above must meet the boundary condition at $x = 0$. From Fig. 5.1, the force that acts on the boundary can be expressed as

$$F(x)|_{x=0} = -F_T \sin\theta \cong -F_T\theta \cong -F_T \left.\frac{\partial U(x)}{\partial x}\right|_{x=0}, \tag{5.39}$$

where force in the upper direction is indicated with a positive sign. According to the boundary condition, namely the boundary impedance, the velocity $V(x)$ can be expressed as

$$F(x)|_{x=0} = -\,F_T \left.\frac{\partial U(x)}{\partial x}\right|_{x=0} = Z_M V(x)\Big|_{x=0}. \tag{5.40}$$

Therefore, by introducing the velocity and force at the boundary such that

$$F(0) = \mathrm{i}kF_T(A-B) \qquad \text{(N)} \tag{5.41}$$

$$V(0) = \mathrm{i}\omega(A+B) \qquad \text{(m/s)}, \tag{5.42}$$

the relationship

$$F_T(A-B) = Z_M c(A+B) \tag{5.43}$$

holds well between the force and velocity at the boundary, where c (m/s) denotes the speed of waves on the string. Consequently, the reflection coefficient B/A is obtained as

$$\frac{B}{A} = -\frac{\hat{Z}_M c - 1}{\hat{Z}_M c + 1} \tag{5.44}$$

in terms of the boundary impedance, where $\hat{Z}_M = Z_M/F_T$.

The boundary impedance can be tentatively represented in terms of the mass, spring constant, or friction coefficient based on vibration of a single degree of freedom of the system. In other words, it can be interpreted as the vibrating string being connected to the rigid boundary via a mass-spring system. Suppose that the boundary impedance can be represented by the mass so that the force is in proportion to the acceleration, or the impedance can be expressed as a pure imaginary number. Then assuming the impedance can be written as

$$\hat{Z}_{MC} = \mathrm{i}\cot k y_{Ec} \qquad (y_{Ec} > 0), \tag{5.45}$$

the reflection coefficient can be rewritten as

$$\frac{B}{A} = -\frac{\mathrm{i}\cot k y_{Ec} - 1}{\mathrm{i}\cot k y_{Ec} + 1} = -\mathrm{e}^{2\mathrm{i}k y_{Ec}}. \tag{5.46}$$

After substituting this reflection coefficient into Eq. 5.38,

$$U(x) = A\mathrm{e}^{-\mathrm{i}kx} + B\mathrm{e}^{\mathrm{i}kx} = A\mathrm{e}^{\mathrm{i}k y_{Ec}}\left(\mathrm{e}^{-\mathrm{i}k(x+y_{Ec})} - \mathrm{e}^{\mathrm{i}k(x+y_{Ec})}\right) \tag{5.47}$$

is derived. This outcome is as if the length of string were shortened by the amount of y_{EC}.

In contrast, if the boundary impedance is a value such that the force is in proportion to the deflection of the string, i.e.,

$$\hat{Z}_{MC} = -\mathrm{i}\cot k y_{Ec} \qquad (y > 0), \tag{5.48}$$

then the reflection coefficient becomes

$$\frac{B}{A} = -\mathrm{e}^{-2\mathrm{i}k y_{Ec}}. \tag{5.49}$$

Therefore, the deflection wave can be rewritten as

$$U(x) = \mathrm{e}^{-\mathrm{i}k y_{Ec}}\left(\mathrm{e}^{-\mathrm{i}k(x-y_{Ec})} - \mathrm{e}^{\mathrm{i}k(x-y_{Ec})}\right). \tag{5.50}$$

This expression can be visualized as if the length of string were extended by y_{EC}.

If the impedance is regarded as resistive, namely the force is in proportion to the vibration velocity, then the impedance is a real function. Therefore, the reflection coefficient is also real with a negative sign whose magnitude is smaller than unity. The results can be intuitively interpreted[22]. For the case of massive impedance, the phase of the reflection wave seems to "lead" the incident wave by $2ky_{Ec}$ at the boundary. This can be interpreted as if the reflection wave is in anti-phase to the incident wave at $x = -2y_{Ec}$, and thus the two waves meet at $x = -y_{Ec}$ so that the "node" might be created by cancellation. This node makes the length of the string effectively shorter by y_{EC}.

In contrast, for the stiffness boundary, the phase of the reflection wave is delayed by $2ky_{Ec}$ from the incident wave at $x = 0$. This can be understood as the reflection from the virtual end at $x = y_{Ec}$. Namely, the string becomes virtually longer than its geometrical length.

Only for the case of resistive impedance is the energy of the incidental wave consumed, i.e., the vibration of a string is transmitted across the boundary. Therefore, the magnitude of the reflective wave is smaller than that of the incidental wave. In a real situation, the boundary impedance can be interpreted as a mixture of the three components.

5.2.4 Effects of Boundary Conditions on Eigenfrequencies

The end correction for a one-dimensional vibrating system such as a vibrating string was described according to the wave theory in the previous section. However, such a boundary effect on the eigenfrequencies can be intuitively understood in terms of vibration for two-degree-of-freedom systems. Namely, it can be described from the view point of transmission or reduction of vibration between vibrating systems.

Consider a coupled vibrating system as shown in Fig. 5.8. Set the exciting force $F_X e^{i\omega t}$ for the mass 2. Then, two equations of motion are written as

$$M_1 \frac{d^2 x_1(t)}{dt^2} = -K_1 x_1(t) - K_2(x_1(t) - x_2(t)) \tag{5.51}$$

$$M_2 \frac{d^2 x_2(t)}{dt^2} = -K_2(x_2(t) - x_1(t)) + F_X e^{i\omega t}. \tag{5.52}$$

By substituting $x_1(t) = U_1(\omega)e^{i\omega t}$ and $x_2(t) = U_2(\omega)e^{i\omega t}$ for the equation above,

$$U_1(\omega) = \frac{K_2 F_X}{(K_1 + K_2 - \omega^2 M_1)(K_2 - \omega^2 M_2) - K_2^2} = \frac{K_2}{D(\omega)} F_X \tag{5.53}$$

$$U_2(\omega) = \frac{K_1 + K_2 - \omega^2 M_1}{D(\omega)} F_X \tag{5.54}$$

are derived. Here system 1 in the figure can be regarded as the system that represents the boundary condition. In other words, if the boundary can be regarded as a completely fixed point, then the vibrating system can be simplified to the single-degree-of-freedom system only.

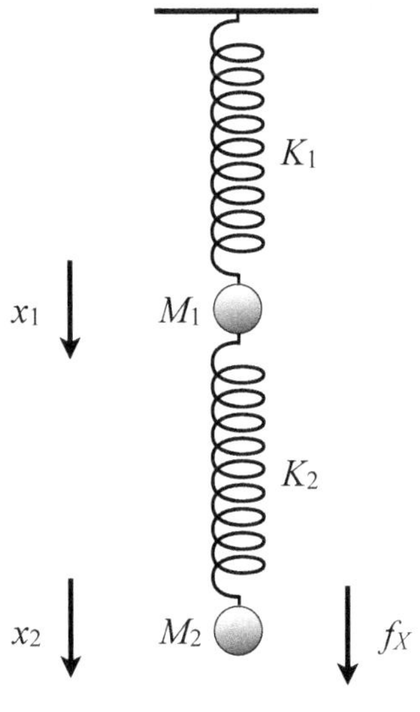

Fig. 5.8 Example of coupled vibrating system

It might be interesting to see the eigenfrequencies of the system including the effect of the "boundary condition". For that purpose, first suppose that the boundary can be assumed to be purely elastic without the mass effect. By assuming such a boundary condition, the denominator of Eq. 5.54 can be written as

$$D(\omega) \cong K_1K_2 - (K_1 + K_2)\omega^2 M_2 \tag{5.55}$$

according to the assumption $M_1 \to 0$. Thus the eigenfrequency can be obtained as

$$\omega^2 = \frac{K_2}{M_2}\frac{K_1}{K_1 + K_2} < \frac{K_2}{M_2}. \tag{5.56}$$

Consequently, it can be seen that the eigenfrequency becomes lower than the original one due to the elastic boundary. This outcome corresponds to the end correction of the elastic boundary for a finite length of a string.

On the other hand, assuming a massive boundary, namely $K_1 \to 0$, then

$$D(\omega) \cong \omega^2(M_1M_2\omega^2 - K_2(M_1 + M_2)). \tag{5.57}$$

Thus the eigenfrequency becomes

$$\omega^2 = \frac{K_2}{M_2}\frac{M_1 + M_2}{M_1} > \frac{K_2}{M_2}. \tag{5.58}$$

The outcome above represents the eigenfrequency becoming higher than the original because of the massive boundary effect. This also corresponds to end correction for the vibration of a finite length string.

5.3 Driving Point Impedance of Vibrating String

Excitation of vibration motion and its travel in a medium can be represented in terms of impedance. The **driving point impedance** is defined by the ratio of the force to vibration velocity at the exciting position.

5.3.1 Driving Point Impedance of Travelling Wave on String

Assume a sinusoidal wave, excited at one end $x = 0$ and travelling on a string, whose displacement $u(x,t)$ is written as

$$u(x,t) = U(x)\mathrm{e}^{\mathrm{i}\omega t} = A\mathrm{e}^{\mathrm{i}(\omega t - kx)}. \qquad \text{(m)} \tag{5.59}$$

Recall that the external force applied at $x = 0$ can be expressed by Eq. 5.39[21]. Thus the displacement and its velocity can be rewritten as

$$u(x,t) = \frac{F_X}{\mathrm{i}kF_T}\mathrm{e}^{\mathrm{i}(\omega t - kx)} = \frac{F_X}{\mathrm{i}\omega\rho_0 c}\mathrm{e}^{\mathrm{i}(\omega t - kx)} \qquad \text{(m)} \tag{5.60}$$

$$v(x,t) = \frac{F_X}{\rho_0 c} e^{i(\omega t - kx)}, \qquad (\mathrm{m/s}) \tag{5.61}$$

where $F_T = \rho_0 c^2$ (N) is the tension, ρ_0 (kg/m) denotes the mass of the string for a unit length, and c (m/s) is the wave speed. Consequently, the **driving point mechanical impedance** of the string becomes

$$Z_{M_{\text{in}}} = \left.\frac{f_X}{v}\right|_{x=0} = \rho_0 c. \qquad (\mathrm{N \cdot s/m}) \tag{5.62}$$

It can be seen that the driving point impedance is real, so a product of the force and velocity is also real such that

$$W_0 = \left.\frac{1}{2}\mathrm{Re}(f_X^* v)\right|_{x=0} = \frac{1}{2}\mathrm{Re}\frac{|F_X|^2}{Z_{M_{\text{in}}}} = \frac{1}{2}\frac{|F_X|^2}{\rho_0 c} \qquad (\mathrm{W}) \tag{5.63}$$

on average over a unit time interval, where Re(*) denotes taking the real part of the corresponding complex quantity, f_X (N) denotes the sinusoidal external source, f_X^* means the complex conjugate of f_X, and 1/2 corresponds to the time-average for a squared sinusoidal function. This means the energy continuously travels from the source into the string.

5.3.2 *Driving Point Impedance of Standing Wave on Finite Length of String*

Instead of the example above, suppose that there is a finite length of string whose ends are fixed ad $x = 0$ and $x = L$. Now set the external sinusoidal source at $x = 0$; the waves travelling on the string must meet the boundary and source conditions. Take the two travelling waves that cause the displacement of vibration such that

$$u(x,t) = Ae^{i(\omega t - kx)} + Be^{i(\omega t + kx)}. \qquad (\mathrm{m}) \tag{5.64}$$

For the boundary and source conditions given by

$$u(L,t) = 0 \tag{5.65}$$

$$-F_T \left.\frac{\partial u(x,t)}{\partial x}\right|_{x=0} = F_X e^{i\omega t} = f_X(0,t), \tag{5.66}$$

$$Ae^{-ikL} + Be^{ikL} = 0 \tag{5.67}$$

$$ikF_T A - ikF_T B = F_X \tag{5.68}$$

hold well.

By solving the simultaneous equations above,

$$A = \frac{F_X e^{ikL}}{i2\rho_0 c\omega \cos kL} \tag{5.69}$$

$$B = \frac{-F_X e^{-ikL}}{i2\rho_0 c\omega \cos kL} \tag{5.70}$$

are obtained as the solutions. After substituting A and B above into Eq. 5.64, the displacement of vibration can be written as

$$u(x,t) = \frac{1}{\rho_0 c\omega} \frac{\sin k(L-x)}{\cos kL} f_X(0,t). \qquad \text{(m)} \tag{5.71}$$

and its velocity is

$$v(x,t) = \frac{\partial u(x,t)}{\partial t} = \frac{i}{\rho_0 c} \frac{\sin k(L-x)}{\cos kL} f_X(0,t). \qquad \text{(m/s)} \tag{5.72}$$

Therefore, the driving point mechanical impedance becomes

$$Z_{M_{\text{in}}} = \frac{f_X(0,t)}{v(0,t)} = -i\rho_0 c \cot kL. \qquad (\text{N} \cdot \text{s/m}) \tag{5.73}$$

Note here that driving point mechanical impedance is purely imaginary. This fact indicates that the power might not be absorbed by the travelling waves on the string. This is an important difference between the **travelling** and **standing** waves. It is the outcome due to the "un-realistic" assumption that there is no damping in the vibration of a string. Namely, the boundary at $x = L$ is ideally fixed without any motion.

The results showing no energy transfer imply that it is impossible to define which is the external source of force for the motion of vibration. Namely, if the system were at the steady state and external source stopped, the motion of the system and its external piston-like source would continue forever. The energy goes back and forth between the piston and the string.

The situation above is not possible in a practical case, and it is only possible from an ideal point of view assuming no energy consumption. Recall that strings are pinned at the ends on the sound board in a piano. If no energy is transferred from the vibrating strings into the sound board, no sound is available from the piano. Such energy transfer through the boundaries is described in terms of the **end correction** for the boundaries as described in subsection 5.2.3.

5.3.3 Driving Point Impedance and Power Injection from External Source

Now reconsider power injection from the external source into a system of interest subject to a small amount of energy loss being assumed in the system. Recall the driving point mechanical impedance $Z_{M_{\text{in}}}$ at $x = 0$ of a finite length of string described in the previous subsection. The power transmitted from the source into the string was written by Eq. 5.73. Here, by introducing the complex variable

$$k = k_r - \mathrm{i}\beta \qquad (k_r >> \beta), \tag{5.74}$$

the driving point impedance can be written as

$$Z_{M_{\mathrm{in}}}(0) = -\mathrm{i}\rho_0 c \frac{\cos(k_r - \mathrm{i}\beta)L}{\sin(k_r - \mathrm{i}\beta)L} \cong -\mathrm{i}\rho_0 c \frac{\cos k_r L + \mathrm{i}\beta L \sin k_r L}{\sin k_r L - \mathrm{i}\beta L \cos k_r L}. \quad (\mathrm{N}\cdot\mathrm{s/m}) \tag{5.75}$$

Therefore, the power injected from the source becomes

$$W_X \cong \frac{1}{2}\frac{|F_X|^2}{\rho_0 c}\frac{\beta L}{\cos^2 k_r L} \cong \frac{1}{2}\frac{|F_X|^2}{\rho_0 c}\frac{\beta L}{\cos^2 kL} \qquad (\mathrm{W}) \tag{5.76}$$

where $k \cong k_r$ is assumed.

Now recall the vibration displacement $u(x,t)$ given by Eq. 5.71 depends on the observation point of x. By taking the spatial average of the squared magnitude response over all the observation points on the string,

$$< u^2 > = \frac{1}{2}\overline{u(x,t)\cdot u^*(x,t)} = \frac{1}{2}\frac{1}{L}\int_0^L |u(x,t)|^2 dx = \frac{1}{4}\frac{|F_X|^2}{\rho_0 c}\frac{1}{\cos^2 kL}\frac{1}{\rho_0 c\omega^2}, \quad (\mathrm{m}^2) \tag{5.77}$$

is obtained. The following relationship

$$W_X = < u^2 > \cdot \rho_0 c\omega^2 \cdot 2\beta L \qquad (\mathrm{W}) \tag{5.78}$$

is found to hold well between the injected power and squared magnitude response averaged over all the space. It can be interpreted as the **energy balance equation** between the injected and consumed power in the system. In other words, the power response from the source located in the system can be estimated by the spatial average of the squared quantity of the vibration magnitude.

5.3.4 Driving Point Impedance and Source Position

The driving point mechanical impedance depends on the source position for standing waves on a finite length of string. Suppose that the external source is located at $x = x' \neq 0$. Set the mechanical impedance for the right hand side of the string to be $Z_{MR_{in}}$. By introducing the velocity at the source position $v(x',t)$, the force for the right hand side string can be written as

$$f_{X_R}(x',t) = Z_{MR_{in}} v(x',t) = -\mathrm{i}\rho_0 c \cot k(L - x') \cdot v(x',t). \qquad (\mathrm{N}) \tag{5.79}$$

where $k \cong k_r$ is assumed and L denotes the length of the string. By similarity taking the left hand side of the string,

$$f_{X_L}(x',t) = Z_{ML_{in}} v(x',t) = -\mathrm{i}\rho_0 c \cot kx' \cdot v(x',t). \qquad (\mathrm{N}) \tag{5.80}$$

is obtained. Consequently, the driven point mechanical impedance is given by

$$Z_{M_{\text{in}}}(x') = \frac{f_{X_R}(x',t) + f_{X_L}(x',t)}{v(x',t)} = -\mathrm{i}\rho_0 c \frac{\sin kL}{\sin kx' \sin k(L-x')} \quad (\mathrm{N \cdot s/m}) \quad (5.81)$$

where $c^2 = F_T/\rho_0$. As shown here, the driving point impedance actually depends on the location of the source. Note that the driving point impedance does not depend on the type of external source. If a velocity type of external source is applied at $x = x'$ on the string instead, then the impedance is the same as that by the equation above.

5.3.5 *Poles and Zeros of Driving Point Impedance*

The impedance function can be formulated by **poles** and **zeros**[23][24]. The poles correspond with the singularities such as the zeros of the denominator, while the zeros indicate zeros of the numerator. Recall the driving point impedance

$$Z_{M_{\text{in}}}(0) = -\mathrm{i}\rho_0 c \frac{\cos kL}{\sin kL} \quad (5.82)$$

$$Z_{M_{\text{in}}}(x') = -\mathrm{i}\rho_0 c \frac{\sin kL}{\sin kx' \sin k(L-x')}. \quad (\mathrm{N \cdot s/m}) \quad (5.83)$$

It can be confirmed that the poles and zeros are interlaced on the frequency axis[23][24]. Suppose that k is a wavenumber of interest. The number of zeros distributed under the wavenumber can be estimated as

$$N_z = \frac{k}{\pi/L}. \quad (5.84)$$

By similarity considering the poles, those numbers can be also estimated as

$$N_p = N_{p_1} + N_{p_2} \quad (5.85)$$

$$N_{p_1} = \frac{k}{\pi/x'} \quad (5.86)$$

$$N_{p_2} = \frac{k}{\pi/(L-x')}. \quad (5.87)$$

Therefore, the difference in the numbers of the poles and zeros is

$$N_z - N_{p_1} - N_{p_2} = \frac{k}{\pi}\left((L-x') - (L-x')\right) = 0. \quad (5.88)$$

Namely, it can be understood that the poles and zeros must be interlaced.

The poles and zeros of driving point mechanical impedance are closely related to the eigenfrequencies of the system. Suppose that the external source can be assumed as **constant force**. The excited velocity on the string by the force can be written as

$$v(x',t) = \frac{f_X(x',t)}{Z_{M_{\text{in}}}(x')} \qquad (\text{m/s}) \tag{5.89}$$

in terms of the driving point mechanical impedance. Then the **zeros of the impedance** create the **resonance for the standing waves** of corresponding frequencies with the zeros, while the **poles** create the **anti-resonance** of the waves. Namely, standing waves of the frequencies corresponding to the poles cannot be excited by the force at the source position. On the other hand,the poles create the resonance when the external source is assumed to be **constant velocity**. The external force can be rewritten as

$$f_X(x = x',t) = v(x',t)Z_{M_{\text{in}}(x')} \qquad (\text{N}) \tag{5.90}$$

according to the driving point mechanical impedance.

In the case of the piano, the external source position (hammer position) is set at around $L/8$ where L is the length of a string in the piano[25]. This is because around 7-9 harmonics might be removed from the higher harmonics. This can be interpreted in terms of driving point impedance. The driving point impedance becomes

$$Z_{M_{\text{in}}}(x')\big|_{x'=L/8} = -i\rho_0 c \frac{\sin kL}{\sin(kL/8)\sin(7kL/8)}. \qquad (\text{N}\cdot\text{s/m}) \tag{5.91}$$

Here the first pole is given by

$$\frac{k_{p_1}L}{8} = \frac{2\pi\nu_{p_1}L}{8c} = \pi. \tag{5.92}$$

Namely,

$$\nu_{p_1} = 8\frac{c}{2L} = 8\nu_1 \qquad (\text{Hz}) \tag{5.93}$$

is obtained. The outcome indicates that the standing wave of the frequency given by ν_{p_1} cannot be excited by the external force, subject to the source being assumed as constant.

5.4 Propagation of Initial Disturbance along Finite Length of String

Propagation of initial disturbance on a finite length of string can also be graphically illustrated according to the boundary conditions. The notion of normal modes or harmonics cannot be explicitly seen in the graphical method for representing wave propagation. However, it can be found that such a wave theoretic concept as normal modes must be hidden in the graphical method by recalling that periodically propagating waves are expressed as odd and spatially periodic functions following the boundary conditions.

5.4.1 Propagation of Initial Displacement

Consider a finite length of string with initial displacement as shown in Fig. 5.9. The initial displacement in the interval between $x = 0$ and $x = L$ can be extended as an odd and periodic function.

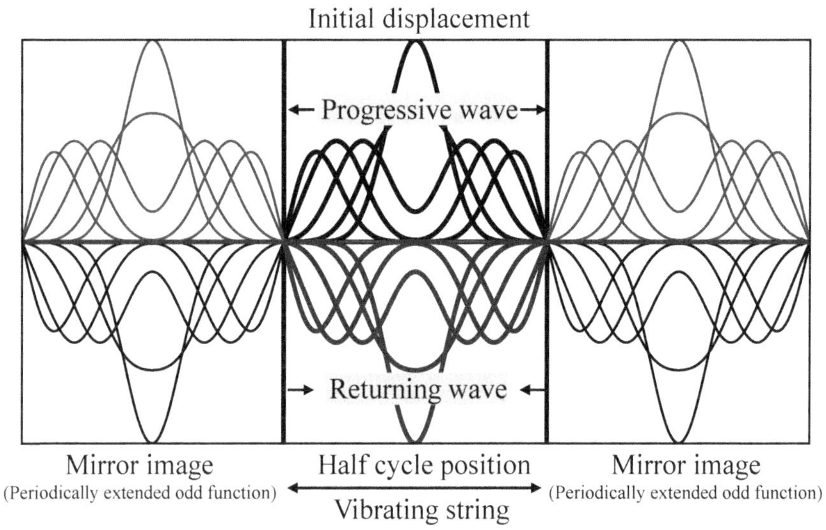

Fig. 5.9 Propagation of initial displacement for finite length of string with clamped ends from [9] (Fig.5.6)

Similar to wave propagation along a long string as described in section 5.1, the initial disturbance is equally separated and propagated to the left and right even on a finite length of string. However, for a finite length of string, such separation and propagation simultaneously occur in every interval between $x = lL$ and $x = (l+1)L$ where l is an integer, as if the disturbance occurred in every interval on an infinitely extending virtual string. Here, every wave entering the interval ($x = 0$ to $x = L$) from outside is interpreted as a reflection wave. It takes a long time for a reflection wave to return to the original interval, as a virtually extended interval from which the reflection comes is far from the original. In other words, a reflection wave coming from the interval between $x = lL$ and $x = (l+1)L$ represents an l-th reflection wave after l collisions into the boundary.

Consequently, waves that are propagated on the string between $x = 0$ and $x = L$ are obtained by periodic superposition of all the reflection waves from outside the interval. The period of vibration is $T_1 = 2L/c = 1/\nu_1$ (s), which actually corresponds to the time interval of successive reflections; namely, reflection waves enter every interval of T_1.

5.4.2 *Propagation of Initial Velocity*

Figure 5.10 illustrates propagation of displacement due to the initial disturbance of velocity instead of displacement. Vibration including build-up and decaying processes is cyclic with a period of $2L/c$ (s), and the vibration pattern itself is extended as an odd and spatially periodic function with an interval of $2L$ (m). Displacement shown in the figure was due to conversion from the initial velocity because of no initial displacement. The displacement might be interpreted as left over after a wave passed in every cycle.

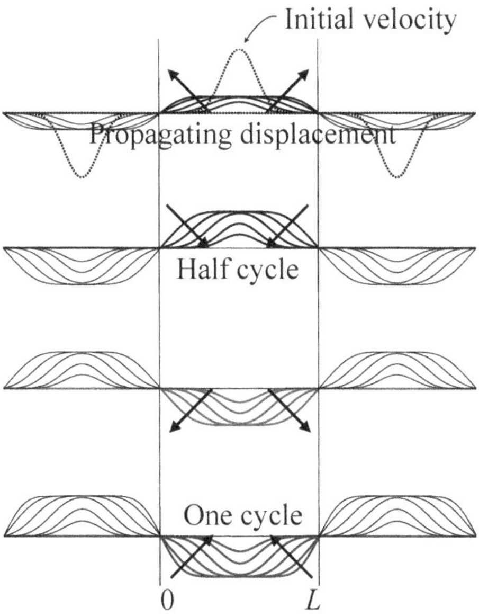

Fig. 5.10 Propagation of displacement due to initial velocity for finite length of string with clamped ends from [9] (Fig.5.7)

5.5 Impulse Response and Its Transfer Function for Vibrating String

A vibrating string is a good example that shows the relationship between an impulse response and its transfer function for a linear system.

5.5.1 *Impulse Response of Finite Length of String*

Suppose that a finite length of string is excited by an impulse-like initial excitation as position S as shown by Fig. 5.11. The response observed at position O is composed of four types of pulse series that start at different initial positions on the time axis, but with the same time intervals of successive pulses as shown in Fig. 5.12.

Here set the time intervals as follows:

$$\tau_{S_R} = (x - x')/c = N_a T_s \tag{5.94}$$
$$\tau_{S_L} = (2x' + (x - x'))/c = N_b T_s \tag{5.95}$$
$$\tau_{O_R} = 2x_o/c = N_\eta T_s \tag{5.96}$$
$$\tau_{O_L} = 2(x)/c = N_\gamma T_s \tag{5.97}$$

and

$$\tau_{SR} + \tau_{OR} = N_c T_s = (N_a + N_\eta) T_s \tag{5.98}$$
$$\tau_{SL} + \tau_{OR} = N_d T_s = (N_b + N_\eta) T_s \tag{5.99}$$
$$\tau_{OR} + \tau_{OL} = \tau_T = N_T T_s \tag{5.100}$$

where T_S denotes the sampling period, c is the wave speed, x', x are shown in Fig. 5.11 and $x + x_o = L$.

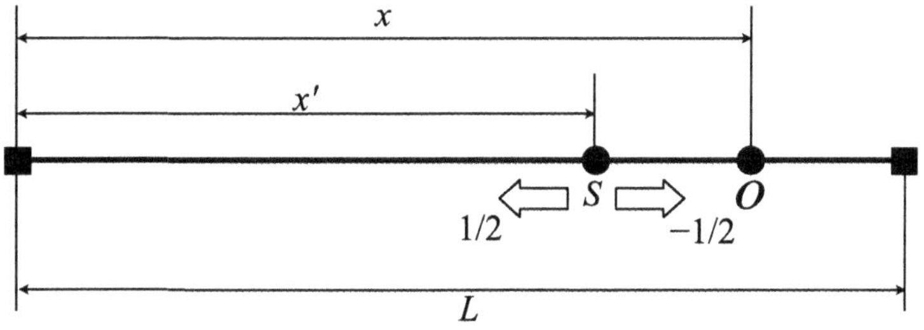

Fig. 5.11 Impulsive source excitation on finite length of string

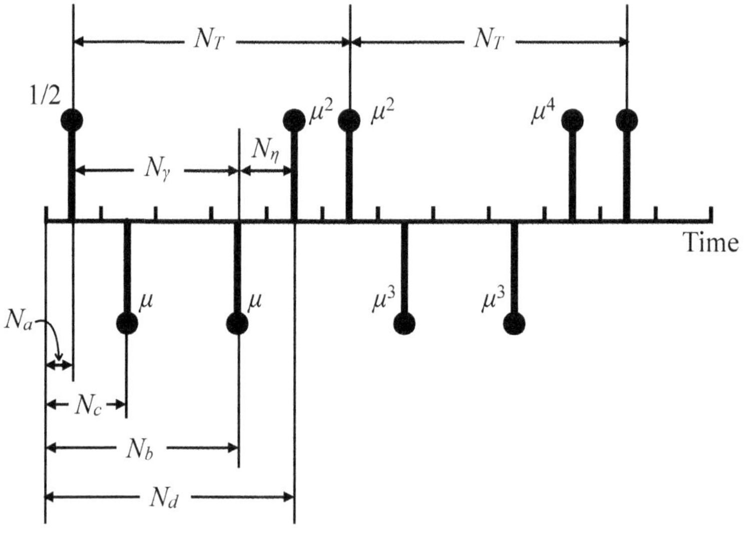

Fig. 5.12 Initial time sequence of impulse response

5.5.2 Transfer Function of Impulse Response

The impulse response can be represented by a sum of the four types of responses:

$$h(n) = h_a(n) + h_b(n) + h_c(n) + h_d(n). \tag{5.101}$$

The **transfer function**, which is defined by

$$H(z^{-1}) = \sum_n h(n) z^{-n}, \tag{5.102}$$

can be written as

$$H(z^{-1}) = \frac{1}{2} \frac{z^{-N_a} - \mu z^{-N_b} - \mu z^{-N_c} + \mu^2 z^{-N_d}}{1 - \mu^2 z^{-N_T}}, \tag{5.103}$$

where μ represents the magnitude of reflection coefficients for the boundaries.

The poles of the transfer function, which corresponds to eigenfrequencies of the string, can be determined by using the resonant characteristics of the string independent of the source and observation positions. In contrast, the zeros are determined by the locations of source or observation points independent of the resonance properties of the vibration of the string. By substituting $z = \mathrm{e}^{\mathrm{i}\Omega}$ for z where $\Omega = \omega T_S$ in the equation above, the **frequency response** of the string can be derived as described in the following section.

5.5.3 Frequency Response of String Vibration

Recall that μ represents the reflection coefficient, and set $\mu = \mathrm{e}^{-\delta}$. Namely,

$$\delta = -\ln \mu \tag{5.104}$$

into Eq. 5.103, the numerator of the equation can be rewritten as

$$\begin{aligned} &(z^{-N_a} - \mu z^{-N_b} - \mu z^{-N_c} + \mu^2 z^{-N_d})\big|_{z=\mathrm{e}^{\mathrm{i}\Omega}} \\ &= \mathrm{e}^{-\mathrm{i}\frac{\Omega}{2}(N_a+N_b+N_\eta)} \mathrm{e}^{-\delta} \left(\mathrm{e}^{\mathrm{i}\frac{\Omega_1}{2}} - \mathrm{e}^{-\mathrm{i}\frac{\Omega_1}{2}}\right) \left(\mathrm{e}^{\mathrm{i}\frac{\Omega_2}{2}} - \mathrm{e}^{-\mathrm{i}\frac{\Omega_2}{2}}\right), \end{aligned} \tag{5.105}$$

where

$$\Omega_1 = \Omega(N_b - N_a) - \mathrm{i}\delta \tag{5.106}$$

$$\Omega_2 = \Omega N_\eta - \mathrm{i}\delta. \tag{5.107}$$

Similarly, the resonant part becomes

$$\left.\frac{1}{1 - \mu^2 z^{-N_T}}\right|_{z=\mathrm{e}^{\mathrm{i}\Omega}} = \frac{\mathrm{e}^{\delta} \mathrm{e}^{\mathrm{i}\frac{N_T \Omega}{2}}}{\mathrm{e}^{\mathrm{i}\frac{\Omega_3}{2}} - \mathrm{e}^{-\mathrm{i}\frac{\Omega_3}{2}}} \tag{5.108}$$

where

$$\Omega_3 = N_T\Omega - \mathrm{i}2\delta. \tag{5.109}$$

Thus the frequency response is obtained such that

$$H\left(\mathrm{e}^{-\mathrm{i}\Omega}\right) = \mathrm{i}\frac{\sin\left(\frac{N_b-N_a}{2}\Omega - \frac{\mathrm{i}\delta}{2}\right)\sin\left(\frac{N_\eta}{2}\Omega - \frac{\mathrm{i}\delta}{2}\right)}{\sin\left(\frac{N_T}{2}\Omega - \mathrm{i}\delta\right)} \tag{5.110}$$

where $N_a + N_b + N_\eta - N_T = 0$.

By substituting $\Omega = \omega T_s$ and $\omega = ck$ into the equation above, the frequency response can be described as

$$\begin{aligned} H(x',x,k) &= \mathrm{i}\frac{\left(\sin kx' - \frac{\mathrm{i}\delta}{2}\right)\sin\left(\sin k(x-x') - \frac{\mathrm{i}\delta}{2}\right)}{\sin(kL - \mathrm{i}\delta)} \\ &\cong -\frac{\sin kx' \sin k(x-x')}{\delta\cos kL + \mathrm{i}\sin kL} \end{aligned} \tag{5.111}$$

for the rigid boundary condition where x' and x are shown in Fig. 5.11. It can be found that the magnitude of the response might take its maxima at around the frequencies of free oscillation. However, such resonant characteristics are lost as the reflection coefficient becomes small.

On the other hand, the magnitude of the response becomes 0 if $\sin kx' = 0$ or $\sin kx = 0$. This indicates that the frequencies of waves cannot be excited by the positions where $\sin kx' = 0$, or the waves that become nodes at the observation point cannot be detected, even if such waves are exited on the string.

Note here also that the initial delay for the direct wave between the source nad receiving positions does not necessarily have to be explicitly represented in the frequency response equation. The phase characteristics due to such a delay can be represented by the **propagation phase**[23][24] for the sinusoidal waves travelling between the source and receiving positions. These are typical characteristics for the waves progressing in a one-dimensional vibrating system.

5.5.4 *Spectral Envelope for Frequency Characteristics*

The numerator of the transfer function in Eq. 5.103 can be understood as the **spectral envelope** for frequency characteristics. Here the spectral envelope can also be interpreted as spectral properties of the source signal that cyclically propagates on the string. From a physical view point, the direct wave without any reflection waves is the first pulse-like wave. However, the periodical properties of the waves are determined by the group of four pulses that periodically travels on the string. Therefore, the group of pulses creates a **spectral envelope** of the signal that is "sampled" by every fundamental frequency in the spectral domain.

Note here that such spectral properties of the original wave depends on the observation position. Therefore, if the observation position is set to be very close to the source position, the spectral properties can be representative of the vibrating string under the condition of the excitation source.

5.5.5 Energy Decay Curve and Impulse Response

Suppose that an impulse response of a vibrating system is given by $h(n)$ that is assumed to be 0 when n is negative. Taking a sum of squared quantity such that

$$E_{Rev}(n) = \sum_{m=n}^{\infty} h^2(m) = \sum_{m=0}^{\infty} h^2(m) - \sum_{m=0}^{n-1} h^2(m) \tag{5.112}$$

is called a **reverberation energy decay curve** representing the reverberation decay of the system vibration energy. For example, set the impulse response as

$$h(n) = \mu^{2n} = (1-\alpha)^n \tag{5.113}$$

following the example of impulse responses in a one-dimensional system shown in the previous section where $\mu^2 = 1 - \alpha = \mathrm{e}^{-2\delta}$.

The energy decay curve can be written as

$$E_{Rev}(n) = \frac{(1-\alpha)^{2n}}{1-(1-\alpha)^2}. \tag{5.114}$$

By introducing

$$(1-\alpha)^{2n} = \mu^{4n} = \mathrm{e}^{-4\delta n} \tag{5.115}$$

$$\ln(1-\alpha) = = -2\delta = -\hat{\alpha} \tag{5.116}$$

the decay curve becomes

$$E_{Rev}(n) = \frac{\mathrm{e}^{-2\hat{\alpha}n}}{1-(1-\alpha)^2} = E(0)\mathrm{e}^{-2\hat{\alpha}n}, \tag{5.117}$$

where $E(0)$ denotes the energy before starting the decay, namely, the **energy at the steady state**. It can be seen that the energy decay follows an exponential function of time. The **reverberation time** is defined as the time required for the energy to decay to 60 dB below the steady state level. Figure 5.13 shows a numerical example of the energy decay curve for a one-dimensional vibrating system. The energy decay follows an exponential function as the decay progresses.

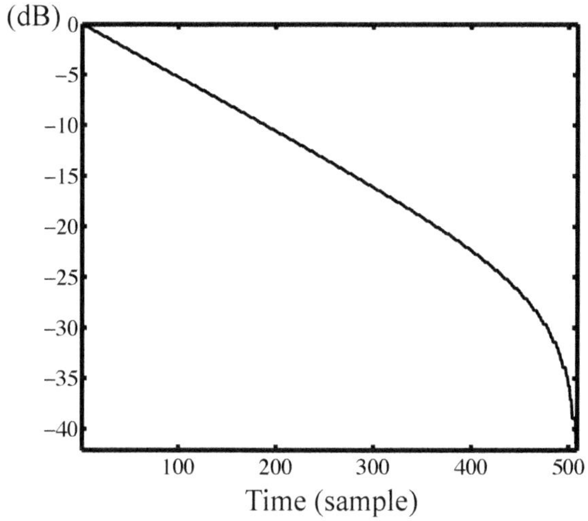

Fig. 5.13 Example of reverberation decay curve for a one-dimensional vibrating system

5.6 Eigenfrequencies and Eigenmodes

Free vibration of a finite length of string can be interpreted in two ways: based on vibration patterns composed of a fundamental and its harmonics, or following graphical representation of expansion of initial disturbance into an odd and periodic function. Eigenfrequencies and eigenmodes are important notions for uniting these ways. This section describes how a fundamental and its harmonics can be interpreted as eigenfrequencies. Consequently it can be confirmed that the period of free vibration is equal to the fundamental period, even if free vibration is composed of many harmonics. This outcome is significant for perception of sounds such as musical sounds.

5.6.1 Eigenfrequencies for Free Vibration of Finite Length of String

As illustrated in Fig. 5.6, vibration patterns with definite periods, namely normal modes of vibration, are expected in free vibration of a finite length of string. Such a vibration pattern is called an **eigenmode**, **eigenfunction**, or **modal function**. Its frequency is called an **eigenfrequency**, and it corresponds to the resonant frequency. These eigenfrequencies and eigenfunctions can be interpreted as eigenvalues and eigenvectors for a linear operator such as a matrix[2][3]. From the fact that vibration corresponding to the n-th modal function is expressed as

$$u_n(x,t) = \sin\frac{n\pi x}{L}\cos\frac{n\pi ct}{L} = \frac{1}{2}\left(\sin\frac{n\pi}{L}(x+ct) + \sin\frac{n\pi}{L}(x-ct)\right), \quad (5.118)$$

a modal function itself can be expressed as a superposition of two progressive waves with a definite frequency. The vibration above can be also rewritten in a general form such as

$$u_\omega(x,t) = \sin\frac{\omega}{c}x\cos\omega t. \tag{5.119}$$

Here, angular frequencies ω, however, must be a series of discrete numbers such that

$$\omega = \omega_n = 2\pi\frac{nc}{2L} \qquad \text{(rad/s)} \tag{5.120}$$

where $n = 1,2,3,....$, because the function representing vibration must be periodic with an interval of $x = 2L$.

The eigenfrequencies are countless but located at every interval of the fundamental frequency. Such a series of frequencies with equal intervals is also called **harmonic structure** in terms of acoustics. The n-th angular eigenfrequency ω_n can be rewritten as

$$\omega_n = n\pi\sqrt{\frac{F_T/L}{\rho_0 L}} = \pi\sqrt{\frac{nF_T/L}{\rho_0 L/n}}, \qquad \text{(rad/s)} \tag{5.121}$$

where ρ_0(kg/m) denotes the density of a string and F_T(N) is tension of the string of interest. If we regard nF_T/L and $\rho_0 L/n$ as the tension density and partial mass of a "divided" string, the n-th eigenfrequency looks like the fundamental for the divided part.

5.6.2 *Superposition of Eigenmodes and Its Period*

Superposition of sinusoidal functions with harmonics is significant from an acoustical point of view. Recall that

$$\sin\omega_1 t = \sin\omega_1(t + lT_1) \tag{5.122}$$

$$\sin n\omega_1 t = \sin n\omega_1(t + lT_1) \tag{5.123}$$

$$T_1 = \frac{2\pi}{\omega_1} \qquad \text{(s)} \tag{5.124}$$

where l is an integer. As shown in Fig. 5.14, superposition

$$u(t) = A_1\sin(\omega_1 t + \phi_1) + A_2\sin(2\omega_1 t + \phi_2) + ... + A_n\sin(n\omega_1 t + \phi_n) \tag{5.125}$$

takes the same value at every T_1 seconds; namely, the signal $u(t)$ is still periodic with the fundamental period. This explains why a listener can perceive the pitch of a compound sound. In other words, the period of a signal composed of harmonic components is equal to the fundamental period regardless of whether the fundamental component is contained or not.

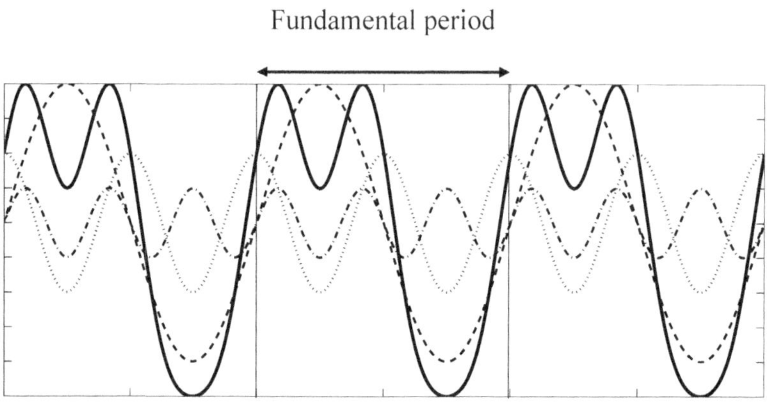

Fig. 5.14 Example of superposition of harmonic functions

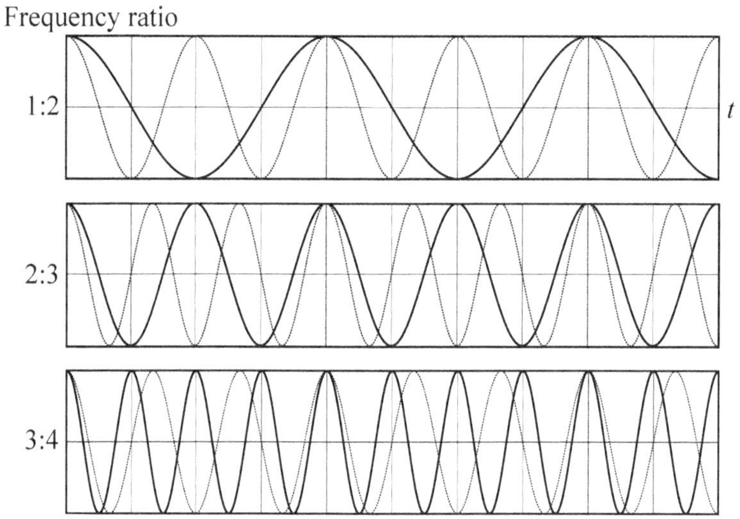

Fig. 5.15 Examples for pairs of two harmonic components from [9] (Fig.5.5)

Figure 5.15 illustrates samples of harmonic pairs of waveforms. The top panel shows waveforms of a fundamental and second harmonics, respectively. The two components return to their initial positions every fundamental period; however, the second harmonic component takes two cycles every fundamental period, while the fundamental one takes only one. Similarly, the middle one illustrates the second and third harmonics. Again both components return to the initial points every fundamental period, but two and three cycles are required for the second and third ones, respectively. The third and fourth harmonics are similarly drawn in the bottom row,

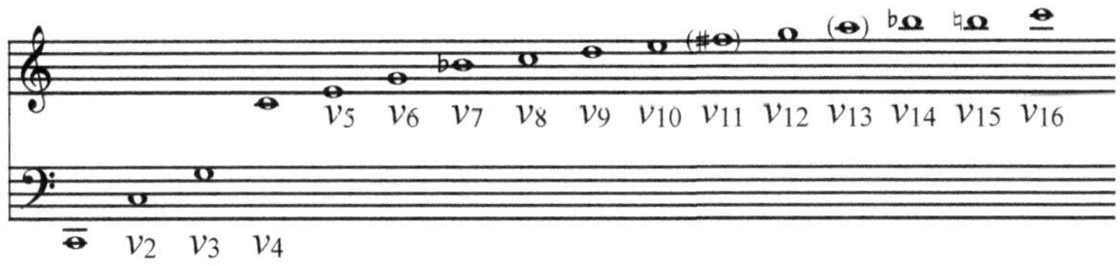

Fig. 5.16 Musical notes following fundamental and its harmonics from [4] (Fig. 2.7)

where the third one needs three cycles, and four cycles are necessary for the fourth one for every fundamental period. These samples indicate that the number of cycles contained in the fundamental period increases with the order of harmonics, and thus a pair composed of higher-order harmonics includes many cycles every fundamental period.

Consequently, a pair of lower-ordered harmonics could be consonant rather than a higher-ordered pair. Figure 5.16 shows musical notes corresponding to a fundamental and its harmonics where the fundamental is set as C[4]. Musical intervals between an adjacent harmonic pair becomes narrow as the order of harmonics increases, and thus tones become dissonant as the order becomes higher.

5.6.3 Expression of Free Vibration under Initial Disturbance

A formal expression can be introduced for free vibration of a string using superposition of modal functions. This formulation provides a physical basis for representation of a function as a periodic expression by using Fourier series in terms of mathematics. An approach to vibration analysis using Fourier series expansion, namely superposition of modal functions, is helpful for analyzing tonal characteristics of sound, which represents for example the differences between a piano and harpsichord. Fourier series expansion of a function can be geometrically interpreted as decomposition of a vector into its orthogonal basis in the space of interest[3]. Fourier coefficients correspond to the coordinates with respect to the basis. Reference [8] might be informative for the theoretical background of mathematical expression of vibration using Fourier series.

Suppose that initial conditions are set to be

$$u(x,0) = a(x) \qquad (0 \le x \le L) \qquad (\mathrm{m}) \tag{5.126}$$

$$v(x,0) = 0. \tag{5.127}$$

Free vibration of a finite string can be expressed as

$$u(x,t) = \sum_{n=1}^{\infty} A_n \sin \frac{n\pi}{L} x \cos(\omega_n t) \qquad (0 \le x \le L) \qquad (m) \tag{5.128}$$

under the boundary conditions such that

$$u(0,t) = 0, \qquad u(L,t) = 0, \qquad \text{(m)} \tag{5.129}$$

where A_n are called modal magnitudes and determined by the initial condition of $a(x)$.

The expression above indicates that the initial disturbance can be expressed as superposition of modal functions (or modal expansion of a periodically extended function) like

$$a(x) = \sum_{n=1}^{\infty} A_n \sin \frac{n\pi}{L} x. \tag{5.130}$$

Following the expression above if the initial disturbance looks like the n-th harmonic function, then vibration could also be made of the n-th mode only. On the other hand, if the n-th harmonic function were not necessary to represent initial disturbance, then vibration might be composed of all the modal functions except the n-th mode. The modal magnitude can be expressed as

$$A_n = \frac{2}{L} \int_0^L a(x) \sin \frac{n\pi}{L} x dx, \tag{5.131}$$

which is called **Fourier coefficients**. Representation of a function as a virtually periodic function by modal expansion is called **Fourier series representation** of a function.

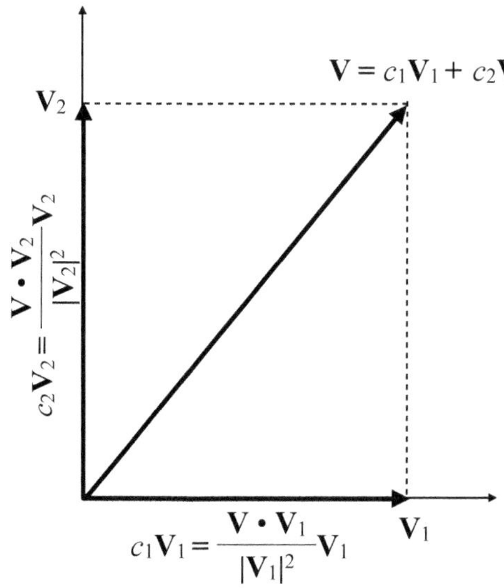

Fig. 5.17 Decomposition of vector into orthogonal vectors

Fourier coefficients, which look like the average of an integrand, normalized by the factor of $L/2$, can be understood as coordinates with respect to the basis of a space of interest, if sinusoidal functions were regarded as the basis[3]. Suppose that two vectors $\mathbf{v}_1$ and $\mathbf{v}_2$ are perpendicular (or orthogonal) and provide the basis of a 2-dimensional space. As shown in Fig. 5.17, vector $\mathbf{v}$ in the space can be written as

$$\mathbf{v} = c_1\mathbf{v}_1 + c_2\mathbf{v}_2 \tag{5.132}$$

where

$$c_1 = \frac{\mathbf{v}\cdot\mathbf{v}_1}{\mathbf{v}_1\cdot\mathbf{v}_1}, \tag{5.133}$$

$$c_2 = \frac{\mathbf{v}\cdot\mathbf{v}_2}{\mathbf{v}_2\cdot\mathbf{v}_2}, \tag{5.134}$$

and the "dot product" of two vectors denotes the **inner product** or the **scaler product** of vectors[2][3].

Therefore, regarding integration for the modal functions as the dot product of modal functions, the Fourier coefficients can be interpreted as decomposition of a vector into the orthogonal vectors. The inner product of identical modal functions is given by

$$\int_0^L \sin^2\frac{n\pi x}{L}dx = \frac{L}{2}, \tag{5.135}$$

which corresponds to the normalized factor of Fourier coefficients.

Figure 5.18 shows an example of wave propagation by initial displacement in a finite length of string with clamped ends. The same type of wave propagation can be seen in Fig. 5.9. This example shows a sample of simplified initial displacement, which is an even function with respect to the center of $x = L/2$. Here, recalling that integration of an odd function over its periodic interval results in 0, the integration over $x = 0$ to $x = L$ of a product of the initial disturbance and an even-ordered modal function results in zero. This is because the integrand could be regarded as an odd function in the interval between $x = 0$ and $x = L$ if the center is shifted to $x = L/2$. Consequently, all the even-ordered harmonic components are removed from the free vibration, if the initial displacement might be a virtually even function as shown by Fig. 5.18. In contrast, if the initial displacement might be an odd function, then all the odd-ordered harmonic components are removed.

According to the outcome above, suppressing definite modes, for example the fifth harmonic component with its harmonics, is possible. Figure 5.19 is an example of initial displacement for removing the fifth mode. Although it cannot be clearly seen in the figure, the fifth (including its odd-ordered higher harmonics) modal functions cannot be included in the propagating waves. This type of consideration was made in the history of musical instruments for removing dissonant harmonics. For that purpose, a position at which to give initial excitation is selected in sound instruments.

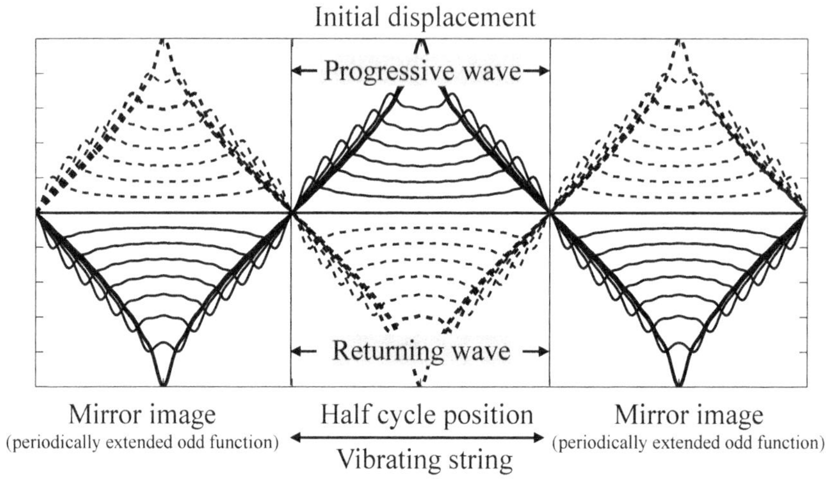

Fig. 5.18 Propagation of initial displacement obtained by modal expansion

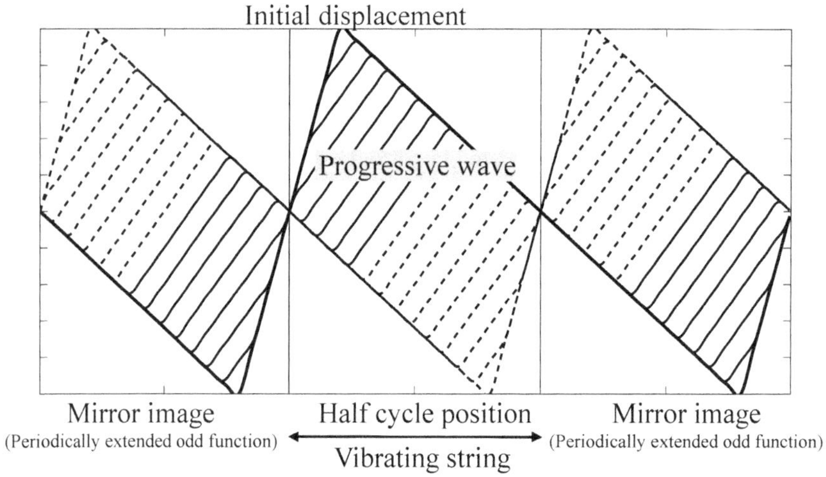

Fig. 5.19 Example of initial displacement and its propagation for removing fifth mode

As shown in Fig. 5.20, propagation of the initial velocity such as that in a piano can be also displayed following the modal expansion. Figure 5.21 is a corresponding example to Fig. 5.19 for removing the fifth modes.

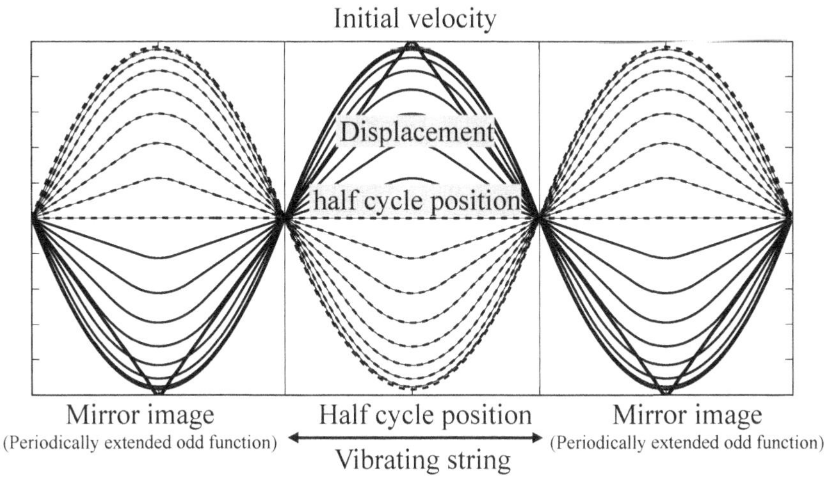

Fig. 5.20 Example of initial velocity and its propagation

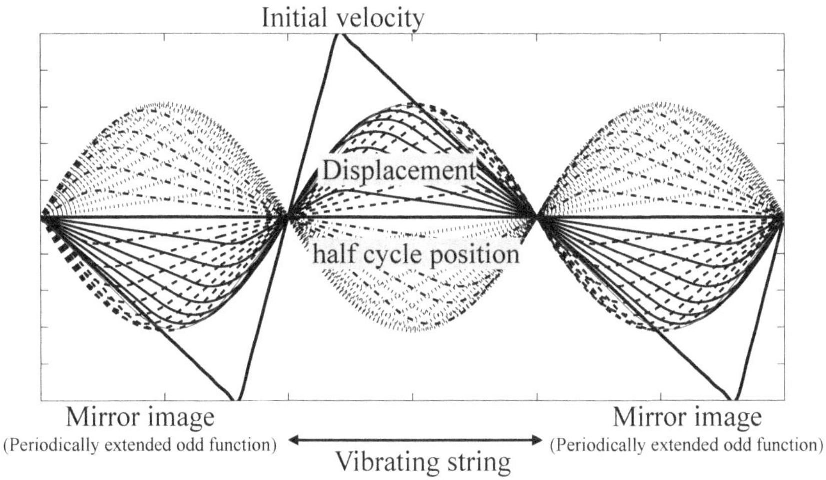

Fig. 5.21 Example of initial velocity and its propagation for removing fifth modal components

Chapter 6
Waves and Speed of Sound in the Air

Sound is a wave that travels in an elastic medium. In particular, sound in the air, such as musical tones, speech, or bird-song, is a longitudinal wave in which change in the local density of the air is propagated. Sound waves in the air can be characterized by sound pressure or particle velocity due to the dilation or condensation of a local area of the air. The direction of the particle velocity or displacement of a **longitudinal wave** is parallel to that in which the wave is travelling. The energy preservation law of sound waves defines the speed of sound. In the last section of this chapter, sound radiated by vibration will also be discussed.

6.1 Propagating Energy and Speed of Sound

6.1.1 Initial Disturbance and Propagating Energy

Suppose that volume $\Delta V(\mathrm{m}^3)$ of a local area of the air changes to $\Delta V(1+s)(\mathrm{m}^3)$ where s denotes the condensation. Such a change of the volume can be interpreted as small displacement of a mass attached to a spring. Therefore, the **potential energy due to change of the density**,

$$E_P = \frac{1}{2}\kappa \Delta V s^2, \qquad (\mathrm{J}) \qquad (6.1)$$

is stored in the compressed or dilated local area as well as in the vibration of a mass-spring system. The compression or dilation of the local area might be a cause of oscillation of the local portion due to the elasticity of the medium. If velocity $v(\mathrm{m/s})$ is yielded due to oscillation, then the kinetic energy of motion

$$E_K = \frac{1}{2}\rho_0 \Delta V v^2 \qquad (\mathrm{J}) \qquad (6.2)$$

is produced, where $\rho_0(\mathrm{kg/m}^3)$ denotes the density of the medium when no oscillation occurs in the medium. Note that change of the density is assumed to be sufficiently small here.

M. Tohyama: *Sound and Signals*, Signals and Communication Technology, pp. 89–102.
springerlink.com

Take an example of a wave travelling as shown in Fig. 5.2. The initial disturbance propagates as longitudinal travelling waves along the horizontal axis. Suppose that the initial potential energy due to the disturbance is unity. The energy-preservation law says that the total energy of the two travelling waves should be unity. Recall that the magnitudes of the equally divided waves are 1/2. The potential energies of the two waves are thus 1/4, respectively. Therefore, the kinetic energy must also be 1/4 for the two waves, and consequently the potential and kinetic energies are the same as each other. This is an important scheme for the propagation of waves. Namely, the two kinds of energy alternate with each other when the wave travels.

Now set both types of energy such that

$$\frac{1}{2}\kappa \Delta V s^2 = \frac{1}{2}\rho_0 \Delta V v^2. \qquad \text{(J)} \tag{6.3}$$

The relationship

$$v = \sqrt{\frac{\kappa}{\rho_0}} s \qquad \text{(m/s)} \tag{6.4}$$

can be derived between the particle velocity v(m/s) and the condensation s. This relationship shows that the particle velocity is in proportion to the condensation. By introducing the relationship into the potential or kinetic energy, the energy density for the unit volume is then written as

$$E_{P_0} = E_{K_0} = \frac{1}{2}\kappa s^2. \qquad (\text{J/m}^3) \tag{6.5}$$

It is proportional to the bulk modulus and square of condensation. The formulation above is the same as that for vibration of a mass-spring system if the bulk modulus and condensation are substituted for the spring constant and displacement, respectively.

6.1.2 *Speed of Sound*

Take a small portion of the medium in which the sound wave travels, and suppose that the cross section of the portion is S(m^2) vertical to the wave-travelling direction. **Volume velocity** is defined as

$$q = Sv = S\sqrt{\frac{\kappa}{\rho_0}} s = Scs \qquad (\text{m}^3/\text{s}). \tag{6.6}$$

Here the volume velocity represents change of volume in a unit time interval due to oscillation as well as particle velocity that indicates temporal variation of the displacement. The expression above shows the volume velocity is in proportion to the length of the portion denoted by c. Namely, the length c is interpreted to be the distance to which the sound is propagated in the unit time interval. Therefore,

$$c = \sqrt{\frac{\kappa}{\rho_0}} \qquad (\mathrm{m/s}) \tag{6.7}$$

represents **sound speed** in the medium. The sound speed in the air is independent of the frequency or magnitude of a wave, but it depends on the volume density and bulk modulus of the air.

Again recall a mass-spring system. The bulk modulus and density correspond to the spring constant and mass, respectively. The ratio between these two quantities defines the eigenfreqeuncy, while the ratio for sound waves in the air shows the sound speed. There is no eigenfrequency for oscillation in the air; instead oscillation in the air yields sound waves traveling with a definite speed.

The sound speed is given by the ratio of the bulk modulus and volume density of the air. It is fast in a medium with a large bulk modulus, even if the volume density is high. The bulk modulus of a gas, however, depends on the variations of the gas state. This means that the sound speed also depends on the dilation and compression processes of a portion of the gas.

Suppose that the process is **isothermal**. As described by Eq. 3.28, the bulk modulus is approximately estimated as

$$\kappa \cong P_a, \qquad (\mathrm{Pa}) \tag{6.8}$$

where P_a denotes the atomic pressure of the air. According to the equation above, sound speed is given by

$$c = \sqrt{\frac{\kappa}{\rho_0}} \cong \sqrt{\frac{P_a}{\rho_0}} \cong 280, \qquad (\mathrm{m/s}) \tag{6.9}$$

where $P_a = 1013(\mathrm{hPa})$ and $\rho_0 = 1.29(\mathrm{kg/m^3})$. This slower speed than the current conventional estimation is equivalent to the historical estimation by Newton[26].

The reason the estimate is a little too slow is due to the assumption that wave propagation might be isothermal. By assuming the **adiabatic process** for sound wave propagation instead, the bulk modulus is then estimated as

$$\kappa \cong \gamma P_a \qquad (\mathrm{Pa}) \tag{6.10}$$

as shown by Eq. 3.32. Consequently, the sound speed is estimated by

$$c = \sqrt{\frac{\kappa}{\rho_0}} \cong \sqrt{\frac{\gamma P_a}{\rho_0}} \cong 330, \qquad (\mathrm{m/s}) \tag{6.11}$$

where $\gamma \cong 1.4$. This estimate by Laplace or Poisson [26] seems reasonable even now[27].

The sound speed, however, also depends on the gas temperature. Recall the squared average of the particle velocity of a gas written as Eq.3.14. The sound speed in a gas can be rewritten as

$$c = \sqrt{\frac{\gamma R_{gas} T_{emp}}{M}} = \sqrt{\frac{\gamma \overline{v^2}}{3}} \qquad (\mathrm{m/s}), \tag{6.12}$$

and thus it is similar to the magnitude of the squared average for the particle velocity in a gas. The sound speed in a gas and particle velocity of the gas increase with the temperature.

6.2 Sound Source and Plane Waves

6.2.1 Sound Pressure and Particle Velocity

A sound wave is excited by a local disturbance or oscillation of a small portion of an elastic medium, similar to the vibration of a local spring being propagated on chained springs with masses. Such a sound wave or elastic wave is characterized by the **sound pressure** and **particle velocity**, both of which travel in the medium as the wave. A wave that one-dimensionally travels in a medium is called a **plane wave**. The pressure and particle velocity can be written using the condensation s as

$$p = \kappa s = \rho_0 c v \qquad (\mathrm{Pa}) \tag{6.13}$$

$$v = cs \qquad (\mathrm{m/s}) \tag{6.14}$$

according to the fact that the potential energy is the same as the kinetic energy. Both the sound pressure and particle velocity are in proportion to the condensation.

The relationship between the sound pressure and particle velocity for a plane wave can also be interpreted following the Newtonian motion law. Suppose a small portion of an elastic medium as shown in Fig. 6.1. The volume that oscillates in a time interval Δt can be estimated by $Sc\Delta t(\mathrm{m}^3)$, where S denotes the area of the cross section (m^2) as shown in the figure and c is the sound speed (m/s). Recall that the acceleration can be expressed as $\Delta v/\Delta t = c\Delta s/\Delta t(\mathrm{m/s}^2)$.

$$\Delta p \cdot S = \rho_0 \cdot (Sc\Delta t) \cdot c\frac{\Delta s}{\Delta t} = \rho_0 c \cdot c\Delta s \cdot S = \rho_0 c v S \qquad (\mathrm{N}) \tag{6.15}$$

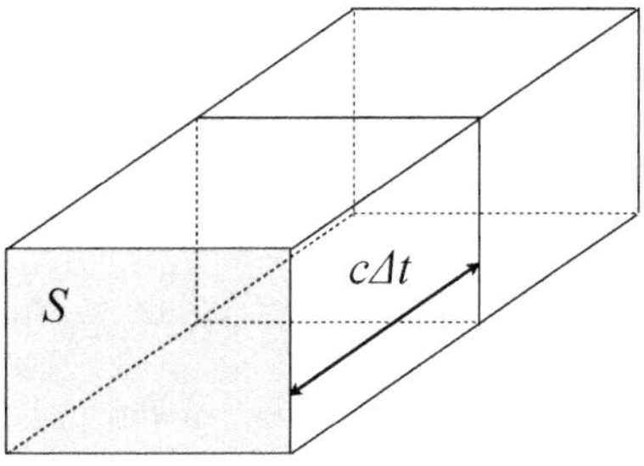

Fig. 6.1 Small portion of elastic medium where sound wave travels

holds well where $\rho_0(\mathrm{kg/m^3})$ is the volume density of the medium. Note here the pressure is in proportion to the condensation, i.e., the particle velocity instead of the acceleration in a plane wave. This is an important characteristic of plane waves. Under constant sound pressure, the volume of a oscillating portion increases with the sound speed, and thus the particle velocity decreases. In contrast, under constant particle velocity, the sound pressure increases with the sound speed.

Figure 6.2 might be a good example of consideration for travelling plane waves and their source. When the plate moves like a piston, a small portion of the medium, which the piston plate faces and is in contact with, moves at the same velocity as the piston plate. Consequently, if the piston goes to the right, then condensation occurs for the medium in the right-hand side of the plate. In contrast, if the plate goes back to the left, then dilation begins in the medium. Namely, the sound pressure $p(\mathrm{Pa})$ yielded by the piston motion of the plate can be estimated as

$$p = \kappa s = \kappa \frac{v}{c}, \qquad (\mathrm{Pa}) \qquad (6.16)$$

where $v = cs(\mathrm{m/s})$ denotes the velocity of the piston plate.

Suppose that the piston plate moves following a sinusoidal function whose angular frequency is $\omega = 2\pi f(\mathrm{rad/s})$. The sound pressure that travels to the right in the tube also follows the sinusoidal function of the same frequency but includes the two variables with respect to the time t and spatial location of x such that

$$p(x,t) = A\sin(\omega t - kx) = A\sin\omega\left(t - \frac{x}{c}\right) = A\sin\omega\,(t-\tau), \qquad (\mathrm{Pa})\,(6.17)$$

where $A(\mathrm{Pa})$ denotes the amplitude of the sound wave. Here $k(1/\mathrm{m})$ is called the **wavenumber**, which is defined by

$$k = \frac{\omega}{c} = \frac{2\pi}{\lambda}, \qquad (1/\mathrm{m}) \qquad (6.18)$$

where $\lambda\,(\mathrm{m})$ denotes the **wavelength**, which is defined by

$$\lambda = cT = \frac{c}{\nu} \qquad (\mathrm{m}) \qquad (6.19)$$

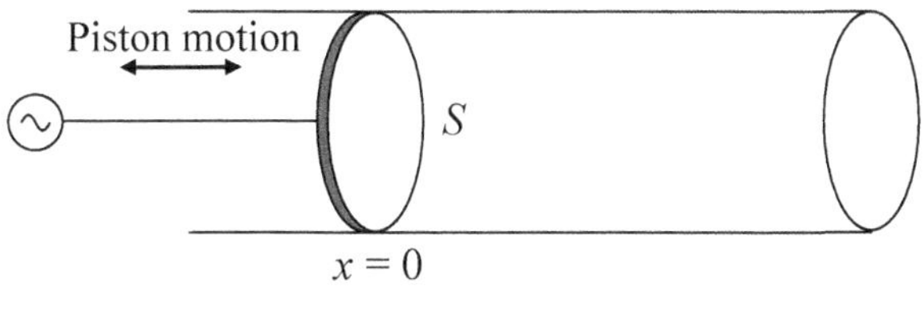

Fig. 6.2 Piston motion for travelling plane waves in tube

$$T = \frac{1}{\nu} = \frac{2\pi}{\omega}. \qquad \text{(s)} \qquad (6.20)$$

In the equation above, τ denotes the time lag that a sound wave requires to travel the distance x. For a sound wave that travels at $t = 0$ to the right-hand side in the figure, the sound wave is observed with **time delay** by τ at a position x. If such a time-delay (or **time difference**) is expressed for a sinusoidal wave, then the difference can be expressed as kx in terms of the **phase difference**. Namely, by recalling that the phase difference is 2π between two points separated by a wavelength,

$$\omega T = 2\pi = k\lambda \qquad \text{(rad)} \qquad (6.21)$$

holds well. The period (s) determines a time interval of a single oscillation cycle, while the wavelength (m) indicates the spatial distance between a pair of oscillating positions in the same phase.

According to Eq. 6.14, the particle velocity is also expressed as a sinusoidal function:

$$v(x,t) = \frac{p(x,t)}{\rho_0 c} = \frac{A}{\rho_0 c} \sin\omega(t - \tau). \qquad \text{(m/s)} \qquad (6.22)$$

The sound pressure is in phase with the velocity for a plane wave. Figure 6.3 displays the phase relationship for the displacement, particle velocity, and sound pressure for a travelling plane wave with normalized equal magnitude. The displacement (broken

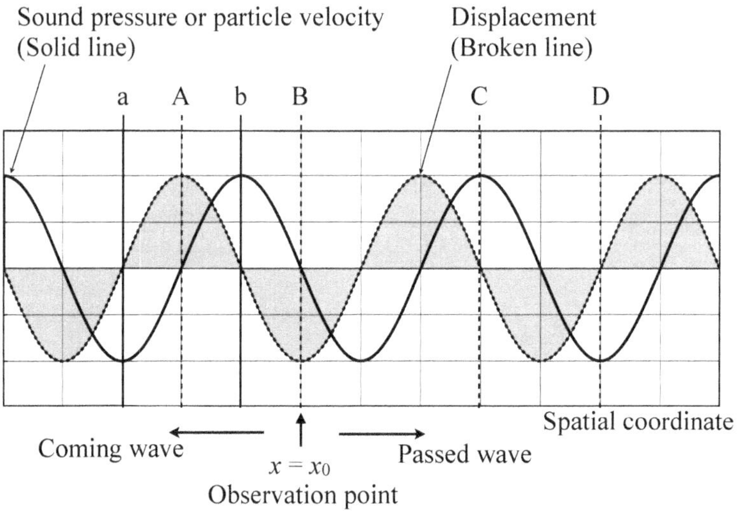

Fig. 6.3 Phase relationship for displacement (broken line), particle velocity (solid line), and sound pressure (solid line) for travelling plane wave from [9] (Fig.4.7)

line) is not in phase with the other two components (solid line). When the absolute of displacement is maximum, i.e., at A in the positive side or at B in the negative side, the particle velocity (solid line) becomes 0. The oscillation instantaneously stops at the instants for taking the maximum displacement with a positive or negative sign. Here the positive sign denotes the displacement to the right-hand side, while the negative one shows the movement to the left-hand side. Note the waves at position $x = x_0$. The waves observed at the right-hand side of $x = x_0$ have already passed the position x_0, while those appearing in the left-hand side will come to the position in the future. In other words, the particle velocity at position A for example, increases in the negative (left-hand) side from now, while that at position B, as another example, increases to the positive (right-hand) side. Similarly, the absolute of the particle velocity takes its maximum at C in the positive side or at D in the negative side, when the displacement is 0 at C or D. Such particle velocity will decrease in the positive (C) and negative (D) sides, respectively.

The outcome of no phase difference between the sound pressure and particle velocity in a plane wave is the same as that between the force and velocity of a mass-spring system at the resonance. There are no eigenfrequencies for a travelling plane wave in free space; however, a travelling plane wave seems just like a resonant wave as far as only the phase relationship is concerned. It might be interpreted as the outcome of balance in a plane wave between the inertia and elastic force, such as that under resonance.

Recall that both the sound pressure and particle velocity are linearly proportional to the condensation. The relationship between the density and displacement can then be understood. Again look at position A or B where the absolute of displacement takes its maximum. The oscillation instantaneously stops at the instants when the volume density comes back to its equilibrium without sound waves. Namely, the portion at A dilates and its density lessens from now, while the density at B is condensed now. On the other hand, the absolute of the condensation takes its maxima at C (positive) and D (negative) when the displacement becomes 0 at C and D. The displacement increases to the positive (C) or negative (D) side to compensate for the condensation or dilation of density. Consequently, the variation of density from the equilibrium decreases through condensation (C) or dilation (D).

The inertia due to the mass of the elastic medium explains why the displacement reaches the maximum (A and B) over its equilibrium (a and b). In contrast, the elastic force restores the displacement from its maxima. Here the scheme of oscillation is similar to that of a mass-spring system. However, there is a difference between the travelling of a plane wave and oscillation of an ideal mass-spring system. Namely, wave travel itself is the outcome of energy consumption through oscillation, which is not taken into account in an ideal oscillator such as a mass-spring system without damping factors. As well as the friction force being in proportion to the oscillation velocity under the balance between the inertia and elastic force at the resonance, the particle velocity and sound pressure are in phase. Thus, the energy of oscillation is consumed to excite oscillation in a surrounding silent portion of the medium. This energy-consuming process itself is wave propagation in a medium. Therefore, a constant sound source that provides the energy is necessary in order to propagate a progressive wave such as a sinusoidal wave in the medium.

6.2.2 Sound Pressure Level

How large (or small) is the **minimum audible sound** for a human listener? This might be a difficult question to answer; however, nowadays it is known that almost 2×10^{-5}(Pa) might be the minimum[28]. It might also be interesting to see how large the expected displacement is in a sound wave in the air for that minimum pressure. Suppose that there is a plane wave that has displacement $u(t)$, velocity $v(t)$, and sound pressure $p(t)$ such that

$$u(t) = A\cos\omega t \qquad (\mathrm{m}) \tag{6.23}$$

$$v(t) = -\omega A \sin\omega t \qquad (\mathrm{m/s}) \tag{6.24}$$

$$p(t) = \rho_0 c v(t). \qquad (\mathrm{Pa}) \tag{6.25}$$

Now the **minimum-audible sound pressure** mentioned above denotes the square-root of the squared average (rms). Then take the rms of sound pressure

$$\sqrt{\overline{p^2(t)}} = \frac{1}{\sqrt{2}}\rho_0 c\omega A, \qquad (\mathrm{Pa}) \tag{6.26}$$

where $\overline{*}$ denotes the time average taken over a single period. Thus, rms of the displacement for the minimum-audible sound can approximately be estimated using

$$\frac{A}{\sqrt{2}} = \frac{1}{1.3 \times 340 \times \pi} \times 10^{-8} \cong 7 \times 10^{-12} \qquad (\mathrm{m}) \tag{6.27}$$

where $\rho_0 = 1.3(\mathrm{kg/m^3})$, $c = 340(\mathrm{m/s})$, and the sound frequency is set to 1000 Hz.

The minimum audible sound is used as the reference for the **sound pressure level** in decibel (dB). The sound pressure level L_p is defined as

$$L_p = 10\log\frac{\overline{p^2(t)}}{P^2_{\mathrm{MA}}}, \qquad (\mathrm{dB}) \tag{6.28}$$

where P denotes the rms of sound pressure in pascal and $P_{\mathrm{MA}} = 2 \times 10^{-5}$(Pa).

6.2.3 Energy of Sound Waves

Suppose that there is a piston plate as a source of sound waves as shown in Fig. 6.2. When the piston plate moves the air in front of the plate, the work done given by the piston to the motion of the air is expressed as

$$W_X = pSv \qquad (\mathrm{W}) \tag{6.29}$$

in a unit time interval, where v(m/s) is the velocity of the piston motion, $S(\mathrm{m}^2)$ is the area of the plate, and p(Pa) is the pressure in front of the plate. A sum $E_0(\mathrm{J/m}^3)$ of the potential E_{P_0} and kinetic E_{K_0} energies for a unit volume of a medium in which a plane wave travels is called the **density of sound energy** or **sound energy density**.

Recall that the velocity of piston motion is just equal to the particle velocity of the medium in front of the piston and that both the potential and kinetic energy are equal to each other such that

$$E_{P_0} = \frac{1}{2}\kappa s^2 = \frac{1}{2}\frac{p^2}{\rho_0 c^2} = \frac{1}{2}\rho_0 v^2 = E_{k_0} = \frac{1}{2}E_0. \qquad (\mathrm{J/m}^3) \qquad (6.30)$$

The work done performed by the piston then can be rewritten as

$$W_X = pSv = \rho_0 v^2 cS = E_0 cS. \qquad (\mathrm{W}) \qquad (6.31)$$

This is equivalent to the energy stored by the plane wave in the volume amount of cS. When a plane wave travels in a medium at the speed of c, the energy is consumed to excite the air with a volume of cS. If such energy is provided by the piston plate, then a plane wave travels constantly in the tube.

By taking the unit area of the cross section or the piston plate,

$$I = pv = E_0 c \qquad (\mathrm{W/m}^2) \qquad (6.32)$$

is called the **density of sound energy flow** or **sound intensity**. The sound energy given by $E_0 c$ flows in the medium across the unit area when the sound wave whose energy density is E_0 travels a distance c in a unit time interval. Namely, the density of sound energy flow indicates the energy flow in a unit time interval across the unit area.

6.2.4 Sound Waves Radiated by Sinusoidal Motion of Piston Plate

Suppose that the thin plate in Fig. 6.2 moves like a piston and its volume velocity is given by a sinusoidal function as

$$q(t) = Q\cos\omega t. \qquad (\mathrm{m}^3/\mathrm{s}) \qquad (6.33)$$

Such a source mentioned above is called a **velocity source**, when the velocity is given independent of environmental conditions surrounding the source.

Two progressive plane waves are excited by the source: the waves travelling in the right-hand or left-hand side of the source in the tube. The plane wave that travels in the right-hand side has the velocity of motion

$$vS = \frac{1}{2}Q\cos(\omega t - kx). \qquad (\mathrm{m}^3/\mathrm{s}) \qquad (6.34)$$

Thus, the sound energy density of the plane wave $E_{0_R}(\mathrm{J/m^3})$ becomes

$$E_{0_R} = \rho_0 \overline{v^2(x,t)} = \frac{1}{8}\rho_0 \frac{Q^2}{S^2} \qquad (\mathrm{J/m^3}) \qquad (6.35)$$

independent of the observation position x, and it is just half the kinetic energy E_0 of the sound source, where

$$E_0 = \frac{1}{2}\rho_0 \overline{v^2(0,t)} = \frac{1}{4}\rho_0 \frac{Q^2}{S^2}. \qquad (\mathrm{J/m^3}) \qquad (6.36)$$

Uniform density of the sound energy independent of spatial positions is a typical characteristic for plane waves. The energy of the sound source is equally divided into two waves, i.e., travelling in the right or left side, and the divided energy is constant independent of the travelling distance or the wave frequency.

6.3 Sound Speed and Radiation of Sound from Vibrating Object

A plane wave can also be observed in three-dimensional space, not only in a one-dimensional tube. The sound speed is a key issue for the condition of whether a plane wave is radiated from a vibrating wall or not.

6.3.1 Radiation of Sound from Vibrating Wall

Suppose that a **transversal** or **bending wave** such as that on a vibrating string travels on the surface of a wall with a speed of $c_b(\mathrm{m/s})$ and wavelength of $\lambda_b(\mathrm{m})$. A plane wave radiated from the vibration, in other words, the sound that is propagated into the surrounding medium from the vibrating wall, travels the space with the speed of $c(\mathrm{m/s})$ and wavelength of $\lambda(\mathrm{m})$ following

$$\lambda_b \sin\theta = \lambda, \qquad (\mathrm{m}) \qquad (6.37)$$

as shown in Fig. 6.4. Assume the vibrating velocity of the wall is $v_b(\mathrm{m/s})$. The particle velocity perpendicular to the wall of the medium close to the wall is equal to v_b, so the velocity might be continuous at the boundary. Thus, the particle velocity $v(\mathrm{m/s})$ and sound pressure $p(\mathrm{Pa})$ of the radiated plane wave are given by

$$v_b = v\cos\theta \qquad (\mathrm{m/s}) \qquad (6.38)$$

$$p = \rho_0 c \frac{v_b}{\cos\theta}, \qquad (\mathrm{Pa}) \qquad (6.39)$$

following Fig. 6.4.

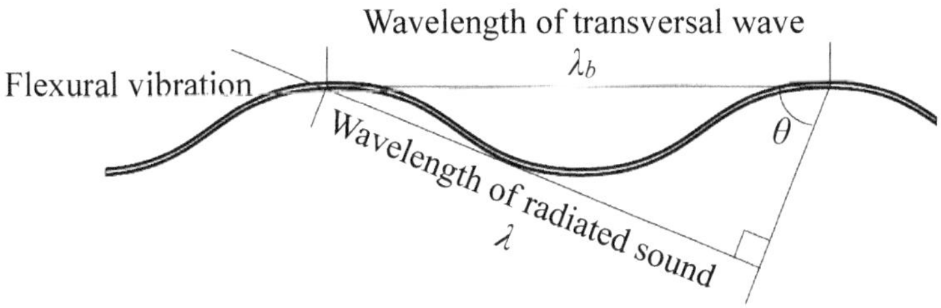

Fig. 6.4 Wavelength of sound wave radiated from vibration of wall

Figure 6.4 geometrically illustrates the relationship between the wavelengths for vibration of the wall and radiation sound in the medium. Note that

$$\cos\theta = \frac{\sqrt{\lambda_b^2 - \lambda^2}}{\lambda_b} = \frac{\sqrt{c_b^2 - c^2}}{c_b}. \tag{6.40}$$

Namely, sound can be radiated into the medium only when the wavelength of wall vibration is longer than that for the propagating sound in the medium. In other words, the speed of the transversal wave must be higher than that of sound in the medium. Otherwise, the right triangle cannot be constructed.

If the speed of the vibration wave is slower than the sound speed in the surrounding medium, then the radiated sound that travels along the boundary in the medium gets ahead of bending waves on the boundary. Consequently, the sound pressure that should have been excited by the wall vibration disappears due to interference by the sound waves coming earlier along the boundary in the medium. The energy of vibration without sound radiation is not dissipated during each oscillation cycle.

6.3.2 *Radiation Impedance and Coincidence Effect*

The condition required for sound radiation into the surrounding medium from the wall vibration can also be formulated in terms of the **radiation impedance**. The ratio of the sound pressure on the surface of the vibrating wall and its volume velocity is called **radiation acoustic impedance**. The radiation acoustic impedance $Z_{A_{\rm rad}}(\mathrm{Pa}\cdot\mathrm{s/m})$ can be expressed as

$$Z_{A_{\rm rad}} = \frac{p}{v_b S} = \frac{\rho_0 c}{S}\frac{1}{\cos\theta} = \frac{\rho_0 c}{S}\frac{\lambda_b}{\sqrt{\lambda_b^2 - \lambda^2}} = \frac{\rho_0 c}{S}\frac{c_b}{\sqrt{c_b^2 - c^2}} \quad (\mathrm{Pa}\cdot\mathrm{s/m}) \tag{6.41}$$

for the example illustrated in the previous subsection, where $S(\mathrm{m}^2)$ denotes the area for the vibrating wall of interest. It indicates that sound can be radiated from the vibration only when the radiation acoustic impedance contains the real part. If the sound speed of vibration is slower than that of the radiated sound, then the radiation impedance is imaginary without the real part. When the speed of vibration is faster than that of sound, the vibration is **supersonic**[29]. Sound radiation is a process of energy conversion from vibration to sound in the medium. Therefore, if the radiation impedance is purely imaginary, then there is no energy consumption due to the radiation of sound into the surrounding medium.

Sound reduction between two adjacent rooms is a fundamental issue for building acoustic problems. A great amount of sound reduction can be expected between the two rooms, as the mass per unit area of a partition increases in general. The **mass law** says that sound reduction increases in proportion to the surface density of a partition[30]. Sound transmission between the space divided by a partition is thus an issue of sound radiation from the vibrating wall. Therefore, the sound speed of the partition vibration is important.

The speed of transversal wave of an elastic body depends on the frequency in general. When the speed of wave is equal to that of sound, the vibration frequency is called the **coincidence frequency** for sound radiation[31]. The radiation acoustic impedance is purely imaginary for the vibration whose frequency is lower than the coincidence frequency, while it is real for the frequencies higher than the coincidence frequencies. The mass law holds well under the coincidence frequency, while sound reduction greatly decreases from the expectation by the mass law for higher frequencies over the coincidence frequency[31].

This is because vibration of a partition is made by supersonic and non-supersonic modal vibration. In other words multi-modal vibration, such as the vibration of a finite string by a stationary external source, can be classified into resonant and non-resonant modal vibration. Here the eigenfrequencies of resonant modes are very close to the vibration frequency, while non-resonant modes have different eigenfrequencies from the frequency. When the vibration frequency is lower than the coincidence frequency, only the non-resonant modal vibration is supersonic. On the other hand, if the frequency is higher than the coincidence, the resonant vibration itself can be supersonic. Therefore, a great amount of energy can be transmitted into the surrounding medium via the vibrating partition. However, magnitudes of the supersonic modes under the coincidence frequency, namely non-resonant modes, decrease in proportion to the unit-area mass of the partition in general. The mass law therefore holds well for the frequencies under the coincidence[31].

As an example, the coincidence frequency is around 1.3 kHz for bending waves due to vibration of a 1-cm thick iron plate. The speed of bending waves c_b(m/s) is in proportion to the square root of the frequency, following

$$c_b = \sqrt{\omega}\left(\frac{B_p}{\rho_0}\right)^{1/4}, \qquad \text{(m/s)} \tag{6.42}$$

where

$$B_p = \frac{Y_M}{1-\sigma}\frac{d_{th}^3}{12}, \tag{6.43}$$

Y_M denotes Young's modulus (N/m^2), d_{th} is the plate thickness, σ is Poisson's ratio, and ρ_0 gives the surface density (kg/m^2)[21][29][30].

Sound reduction through a partition is not always possible by reducing vibration of the partition of interest. It is necessary to analyze modal vibration of the partition according to the radiation condition or radiation impedance so that the supersonic modes might be well damped.

6.3.3 Sound Radiation from Moving Object

The speed of waves along the boundary can be interpreted as the speed of a moving object. By applying the radiation condition to the sound radiated from a moving object, supersonic waves can be understood as the sound when an object moves faster than the speed of sound. Taking an example of water waves excited by a boat, it can be imagined that a wavefront regularly progresses on the water when a boat runs faster than the speed of waves in the water.

Suppose that there is a sound source moving faster than the speed of sound[32], and take an example of spherical waves instead of plane waves. A progressive wave traveling and making a spherical wavefront is called a **spherical wave**. Figure 6.5(a) shows the image of a wavefront when the object moves faster than the speed of sound. A spherical wave radiated from an object at x_1 reaches the circle with radius r_1, while the object moves to x_2. Here the radius r_1 is shorter than the moving distance of the object. Suppose another spherical wave is radiated from the object at x_2. It comes to the circle with radius r_2, while the object reaches x_3. Again the radius r_2 is shorter than the moving distance. These progressive spherical waves build

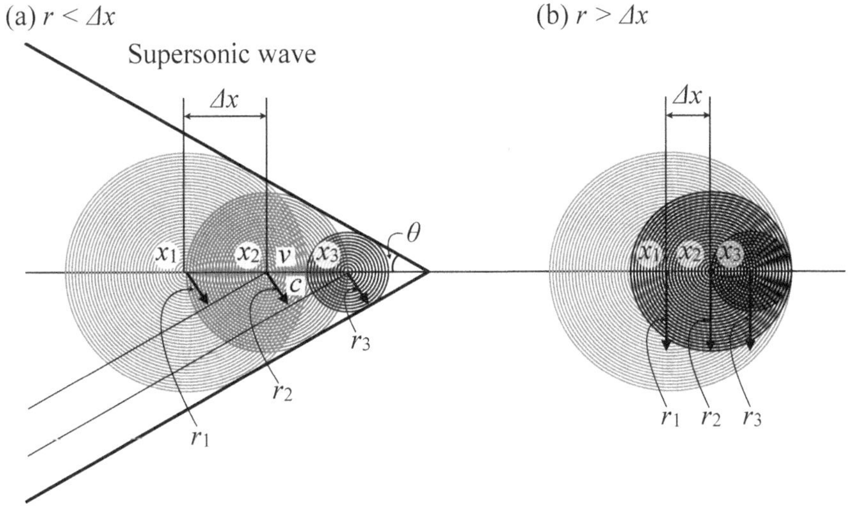

Fig. 6.5 Image of wavefront excited by object moving faster (a) or slower (b) than speed of sound[9] (Fig.4.10)

the wavefronts that depict the common tangent for spherical waves as shown in Fig. 6.5(a). Consequently, the sound is propagated in the wide area behind the moving object. In other words, there is no sound in front of a moving object that runs faster than the speed of sound. From this illustration,

$$c = v \sin\theta \qquad (\mathrm{m/s}) \tag{6.44}$$

holds well, where c and v show the speed of sound and the moving object, respectively.

On the other hand, Fig. 6.5(b) illustrates the sound from a moving object that moves slower than the speed of sound. Suppose spherical waves are radiated from the object. The spherical waves travel further than the distance the object moves. Therefore, these spherical waves build wavefronts that are densely located in front of the object but are sparse behind the object. Consequently, there are no widely spread wavefronts of travelling waves around the object; instead, the sound energy is concentrated only in front of the object.

The distribution of energy or wavefronts also changes the sound frequency. This is called the **Doppler effect** by a moving sound source. The sound frequency for an approaching object increases; it decreases after the object passes. Namely, the frequency $\nu(\mathrm{Hz})$ is expressed as

$$\nu = \frac{1}{1 - \frac{v}{c}} \nu_o, \qquad (\mathrm{Hz}) \tag{6.45}$$

where v denotes the speed of a moving object taking a positive (negative) sign when (after) it approaches (passes) the position of interest, c is the sound speed of the medium, and ν_o is the frequency of sound radiated from the object if it is not moving. The denominator of the equation above approaches 0 when the speed of the object is close to the speed of sound. Consequently, the frequency changes rapidly, and thus the concentration of wavefronts demonstrated by Fig. 6.5(b) reaches a maximal limit. This energy concentration reaching its maximal limit causes a **sonic boom**. It might also be interpreted that divergence of the radiation impedance causes the sonic boom.

The sonic boom is made when the speed of a moving object exceeds the speed of sound. The equation above indicating Doppler's effect cannot be applied to an approaching object with supersonic sound. There is no sound in front of the object with the speed higher than that of sound. In contrast, the supersonic sound travels widely behind the object, but the frequency rapidly decreases as shown by the equation with a negative sign of v. Sound radiation is an issue of interaction between a vibration source and environment surrounding the source. As it can be inferred from the sound radiated by a moving object, the vibrating speed (or wavelength) must be higher (or longer) than that of the sound in the medium so that the radiated sound might widely travel around the source. The interaction can be formulated in terms of the radiation impedance of a complex number. Such the impedance will be described in detail later.

Chapter 7
Sound in Tube

Sound waves traveling in an acoustic tube are a typical example of one-dimensional sound waves in the air. An acoustic tube is a basic model for musical wind instruments, just as vibration of a string is the fundamental basis of string instruments. Similar to vibration of a string, there are eigenfrequencies of an acoustic tube due to waves travelling between both ends. Resonance, which makes the acoustic tube an musical instrument, occurs at the eigenfrequencies. In this chapter, basic properties are described, related to sound in an acoustic tube.

7.1 Schematic View of Wave Propagation in Tube

7.1.1 Reflection Waves at Ends of Tube

Suppose that there is an acoustic tube in which a sound wave travels. A progressive longitudinal wave of sound is reflected back into the tube at both ends of the tube. This is similar to a travelling wave on a finite string being a transversal wave.

Figures 7.1 and 7.2 are typical examples of the end conditions for acoustic tubes[33]. Following Fig. 7.1, suppose that a compressed wave is fed into the tube from the left end. When the pulse-like compressed wave reaches the right open end, it comes back to the left as a reflected wave. Note here however that the reflected wave is not compressed but dilated. Namely, the sign of the wave is changed at the open end. This is because the sound pressure becomes lower at the open end than inside the tube, when the progressive wave reaches the end. Thus, particles in the medium move to the right to compensate for the excess decrease in pressure. Consequently, a dilated wave comes back to the left as a reflected wave.

In contrast, following Fig. 7.2, a compressed wave again comes back to the left as a reflected wave at the right closed end. Namely, the reflected wave is of the same sign as that for the wave approaching the closed end. This is because the sound pressure becomes higher at the closed end than inside the tube, when the approaching wave reaches the end. Thus, particles in the medium move to the left to compensate for that the excess increase in pressure. Consequently, a compressed wave comes back to the left.

M. Tohyama: *Sound and Signals*, Signals and Communication Technology, pp. 103–120.
springerlink.com

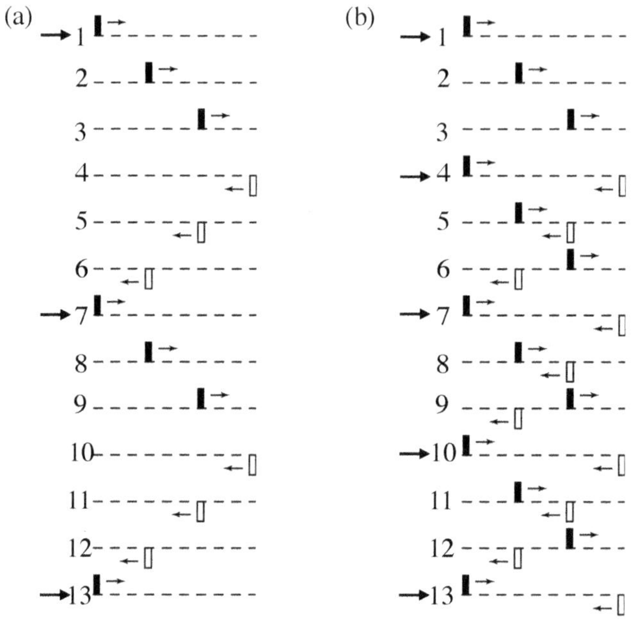

Fig. 7.1 Schematic illustration of sound propagation in open-open tube from [33] (Fig.4.1, Fig.4.2)

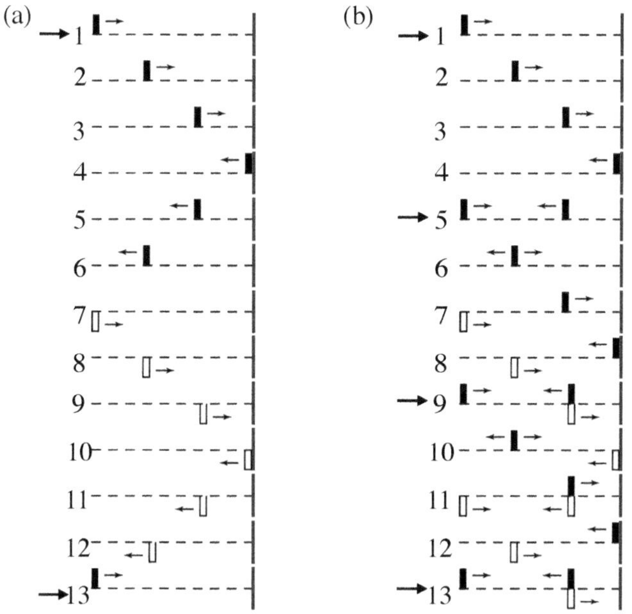

Fig. 7.2 Schematic illustration of sound propagation in open-closed tube from [33] (Fig. 4.5, 4.6)

7.1.2 *Fundamentals and Harmonics*

Resonance also occurs for the travelling waves in a tube. Figure 7.1(a) illustrates that when the reflected wave comes back to the left (open) end as a dilated wave, it again changes into a compressed wave that travels to the right in the tube. This indicates a good moment for a compressed wave to be fed into the tube. In other words, if a compressed wave is fed into the tube every time the dilated reflection wave comes back to the left end, then a travelling resonant wave might build up to steady motion in the tube. The period $T_{o1}(\mathrm{s})$ of a wave travelling in an open-open tube is

$$T_{o1} = \frac{2L}{c} \qquad (\mathrm{s}) \tag{7.1}$$

$$\nu_{o1} = 1/T_{o1}, \qquad (\mathrm{Hz}) \tag{7.2}$$

where $L(\mathrm{m})$ is the length of the tube, $c(\mathrm{m/s})$ is the sound speed, and $\nu_{o1}(\mathrm{Hz})$ is called the **fundamental** for an open acoustic tube.

The **harmonics** can also be built up as well as the fundamental. As illustrated by Fig. 7.1(b), if a compressed wave is fed into the tube every time the dilated reflection wave starts at the right end, the second harmonic might build up to steady motion. Here the period of the second harmonic is 1/2 the period of the fundamental. In other words, the frequency of the second harmonic is two times higher than that for the fundamental. From the figure, the sound wave is always in-phase inside the tube for the fundamental mode of a travelling wave, while for the second harmonic the sound is in anti-phase between the right and left parts of the tube. This recalls the standing wave of the second mode for vibration of a string.

Similarly, Fig. 7.2 indicates the fundamental and its harmonics for a open-closed tube. As shown in the figure, dilated and compressed waves come back into the tube at the left-open and right-closed ends, respectively. This combination makes the period of the fundamental two times longer than that in an open-open tube. Namely, the period $T_{c1}(\mathrm{s})$ of a wave travelling in an open-closed tube is

$$T_{c1} = \frac{4L}{c} \qquad (\mathrm{s}) \tag{7.3}$$

$$\nu_{c1} = 1/T_{c1}. \qquad (\mathrm{Hz}) \tag{7.4}$$

Figure 7.2 also shows the harmonics such that

$$\nu_{cn} = n\nu_{c1}, \qquad (\mathrm{Hz}) \tag{7.5}$$

where $n = 2m - 1$ and m is a positive integer.

The fundamentals with their harmonics for open-open and open-closed tubes explain the difference between the flute and clarinet. Although there are no remarkable differences in the lengths of the flute and clarinet, the fundamental frequency of the clarinet is lower than that of the flute. The difference of fundamentals between the two types of instruments explain the difference in pitch. In addition, the difference in the harmonics also indicates the timbre difference between the flute and clarinet.

The fundamental and its harmonics also indicate that resonance occurs for the travelling waves in an open-closed tube. Figure 7.2 illustrates that there is a good moment for building up the fundamental. If a compressed wave is fed into the tube at the left end every two times the reflection wave comes back to the end, the resonant fundamental mode might build up to steady motion. It can also be understood that the second harmonic cannot be built up by cancellation, even if the compressed wave is fed every time the reflected wave comes back to the left end.

However, it might be possible to excite the third harmonic if a compressed wave is fed with a period 1/3 that of the fundamental one. Sound travelling of the third harmonic for the open-closed tube is considered as well as that for the harmonics for the open-open tube. The compressed and dilated waves are always cancelled at the positions inside $L/3$ from both ends. A position where the waves are always cancelled to be null is called the **node** of a wave, such as the center for the second harmonic in the open-open tube or the $L/3$ positions for the third harmonic in the open-close tube. The sound is in anti-phase between both sides of a node. The fundamental and harmonics are related to the speed of sound traveling in a tube. If the air is replaced by gas with a higher speed of sound, then all the fundamentals and harmonic frequencies are shifted to the higher ones.

7.2 Eigenfrequencies in Acoustic Tubes

Wave travel with the fundamental and harmonic frequencies can be formulated following the wave equation.

7.2.1 Eigenfrequencies for Open-Open Tube with Constant Pressure Source

Sound travelling in a thin acoustic tube can be analyzed following propagation of a sinusoidal plane wave. Again suppose that there is an acoustic pipe of length L(m) in which a pressure source is located at the left end such as in Fig. 7.3.

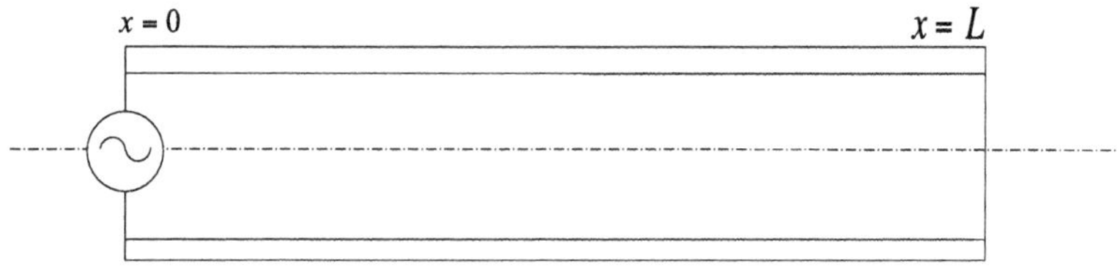

Fig. 7.3 Pressure source in open pipe

The sound waves traveling in the acoustic tube can be expressed as the sum of two progressive waves such that

$$p(x,t) = P(x,\omega)\mathrm{e}^{\mathrm{i}\omega t} = \left(A\mathrm{e}^{-\mathrm{i}kx} + B\mathrm{e}^{\mathrm{i}kx}\right)\mathrm{e}^{\mathrm{i}\omega t}, \tag{7.6}$$

where the first and second terms represent the plane waves propagating to the right (positive direction) and left (negative direction), respectively, and $P(x,\omega)$ denotes magnitude of the sound pressure (Pa) of the angular frequency ω observed at position x. Here A and B have to be determined following the boundary conditions such that there is an open right end and a pressure source at the left end.

First the sound pressure should be null at the open right end. Namely, the general form of the solution must follow the condition that

$$p(x = L, t = 0) = 0 \tag{7.7}$$

at the right open end. In contrast, the sound waves traveling in the tube are excited by the pressure source at the left end as illustrated by Fig.7.3. This type of condition can be a simplified model for the sound source of a flute. Following this condition, the sound pressure must be expressed as

$$p(x = 0, t) = P_X \mathrm{e}^{\mathrm{i}\omega t} \qquad \text{(Pa)} \tag{7.8}$$

at the left end, where P_X denotes the constant magnitude for the sinusoidal pressure source. From the conditions above, simultaneous equations are derived as

$$P(x = L, \omega) = P(L,\omega) = A\mathrm{e}^{-\mathrm{i}kL} + B\mathrm{e}^{\mathrm{i}kL} = 0 \tag{7.9}$$

$$P(x = 0, \omega) = P(0,\omega) = A + B = P_X \tag{7.10}$$

for the two variables A and B to be determined.

According to the equations above, the sound pressure $p(x,t)$(Pa) observed at x is written as

$$P(x,\omega)\mathrm{e}^{\mathrm{i}\omega t} = \left(A\mathrm{e}^{-\mathrm{i}kx} + B\mathrm{e}^{\mathrm{i}kx}\right)\mathrm{e}^{\mathrm{i}\omega t} = P_X \frac{\sin k(L-x)}{\sin kL}\mathrm{e}^{\mathrm{i}\omega t}. \quad \text{(Pa)} \tag{7.11}$$

Here the frequency ν_{Pn}(Hz),

$$\nu_{Pn} = \frac{ck_n}{2\pi} = \frac{nc}{2L} = \nu_{on}, \qquad \text{(Hz)} \tag{7.12}$$

at which the denominator of the equation above becomes zero is called the **eigenfrequency for an open pipe** excited as a constant pressure source, and n is the positive integer. The eigenfrequencies are composed of the fundamental $c/2L$(Hz) and its harmonics.

Figure 7.4 illustrates the samples of sound pressure distribution that are calculated by the numerator of the magnitude in an open tube under pressure excitation [34]. The horizontal axis shows the normalized position by x/L between 0 (the left

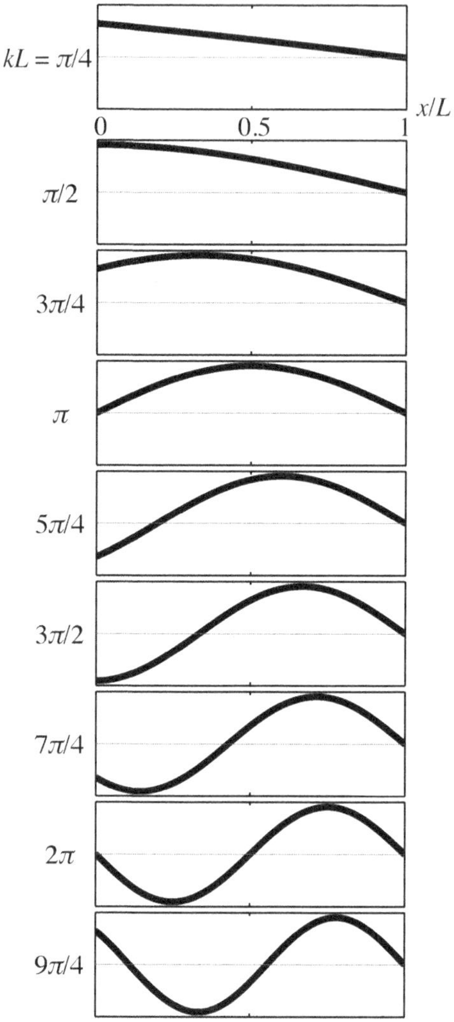

Fig. 7.4 Patterns of sound pressure distribution in open pipe under sound pressure source from [34] (Fig. 3.6)

end) and unity (the right end). The frequencies at which the numerator becomes zero are called the **zeros** or **nodes**, while the frequencies at which the denominator becomes zero are called the **poles** or the eigenfrequencies. The nodes move to the right as the frequency of the external source increases. Here the fundamental corresponds to the condition that $kL = \pi$, and the second harmonic corresponds to that for $kL = 2\pi$.

The dotted line in Fig. 7.5 shows the denominator ($\sin kL$). The denominator is independent of the observation point of x/L, and thus the poles are also

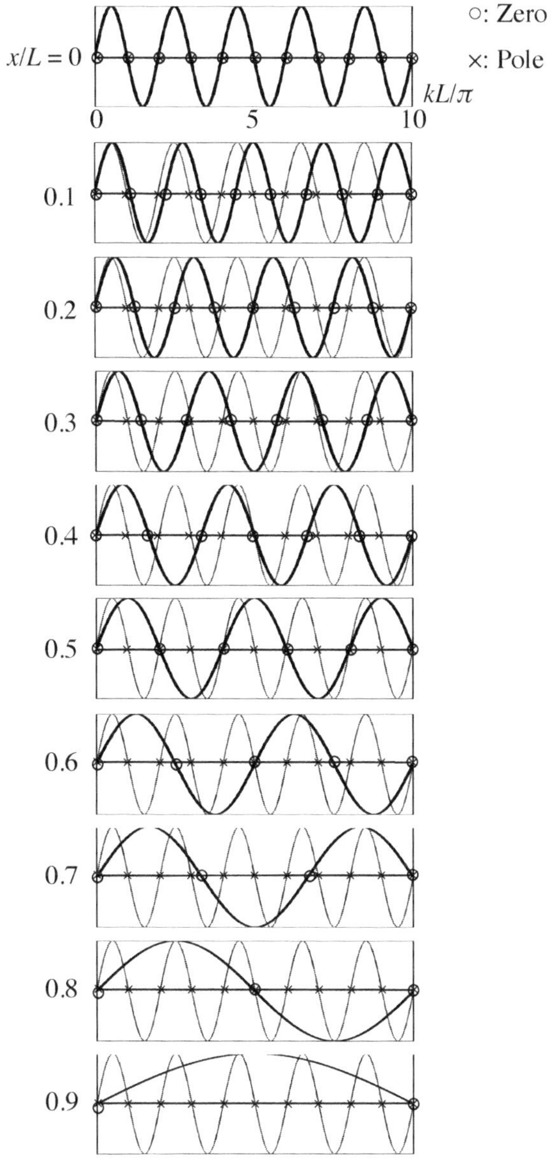

Fig. 7.5 Poles (cross mark) and zeros (open circle) for sound pressure from [34] (Fig. 3.7), solid line: numerator, dotted line: denominator

independent of the observation locations. Sign changes in the denominator due to the poles are cancelled out by the sign changes of the numerator due to the zeros at the positions close to the sound source at the left end, namely where $x/L \to 0$. As the observation point moves further from the source, namely as x/L becomes larger, the nodes (zeros for the numerator) move to the high frequencies ($kL \to$ large) and the number of nodes decreases; consequently, the sign changes due to the poles are left uncancelled.

By recalling the relationship between the sound pressure and particle velocity for a plane wave, the particle velocity of the medium in the pipe can be expressed as the complex-function form

$$v(x,t) = \frac{-\mathrm{i}P_X}{\rho_0 c}\frac{\cos k(L-x)}{\sin kL} = \frac{\mathrm{i}}{\omega\rho_0}\frac{\partial p(x,t)}{\partial x}. \quad \text{(m/s)} \tag{7.13}$$

Although the sound pressure and particle velocity are in-phase and the ratio between them is $\rho_0 c$ for each progressive wave to the right or left, this is not the case for the sum of the two waves. Take the ratio of the sound pressure and particle velocity

$$\frac{p(x,t)}{v(x,t)} = \mathrm{i}\rho_0 c\frac{\sin k(L-x)}{\cos k(L-x)}. \quad (\mathrm{Pa}\cdot\mathrm{s/m}) \tag{7.14}$$

The ratio depends on the observation position and the frequency. Figure 7.6 is a sample of comparison between the sound pressure and particle velocity in a pipe. The particle velocity is in proportion to the sound pressure gradient, and thus the velocity is zero at the position where the sound pressure takes the maximum.

The ratio of the sound pressure and particle velocity is no longer real but is purely imaginary instead. There is the phase difference of $\pi/2$ between the sound pressure and the particle velocity, and thus sound energy is not consumed but stored in the tube during each cycle. In other words, any sound is not radiated outside the pipe. This outcome, which is against our intuition, is due to the unrealistic assumption of the open end, i.e., no sound pressure is yielded at the open end. Indeed, the energy flow defined by the product of the pressure and particle velocity is always zero at the open end when the unrealistic boundary condition is followed.

7.2.2 Eigenfrequencies for Open Tube with Constant Velocity Source

The eigenfrequencies for an acoustic pipe depend on the boundary conditions or external sources. Suppose that a velocity source is located at the left end in Fig. 7.3. This is understood as a simplified model for the sound source of a clarinet. According to this condition, the sound pressure must be expressed as

$$\left.\frac{\partial p(x,t)}{\partial x}\right|_{x=0} = -\mathrm{i}\omega\rho_0 V_X e^{\mathrm{i}\omega t} \tag{7.15}$$

n	Pattern of distribution	Wavelength: λ_n	Eigenfrequency: ν_n
$n=1$	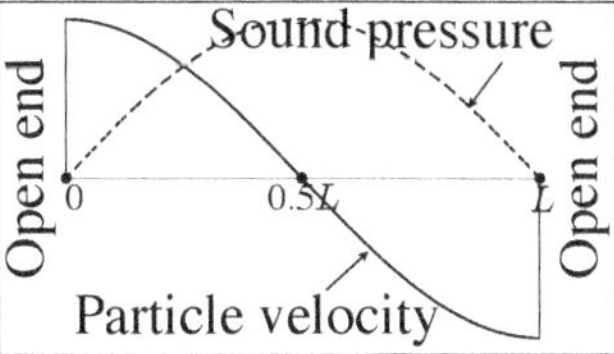	$\lambda_1 = 2L$	$\nu_1 = c/2L$
$n=2$	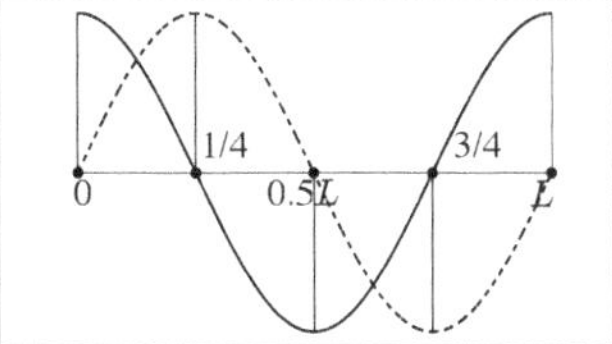	$\lambda_2 = \lambda_1/2$	$\nu_2 = 2\nu_1$
$n=3$	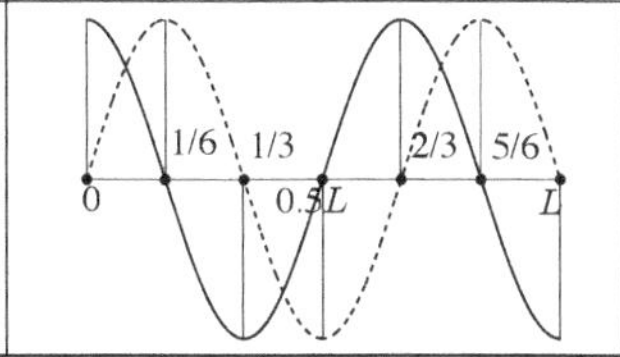	$\lambda_3 = \lambda_1/3$	$\nu_3 = 3\nu_1$

Fig. 7.6 Distribution patterns of sound pressure (dotted line) or particle velocity (solid line) from [19] (Fig. 2.3)

where V_X(m/s) denotes the magnitude of the sinusoidal velocity source at the left end. The condition at the right end is the same as in Eq. 7.7, and thus the simultaneous equations

$$P(L,\omega) = Ae^{-ikL} + Be^{ikL} = 0 \tag{7.16}$$

$$i\omega\rho_0 V_X = ik(A - B) \tag{7.17}$$

hold well.

By solving the simultaneous equations for the complex variables A and B,

$$P(x,\omega)e^{i\omega t} = \left(Ae^{-ikx} + Be^{ikx}\right)e^{i\omega t} = i\rho_0 c V_X \frac{\sin k(L-x)}{\cos kL} e^{i\omega t} \tag{7.18}$$

is obtained as an expression for the sound pressure in the open pipe under constant velocity excitation. Therefore, the eigenfrequency ν_{Vn} (Hz), or the pole at which the denominator becomes zero, is given by

$$\nu_{Vn} = \frac{nc}{4L} - \nu_{cn}, \tag{7.19}$$

where $n = 2m - 1$ and m is a positive integer. The series of eigenfrequencies shows the fundamental ν_{V1} for $n = 1$ is lower by one octave than that under pressure. In addition, only the odd-ordered for such as ν_{V3} and ν_{V5} are available. The differences in eigenfrequencies between the two types of source conditions imply the tonal differences between the flute and clarinet.

7.2.3 Driving-Point Acoustic Impedance

Suppose that the sound pressure $p(x',t)$ (Pa) and the volume velocity $q(t)(\mathrm{m}^3/\mathrm{s})$ are observed at a position very close to the sound-source position x' in an acoustic pipe. Then the ratio between the sound pressure and volume velocity is called the **driving-point acoustic impedance**. Very interestingly, the driving-point acoustic impedance at the end in an open pipe can be expressed as

$$Z_{A_{\mathrm{in}}}(x'=0,\omega) = \frac{\mathrm{i}\rho_0 c}{S}\tan kL \tag{7.20}$$

independent of the external source conditions, where $q(t) = S\cdot V_X \mathrm{e}^{\mathrm{i}\omega t}(\mathrm{m}^3/\mathrm{s})$, $S(\mathrm{m}^2)$ denotes the cross section of area of the pipe, $V_X\mathrm{e}^{\mathrm{i}\omega t}(\mathrm{m/s})$ is the velocity of the source, and the relationship expressed by Eq. 7.13 is used between the sound pressure and velocity. The driving point acoustic impedance defined above can be characterized in terms of **poles** and **zeros**. The zeros are the frequencies that make $\sin kL = 0$, while the poles are the frequencies such that $\cos kL = 0$. Thus, the poles and zeros alternate with each other in the driving-point acoustic impedance. The zeros correspond to the eigenfrequencies of the open pipe under pressure excitation, while the poles indicate the eigenfrequencies for a velocity source.

The driving-point acoustic impedance depends on the position of the source. Suppose that there is a velocity source whose volume velocity is $q(t)(\mathrm{m}^3/\mathrm{s})$ at a position $x'\neq 0$ inside an open pipe as shown in Fig. 7.7. The driving-point acoustic impedance to the right-hand side of the pipe, $Z_{AR_{\mathrm{in}}}(\mathrm{Pa\cdot s/m^3})$, can be expressed as

$$Z_{AR_{\mathrm{in}}}(x',\omega) = \frac{p(x',t)}{q_R(t)} = \frac{\mathrm{i}\rho_0 c}{S}\tan k(L-x'), \tag{7.21}$$

where $p(x',t)(\mathrm{Pa})$ denotes the sound pressure yielded at the sound source position, and $q_R(t)(\mathrm{m}^3/\mathrm{s})$ shows the volume-velocity component to the right-hand side of the pipe. Similarly, the impedance to the left side, $Z_{aL}(\mathrm{Pa\cdot s/m^3})$, is given by

$$Z_{AL_{\mathrm{in}}}(x',\omega) = \frac{p(x',t)}{q_L(t)} = \frac{\mathrm{i}\rho_0 c}{S}\tan kx', \tag{7.22}$$

where $q_L(t)(\mathrm{m}^3/\mathrm{s})$ is the velocity component to the left-hand side of the pipe. Consequently, the **driving-point acoustic impedance** $Z_{A_{\mathrm{in}}}(\mathrm{Pa\cdot s/m^3})$ becomes

$$Z_{A_{\mathrm{in}}}(x',\omega) = \frac{p(x',t)}{q(t)} = \frac{\mathrm{i}\rho_0 c}{S}\frac{\sin kx'\cdot\sin k(L-x')}{\sin kL} \tag{7.23}$$

where $q(t) = q_L(t) + q_R(t)(\mathrm{m}^3/\mathrm{s})$.

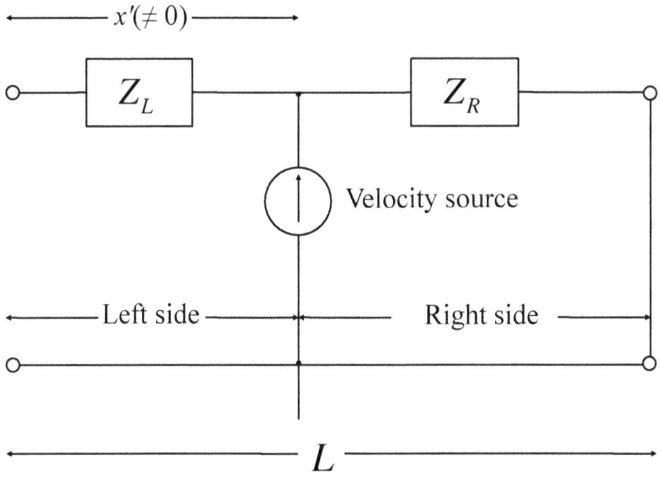

Fig. 7.7 Schematic of velocity source inside pipe

Here it can be reconfirmed that no sound is radiated from the pipe because the impedance is purely imaginary without a real part. The sound pressure and velocity are out-of-phase with each other at the source position, and thus there is no work done by the external source. This is the outcome according to the idealized boundary condition for the open pipe such that there is no sound pressure yielded at the open end.

7.2.4 Transfer Acoustic Impedance

When the ratio defined above is taken at a point x away from the source position x', it is called the **transfer acoustic impedance** instead of the driving-point acoustic impedance. By taking the source position at the left end, such as $x' = 0$, the transfer acoustic impedance $H_{A_{\rm tf}}(x' = 0, x = x, \omega)(\mathrm{Pa} \cdot \mathrm{s/m^3})$ can be expressed as

$$H_{A_{\rm tf}}(x' = 0, x = x, \omega) = \frac{i\rho_0 c}{S} \frac{\sin k(L - x)}{\cos kL} \tag{7.24}$$

for an open pipe. Figure 7.8 presents the outline of the transfer acoustic impedance for an open pipe excited at the left end. The poles are independent of the observation point, while the zeros move to higher frequencies as the observation point moves further from the source[23][24].

When a velocity source is located inside the pipe, the sound pressure $p(x', t)$ (Pa) yielded at the source position can be estimated by

$$p(x', t) = Z_{A_{\rm in}} \cdot q(t) = \frac{i\rho_0 c}{S} \frac{\sin kx' \cdot \sin k(L - x')}{\sin kL} \cdot q(t). \tag{7.25}$$

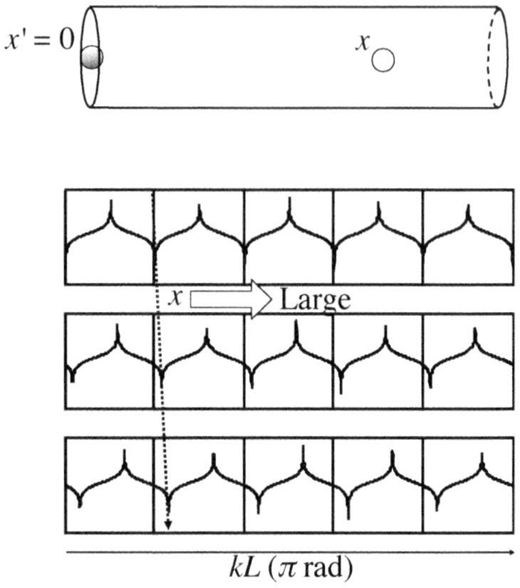

Fig. 7.8 Transfer acoustic impedance for open pipe excited at left end from [34] (Fig.3.11)

Thus the sound pressure $p_R(x = x,t)$(Pa) such that

$$p_R(x = x,t) = p(x',t)\frac{\sin k(L-x)}{\sin k(L-x')} = \frac{\mathrm{i}\rho_0 c}{S}\frac{\sin kx' \cdot \sin k(L-x)}{\sin kL}q(t), \quad (7.26)$$

which corresponds to that for the pipe with the length of $L - x'$ excited by the pressure $p(x',t)$ at the left end, is observed in the right-hand side of the pipe from the source. Here x denotes the distance from the left end of the pipe. Therefore, the **transfer acoustic impedance** $H(\omega,x',x)(\mathrm{Pa}\cdot\mathrm{s/m^3})$ is described as

$$H_{A_{\mathrm{tf}}}(x',x = x,\omega) = \frac{\mathrm{i}\rho_0 c}{S}\frac{\sin kx' \sin k(L-x)}{\sin kL} = \frac{N(\omega)}{D(\omega)}. \quad (7.27)$$

Note here again that the poles of the transfer acoustic impedance are independent of the locations for the source and observation positions, as long as the source is located inside the pipe. However, they are different from those defined by Eq.7.24 when the source is located at the left or right end. This is because not only the location of the source but also the boundary condition are different, when the source is situated at the end of a pipe.

Figure 7.9 shows examples of the poles and zeros for the acoustic transfer impedance by Eq.7.27 where $x' = L/7$. The solid line shows the curve calculated by the numerator of the equation such that

$$N(x',x,\omega) = \sin kx' \sin k(L-x), \quad (7.28)$$

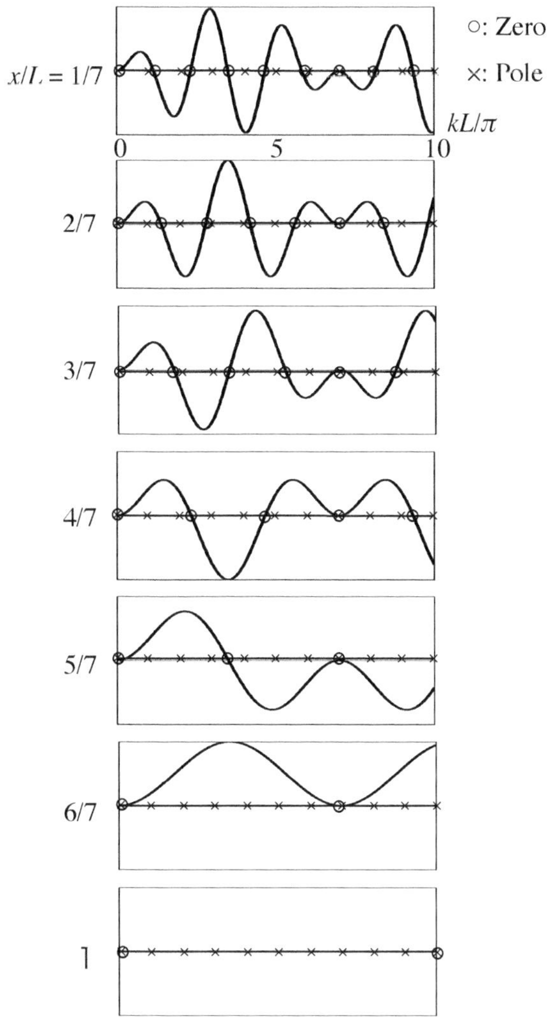

Fig. 7.9 Examples of poles and zeros for transfer acoustic impedance given by Eq. 7.27 where $x' = L/7$ from [34] (Fig. 3.11), open circle: zero, cross: pole

while the crosses plot the poles, which are the zeros of the denominator. The zeros move to the right (higher frequencies) as the position of observation point x becomes moves further the source location.

7.2.5 Sound Radiation from Open End of Acoustic Pipe

When the driving-point impedance is purely imaginary, sound is not radiated from the open-end pipe. The sound energy radiated by a source in a unit time interval is called the **sound power output of a source**. It can be formulated using the driving-point acoustic impedance such that

$$W_X = \mathrm{Re}\frac{|Q|^2}{2}Z_{A_{\mathrm{in}}} = \frac{|Q|^2}{2}R_{A_{\mathrm{in}}}, \qquad \text{(W)} \tag{7.29}$$

where the power output W_X (W) of a source is in proportion to the real part $R_{A_{\mathrm{in}}}$ of the driving-point acoustic impedance $Z_{A_{\mathrm{in}}}(\mathrm{Pa}\cdot\mathrm{s/m^3})$ or the magnitude of volume velocity of the source $Q(\mathrm{m^3/s})$. If there is no real component of the driving-point acoustic impedance, sound radiation cannot be expected from the source independent of the volume velocity. This is the same as when no sound is radiated from a moving source that runs slower than the speed of sound in the medium as described in the previous chapter.

Recall that there is no real part of the driving-point acoustic impedance for an open-end pipe. This is because the boundary condition for an open end was unrealistically idealized so that there is no sound pressure yielded at the open end. To introduce a more realistic condition for the open end, assume the acoustic impedance Z_A the boundary such that

$$p(x=L,t) = Z_A v(x=L,t)\cdot S \tag{7.30}$$

$$v(x=0,t) = V_X \mathrm{e}^{\mathrm{i}\omega t}. \tag{7.31}$$

Recall that the particle velocity can be express following Eq. 7.15. The relationship such that

$$A\mathrm{e}^{-\mathrm{i}kL} + B\mathrm{e}^{\mathrm{i}kL} = \frac{Z_A S}{\rho_0 c}\left(A\mathrm{e}^{-\mathrm{i}kL} - B\mathrm{e}^{\mathrm{i}kL}\right) \tag{7.32}$$

holds well at the open end ($x = L$) subject to the acoustic impedance Z_A at the boundary. Thus the **reflection coefficient** is written as

$$\frac{B}{A} = \mathrm{e}^{-2\mathrm{i}kL}\left(\frac{Z_A - Z_0}{Z_A + Z_0}\right), \tag{7.33}$$

where $p(x,t)$ is given by Eq. 7.6 and $Z_0(\mathrm{Pa}\cdot\mathrm{s/m^3})$ is defined by

$$Z_0 = \frac{\rho_0 c}{S} = \frac{z_0}{S}, \tag{7.34}$$

$S(\mathrm{m^2})$ is the cross-sectional area of the tube, $z_0 = \rho_0 c(\mathrm{Pa}\cdot\mathrm{s/m})$ is unique for the medium of interest. The is called the **specific impedance of a medium**, which is given by a ratio of the sound pressure and particle velocity for a plane wave in the medium.

Similarly by recalling the particle velocity for the source position at the left end ($x' = 0$) is expressed as

$$v(x'=0,t) = \frac{A-B}{\rho_0 c}\mathrm{e}^{\mathrm{i}\omega t}, \tag{7.35}$$

the driving-point acoustic impedance $Z_{A_{\mathrm{in}}}(x'=0,\omega)(\mathrm{Pa}\cdot\mathrm{s/m^3})$ can be obtained by

$$Z_{A_{\mathrm{in}}}(x'=0,\omega) = Z_0\frac{A+B}{A-B} = Z_0\frac{Z_A\cos kL + \mathrm{i}Z_0\sin kL}{Z_0\cos kL + \mathrm{i}Z_A\sin kL}. \tag{7.36}$$

Therefore, by taking the real part of the impedance above,

$$R_{A_{\mathrm{in}}} = \frac{R_A\cdot Z_0}{(Z_0\cos kL - X_A\sin kL)^2 + R_A^2\sin^2 kL}\cdot Z_0 \tag{7.37}$$

is derived where

$$Z_A = R_A + \mathrm{i}X_A. \tag{7.38}$$

The sound power output of the source located at the left end of the open pipe indicates the acoustic-power flow from the source to the right open end. Therefore, the sound power radiated from the pipe to the outside increases as the real part of the impedance at the right end increases. In other words, more power is required to be transmitted to the pipe from the source in order to keep a wave travelling in the tube, as the real part of the impedance becomes larger and greater sound is radiated from the pipe. Consequently, free vibration might decrease more rapidly as the radiation sound becomes louder. It seems that the sound energy goes back and forth between the pipe and source when the open end approaches the ideal one, where no sound pressure is yielded. Thus, if the driving-point impedance is purely imaginary, it might be impossible to define which is the exciter between the source and tube.

By taking the ratio of the force and velocity, the **driving point mechanical impedance** $Z_{M_{\mathrm{in}}}(\mathrm{N}\cdot\mathrm{s/m})$, is obtained instead of the driving-point acoustic impedance stated above

$$Z_{M_{\mathrm{in}}} = \frac{f_X(t)}{v(t)} = \mathrm{i}\omega\frac{f_X(t)}{\dot{v}(t)} = R_{M_{\mathrm{in}}} + \mathrm{i}X_{M_{\mathrm{in}}}, \tag{7.39}$$

is obtained instead of the acoustical driving point impedance. By rewriting the equation as

$$\frac{Z_{M_{\mathrm{in}}}}{\mathrm{i}\omega} = \frac{f_X(t)}{\dot{v}(t)} = \frac{R_{M_{\mathrm{in}}} + \mathrm{i}X_{M_{\mathrm{in}}}}{\mathrm{i}\omega} = \frac{R_{M_{\mathrm{in}}}}{\mathrm{i}\omega} + \frac{X_{M_{\mathrm{in}}}}{\omega}, \tag{7.40}$$

it can be seen that the real part "$X_{M_{\mathrm{in}}}$" is in-phase with the acceleration, and the imaginary part "$R_{M_{\mathrm{in}}}$" is in-phase with the velocity. Here the component in-phase with the acceleration represents the force required to oscillate the sound source, while the component in-phase with the velocity expresses the energy consumed for sound propagation in the medium.

This can be understood by taking an example of piston-like motion in a tube as shown in Fig.7.10. The equation of motion of the piston plate can be written as

$$M\ddot{x}(t) + Kx(t) = F_X e^{\mathrm{i}\omega t} - pS, \tag{7.41}$$

where M denotes the mass of the plate, $x(t) = Ae^{\mathrm{i}\omega t}$ shows the displacement due to sinusoidal motion of the plate, $f_X(t) = F_X e^{\mathrm{i}\omega t}$ is the external force to make the

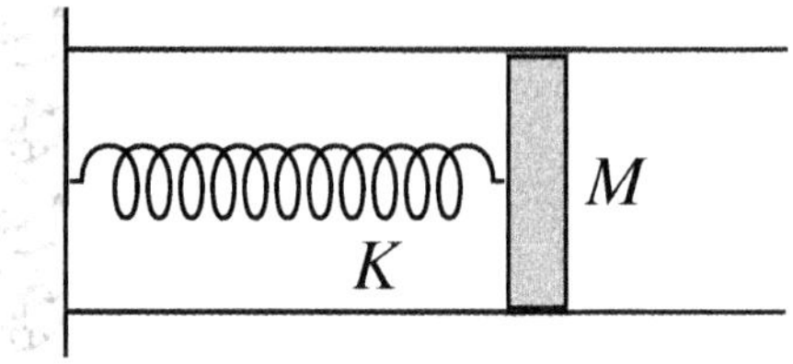

Fig. 7.10 Piston-like motion of plate in acoustic pipe

sinusoidal motion, $\ddot{x}(t) = \frac{\partial^2 x(t)}{\partial t^2}$, S is the area of the plate, $p(t)$ is the sound pressure on the plate due to the piston motion, and K corresponds to the spring constant. By following the equation of motion,

$$\frac{f_X(t)}{\ddot{x}} = \frac{\rho_0 c S}{\mathrm{i}\omega} + M\left(1 - \frac{\omega_0^2}{\omega^2}\right) \tag{7.42}$$

is derived where $p = \rho_0 c \dot{x}(t)$, $\dot{x}(t) = \frac{\partial x(t)}{\partial t}$, and $\omega_0^2 = K/M$.

The real part, the second term of the right-hand side, which is in-phase with the acceleration, expresses resonance of the mass-spring system. On the other hand, the first term, the imaginary component, which is in-phase with the velocity of the sinusoidal motion, indicates the force that is converted to the energy consumed for sound propagation in the tube. All the external force can be used as the energy required for sound to travel in the tube, when the frequency of the external force is equal to the eigenfrequency $(\omega_0/2\pi)$ of the mass-spring system and the second term becomes zero. The resonance at the eigenfreqeucies of the pipe should be noticeable only when the real part of the impedance and the radiated sound become small.

7.2.6 End-Correction for Acoustic Open Pipe

The effect of boundary conditions on the traveling waves or eigenfrequencies in a tube can be formulated by the reflection coefficient. Such effects of the end condition are called **open-end corrections**[7][19]. According to the formulation of sound pressure waves by Eq. 7.6, the volume velocity $Q(x=0,\omega)\mathrm{e}^{\mathrm{i}\omega t}$ (m^3/s) and sound pressure $P(x=0,\omega)\mathrm{e}^{\mathrm{i}\omega t}$ (Pa) are written respectively as

$$Q(x=0,\omega) = SV_a(x=0,\omega) = S\frac{A-B}{\rho_0 c} \tag{7.43}$$

$$P(x=0,\omega) = A+B = Z_A Q = Z_A S \cdot \frac{A-B}{\rho_0 c}. \tag{7.44}$$

where S(m^2) denotes the cross-sectional area of the tube. By recalling Eq. 7.33, the reflection coefficient for the reflected wave from the open end at $x=0$ is derived as

$$\frac{B}{A} = -\frac{1 - \frac{Z_A S}{\rho_0 c}}{1 + \frac{Z_A S}{\rho_0 c}}. \tag{7.45}$$

The reflection coefficient depends on the impedance at the open end. If the open end is closed by a rigid wall, the acoustic impedance $Z_A(\mathrm{Pa}\cdot\mathrm{s/m^3})$ approaches infinity, and then the reflection coefficients approach unity (**perfect in-phase reflection**). In contrast, when the open end can be idealized such that there is no pressure at the boundary (a so-called **free boundary**), the acoustic impedance becomes 0 and thus the reflection coefficient is -1 (**perfect anti-phase reflection**). In particular, there can be no reflection waves under the condition that $Z_A S = \rho_0 c$.

Similarly to subsection 5.2.3 the **open-end correction** is an approximated expression of the acoustic impedance of an open end[19]. Suppose the acoustic impedance can be mostly purely imaginary such that

$$\frac{Z_A S}{\rho_0 c} = \mathrm{i}\tan k y_{Ec}, \tag{7.46}$$

recalling the driving-point acoustic impedance by Eq. 7.20, where $y_{Ec} > 0$. The reflection coefficient can then be rewritten as

$$\frac{B}{A} = -\frac{1 - \mathrm{i}\tan k y_{Ec}}{1 + \mathrm{i}\tan k y_{Ec}} = -\mathrm{e}^{-\mathrm{i}2k y_{Ec}}. \tag{7.47}$$

Therefore, the sound pressure waves $P(\mathrm{Pa})$ composed of incident and reflected waves are

$$P(x,\omega) = A\mathrm{e}^{-\mathrm{i}kx} + B\mathrm{e}^{\mathrm{i}kx} = A\mathrm{e}^{-\mathrm{i}k y_{Ec}}(\mathrm{e}^{-\mathrm{i}k(x-y_{Ec})} - \mathrm{e}^{\mathrm{i}k(x-y_{Ec})}). \tag{7.48}$$

Consequently, the incident wave is not reflected at $x = 0$ but at $x = y_{Ec}$ instead, as if there is a free boundary at $x = y_{Ec}$, which perfectly reflects the anti-phase wave. The extension denoted by $y(\mathrm{m})$ above is called **open-end correction** for an acoustic open pipe with a small cross-sectional radius as considered here. The open-end correction can be approximately written as

$$y_{Ec} \cong \frac{8a}{3\pi}, \qquad (\mathrm{m}) \tag{7.49}$$

where $a(\mathrm{m})$ denotes the radius of the cross section of the pipe. As described in subsection 9.4.1, the correction above can be derived from the imaginary part of the mechanical radiation impedance for a circular thin plate or membrane on an infinitely extending rigid wall such that

$$X_{M_{\mathrm{rad}}} \cong \frac{8ka}{3\pi}\rho_0 c\pi a^2 = \omega\rho_0\frac{8a}{3\pi}(\pi a^2) = \omega\rho_0 y_{Ec}\cdot S \tag{7.50}$$

$$y_{Ec} = \frac{8a}{3\pi} \tag{7.51}$$

$$S = \pi a^2, \tag{7.52}$$

where $ka < 1/2$[5][6].

The end correction indicates that the length of a tube must be virtually longer than the real length. Namely, the eigenfrequencies become lower than those estimated according to the real length of a pipe. Such an effect of the boundaries can be interpreted as **incompressibility of a medium**. Namely, an amount of the outer medium might move together at the open end with the medium inside the tube, and thus the medium is not as compressed as that inside the tube.

Chapter 8
Sound in Space as Plane and Spherical Waves

This chapter gives an overview of the nature of sound radiation and propagation in space. It begins with plane-wave propagation such as reflection and refraction between two media across their boundary, and spherical waves and their simple sound sources are introduced as well as plane waves. The most fundamental nature of spherical waves is the phase difference between the sound pressure and particle velocity.

8.1 Incidence and Reflection of Waves at Boundary

When a progressive plane wave comes to the boundary between different media, reflection and refraction waves are yielded in the media. Suppose that there is an incident wave as shown in Fig. 8.1. Three waves, the incident, **reflection**, and **refraction** waves, are shown.

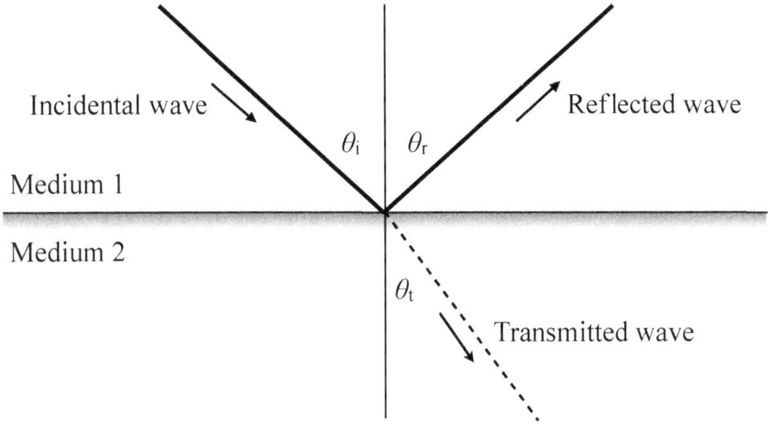

Fig. 8.1 Incident, reflection, and refraction waves at boundary of two media

M. Tohyama: *Sound and Signals*, Signals and Communication Technology, pp. 121–138.
springerlink.com

8.1.1 Reflection Waves and Huygence Principle

A reflection wave is yielded when an incidental wave comes to the boundary. There is a relationship between the two waves such that $\theta_i = \theta_r$. This is called the **law of reflection**. The reflection law can be graphically interpreted according to the **Huygence principle** as shown in Fig. 8.2[35]. By following the Huygence principle, the

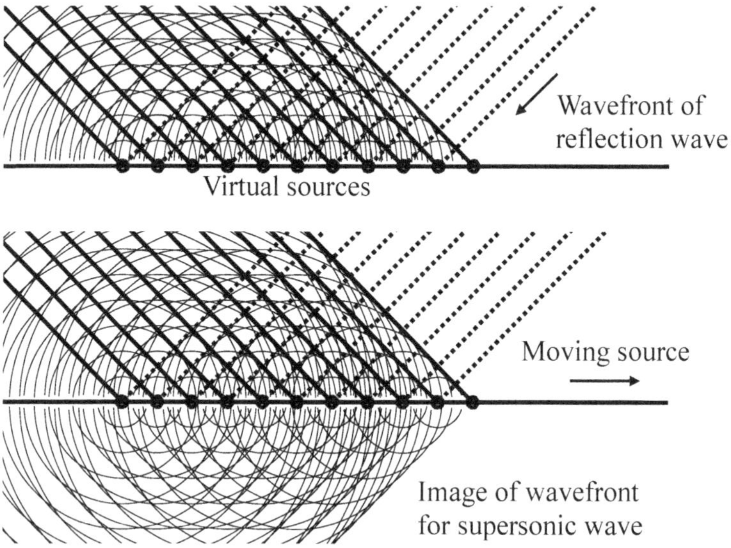

Fig. 8.2 Reflection waves according to Huygence principle from [9] (Fig. 7.2)

reflected wave can be interpreted as superposition of spherical waves (so-called **secondary waves**) that are radiated from the **virtual sources** assumed on the boundary between the two media. Such virtual sources are excited in location order by an incident wave approaching the boundary. Consequently, by tracing the equi-phase contour of the secondary waves, the wavefront can be seen as shown in the figure. The wavefront built up by the secondary waves follows the law of reflection. In particular, if the "lower" part were added symmetrically as shown in the figure, it would look like the supersonic wave as shown in Fig. 6.5. In other words, if the phase difference of virtual sources on the boundary is compared to propagation of the waves on the boundary, the propagation speed is faster than the speed of sound in the medium[5].

8.1.2 Fermat's Principle on Reflection Wave

The law of reflection is shown in the **image theory** of mirror image sources. Suppose that there is a sound source that emanates sound waves above the rigid wall.

The effect of a rigid wall on the sound field can be represented by a single mirror image source as shown in Fig. 8.3. The sound field can be interpreted as if there were two sound sources in a free field where there is no rigid wall.

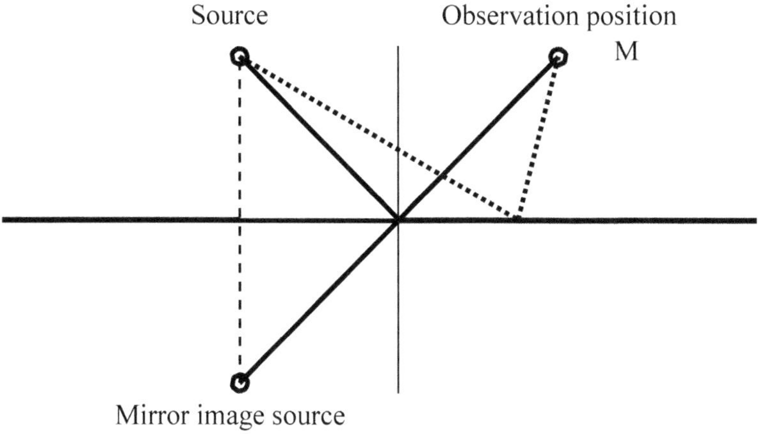

Fig. 8.3 Single mirror image source by rigid wall

The mirror-image theory can also be understood based on **Fermat's principle**[35][36]. Fermat's principle states that the total length of a sound path, which starts at the source, touches a point on the wall, and reaches the receiving position, must be the minimum of all the possible routes. Such a path of minimum length can be achieved by arranging the mirror source as shown in the figure.

8.1.3 Boundary Conditions for Reflection Wave

The law of reflection can be formulated by assuming conditions on the boundary[4]. Suppose that a progressive plane coming into the boundary is expressed as

$$p_i(\mathbf{r},t) = P_i e^{i\omega t} e^{i\mathbf{k}_1 \cdot \mathbf{r}_i} \qquad \text{(Pa)} \tag{8.1}$$

in a two-dimensional plane as shown in Fig. 8.1, where the vector $\mathbf{r}_i$ represents the position vector from the origin, and thus

$$\mathbf{k}_1 \cdot \mathbf{r}_i = k_{1_x} r_{i_x} + k_{1_z} r_{i_z} = k_1 \sin\theta_i \cdot (-x) + k_1 \cos\theta_i \cdot (z), \tag{8.2}$$

and k_1 denotes the wavenumber in medium 1. Similarly, the reflected and transmitted waves are written as

$$p_r(\mathbf{r},t) = P_r e^{i\omega t} e^{-i\mathbf{k}_1 \cdot \mathbf{r}_r} \qquad \text{(Pa)} \tag{8.3}$$

$$p_t(\mathbf{r},t) = P_t e^{i\omega t} e^{-i\mathbf{k}_2 \cdot \mathbf{r}_t} \qquad \text{(Pa)} \tag{8.4}$$

$$\mathbf{k}_1 \cdot \mathbf{r}_r = k_1 \sin\theta_r \cdot (x) + k_1 \cos\theta_r \cdot (z) \tag{8.5}$$
$$\mathbf{k}_2 \cdot \mathbf{r}_t = k_2 \sin\theta_t \cdot (x) + k_2 \cos\theta_t \cdot (-z). \tag{8.6}$$

Now introduce the boundary conditions such that

$$p_i(z=0,t) + p_r(z=0,t) = p_t(z=0,t) \qquad \text{(Pa)} \tag{8.7}$$
$$v_{i_z}(z=0,t) + v_{r_z}(z=0,t) = v_{t_z}(z=0,t) \qquad \text{(m/s)} \tag{8.8}$$

on the boundary $(z=0)$, where $v_{i_z}, v_{r_z}, v_{t_z}$ represent the z-component of the velocities, respectively. Following the boundary conditions above,

$$P_i e^{-ik_1 x \sin\theta_i} + P_r e^{-ik_1 x \sin\theta_r} = P_t e^{-ik_2 x \sin\theta_t} \tag{8.9}$$
$$k_1 \cos\theta_i P_i e^{-ik_1 x \sin\theta_i} - k_1 \cos\theta_r P_r e^{-ik_1 x \sin\theta_r} = k_2 \cos\theta_t P_t e^{-ik_2 x \sin\theta_t} \tag{8.10}$$

must hold well on the boundary independent of the position coordinate x. The law of reflection, which can be interpreted as

$$k_1 \sin\theta_i = k_1 \sin\theta_r = k_2 \sin\theta_t, \tag{8.11}$$

must be satisfied for that purpose. By substituting $k_1 = 2\pi/\lambda_1$ for k_1 above,

$$\lambda_T = \frac{\lambda_1}{\sin\theta_i} \geq \lambda_1, \tag{8.12}$$

which represents the **trace wavelength on the boundary**, is derived[5]. This outcome corresponds to the supersonic wave when the wavelength of a vibrating boundary is longer than that for the sound in the surrounding medium.

8.1.4 Reflection and Transmission Coefficients

Following the law of reflection, the reflection and transmission coefficients of sound pressure are given by

$$\mu_{12} = \frac{P_r}{P_i} = -\frac{z_{0_1}\cos\theta_t - z_{0_2}\cos\theta_i}{z_{0_1}\cos\theta_t + z_{0_2}\cos\theta_i} \tag{8.13}$$
$$\eta_{12} = \frac{P_t}{P_i} = \frac{z_{0_2}}{z_{0_1}} \frac{2 z_{0_1}\cos\theta_i}{z_{0_1}\cos\theta_t + z_{0_2}\cos\theta_i}, \tag{8.14}$$

where the sound pressure (Pa) and particle velocity (m/s) are expressed as

$$p_i = z_{0_1} v_i \qquad p_r = -z_{0_1} v_r \qquad p_t = z_{0_2} v_t, \tag{8.15}$$

and

$$\mu_{21} = -\mu_{12} \tag{8.16}$$
$$\eta_{21} = \frac{z_{0_1}}{z_{0_2}} \eta_{12}. \tag{8.17}$$

Here it should be noted that $\mu_{12}+\eta_{12}\neq 1$. Recall the energy flux density per a unit area during a unit time interval for the incident, reflected, and transmitted waves such that

$$I_i = \frac{1}{2}\frac{|P_i|^2}{\rho_1 c_1} = \frac{1}{2}\frac{|P_i|^2}{z_{0_1}} \qquad (\mathrm{W/m^2}) \tag{8.18}$$

$$I_r = \frac{1}{2}\frac{|P_r|^2}{z_{0_1}} \tag{8.19}$$

$$I_t = \frac{1}{2}\frac{|P_t|^2}{\rho_2 c_2} = \frac{1}{2}\frac{|P_t|^2}{z_{0_2}} \tag{8.20}$$

The energy preservation law states that

$$I_i = I_r + I_t. \qquad (\mathrm{W/m^2}) \tag{8.21}$$

Therefore, the equation above can be rewritten as

$$\mu_{12}^2 + \frac{z_{0_1}}{z_{0_2}}\eta_{12}^2 = 1 \tag{8.22}$$

in terms of the reflection and transmission coefficients.

For the normal incidence (perpendicular incidence) from medium 1 to 2,

$$\mu_{12} = \frac{P_r}{P_i} = -\frac{1-z_{0_{12}}}{1+z_{0_{12}}} \tag{8.23}$$

$$\eta_{12} = \frac{P_r}{P_i} = \frac{2z_{0_{12}}}{1+z_{0_{12}}} \tag{8.24}$$

$$z_{0_{12}} = \frac{z_{0_2}}{z_{0_1}} \tag{8.25}$$

show the reflection and transmission coefficients. The ratio of specific impedance between the two media close to the boundary is a key for determining the wave reflection on the boundary or traveling across the boundary. The reflection coefficient comes close to unity as the ratio $z_{0_{12}}$ increases. However, the transmission coefficients cannot be estimated as a limit when $z_{0_{12}}$ increases. Following the energy preservation law stated by Eq. 8.22, the transmission coefficient should be zero, as the reflection coefficient becomes unity because of the large $z_{0_{12}}$. For example, the volume density is about 1.3 $(\mathrm{kg/m^3})$ for air in which sound travels at around 340 (m) every unit time interval, while the corresponding density and sound speed are almost $10^3(\mathrm{kg/m^3})$ and $1500(\mathrm{m/s})$, respectively in water. Most of the sound energy for a coming wave from the air is reflected into the air again.

8.2 Refraction of Transmitted Waves from Boundary

The law of reflection was described in the previous section in terms of incident and reflection waves. The nature of refraction waves are mentioned in this section in terms of the transmitted waves.

8.2.1 Incident and Transmitted Angles

Again look at the incident and transmitted waves as shown by Fig. 8.4, where the sound speeds are c_1 and c_2 respectively in mediums 1 and 2. As illustrated in the figure when the wavefront progresses by $c_1\Delta t$ in medium 1, the transmitted wave into medium 2 travels by $c_2\Delta t$. Namely, if $c_1 < c_2$, the sound wave travels longer

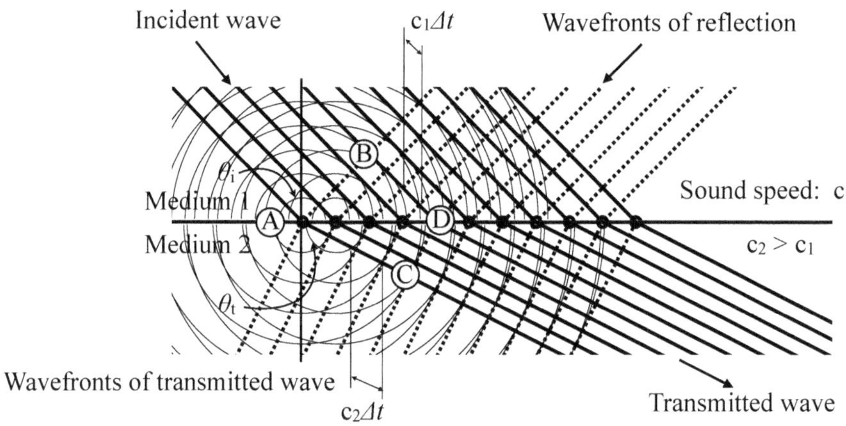

Fig. 8.4 Incident and refraction waves at boundary of two media where sound speed of medium 2 is faster than that for medium 1 from [9] (Fig. 7.4)

than that in medium 1, and vice versa. Consequently, in each case the sound changes its direction at the boundary. This is called **refraction** of a wave. The refraction depends on the ratio of the sound speeds in the two media, while the reflection depends on the ratio of specific impedance between the two media.

Recalling the reflection law in Eq. 8.11 in the previous section,

$$\sin\theta_t = \frac{k_1}{k_2}\sin\theta_i = \frac{c_2}{c_1}\sin\theta_i. \tag{8.26}$$

This relationship, the **law of refraction**, can also be understood by looking at Fig. 8.4. In the figure, the triangles ABD and ACD are right angles where AD is the common hypotenuse. The equation above shows the relationship between the geometrical lengths of the triangles; the angle for the refracted (transmitted) wave θ_t becomes wider (narrower) than the incident one (θ_i) when the sound speed is faster (slower) in medium 2 than in medium 1. Wave refraction occurs due to the difference in the sound speed between the two media. It occurs even in a medium without a boundary between different media, if there is a change of sound speed in the medium due to variances in the temperature.

8.2.2 *Critical Angle of Incidence*

The law of reflection indicates that there is no possibility of wave transmission into medium 2 if there is no transmission angle to match the requirement specified by Eq. 8.26. Namely, it is necessary for the two media to meet the condition that

$$0 < \frac{c_2}{c_1} \sin\theta_i \leq 1 \tag{8.27}$$

when there is a refracted wave into medium 2. If the condition above does not hold well, such a case being called **total reflection**, no refracted waves travel into medium 2. It can be intuitively understood that the length of AD cannot be shorter than that for AC, shown in Fig. 8.4. In other words, there are no supersonic waves in the case of total reflection. The relationship between the incident and refracted waves coming from medium 1 into 2 can also be interpreted as wave transmission from medium 2 into 1. In such a case, the incident (refraction) angle in Fig. 8.4 can be interpreted as the refraction (incident) angle.

8.2.3 *Refraction Waves and Law of Snellious*

As shown in Fig. 8.5 refraction of a plane wave at the boundary can be represented by the ratio of the sound speed in both media such that

$$\frac{x_i}{x_t} = \frac{\sin\theta_i}{\sin\theta_t} = \frac{c_1}{c_2}, \tag{8.28}$$

where c_1 and c_2 denote the speed of sound (m/s) in medium 1 and 2 respectively. The equation above, which says the ratio between the incident and refracted angles is independent of the incident angle but depends instead on the combination of the media, is called the **refraction law of Snellious**.

Recall the reflection coefficient from the boundary can be rewritten as

$$\mu_{12} = \frac{P_r}{P_i} = \frac{z_{0_2}\dfrac{\cos\theta_i}{\cos\theta_t} - z_{0_1}}{z_{0_2}\dfrac{\cos\theta_i}{\cos\theta_t} + z_{0_1}}. \tag{8.29}$$

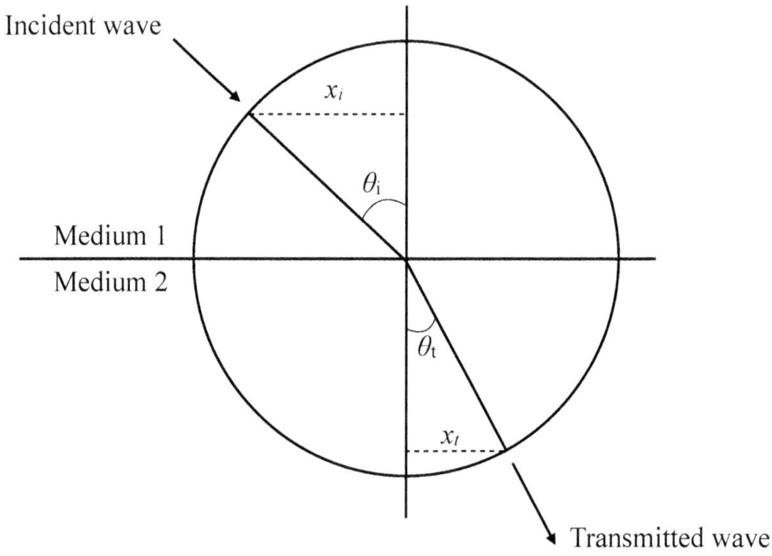

Fig. 8.5 Refraction of plane wave

If the impedance of the boundary (**wall impedance**)[5] is defined by the ratio of the sound pressure and normal velocity of the particle, it can be written as

$$z_{wall}(\theta_i) = \frac{p_t(z=0,t)}{v_t(z=0,t)\cos\theta_t} = \frac{z_{wall}(\theta_i=0)}{\cos\theta_t} \tag{8.30}$$

at the boundary as a function of the incident angle θ_i, where $\cos\theta_t$ is given by

$$\cos\theta_t = \sqrt{1-\left(\frac{c_2}{c_1}\right)^2\sin^2\theta_i} \tag{8.31}$$

following the law of Snellious on refracted waves. By introducing the wall impedance, the reflection coefficients can be formulated as

$$\mu_{12} = \frac{z_{wall}(\theta_i)\cos\theta_i - z_{0_1}}{z_{wall}(\theta_i)\cos\theta_i + z_{0_1}}, \tag{8.32}$$

again as a function of the incident angle and the wall impedance instead of the specific impedance of the medium behind the boundary wall.

8.2.4 *Fermat's Principle on Refraction of Wave*

Refraction of a wave can also be interpreted as a result of **Fermat's principle**[35][36]. Now consider the time τ_{SP} that a sound wave takes to travel from S to P, following Fig. 8.6. The time τ_{SP} is represented as

$$\tau_{SP} = \frac{L_{SO}}{c_1} + \frac{L_{OP}}{c_2}, \tag{8.33}$$

where c_1 and c_2 are the respective sound speeds of the media, and L_{SO} and L_{OP} denote the length of the corresponding path. Fermat's principle on the refraction of a wave states that the law of refraction can be interpreted as a traveling root which minimizes the traveling time from S to P between the two media.

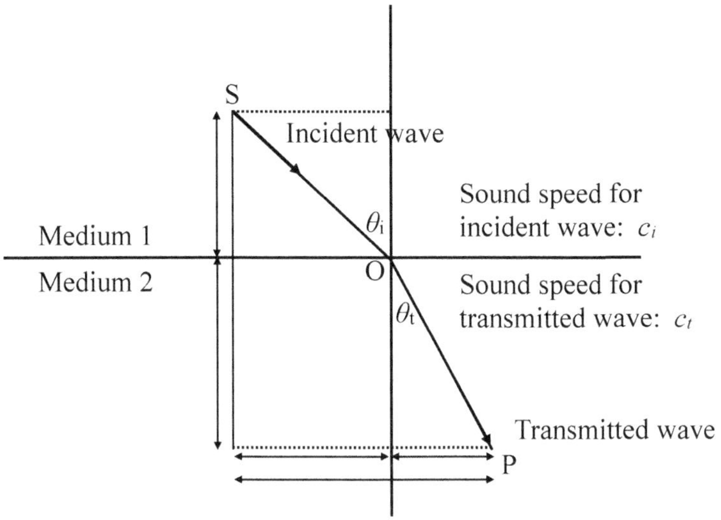

Fig. 8.6 Fermat's principle for refraction of plane wave

8.3 Radiation of Waves from Spherical Source

The magnitude of a plane wave is independent of the distance from the source. However, this goes against our intuition because most daily sounds become louder as the distance to the source decreases. Such experiences suggest that there is another wave mode for sound traveling in space. Indeed, **spherical waves** propagate in three-dimensional space. In particular, spherical waves are significant in the area close to a sound source because the nature of spherical waves depends on the distance from the source. This section describes the fundamental nature of the pressure and particle velocity as a function of the distance from a source when sound travels as a spherical wave in three-dimensional space.

8.3.1 Radiation of Waves from Ideal Point Source

An ideal source from which a spherical wave is radiated is a **spherical source** that uniformly dilates or contracts. Such a spherical source is called a **point source**, if the radius of the spherical source is sufficiently smaller than the sound wavelength. A spherical wave from such a spherical source is called a symmetric spherical source. The wavefront of a symmetrical spherical wave spreads over a spherical surface of radius r as r increases. Consequently, the energy-flux density across the wavefront decreases in inverse proportion to the surface area of the spherical wavefront. This explains our daily experience that sound becomes softer as the distance from the source increases.

8.3.2 Particle Velocity and Sound Pressure from Ideal Point Source

For a plane wave, the magnitude of the sound pressure is independent of the distance from the source, and it is in-phase with the particle velocity. It is not the case for a spherical wave radiated from an ideal point source. Now consider the equation of motion for a small portion of a medium in which a spherical sound wave travels as shown in Fig. 8.7. The force

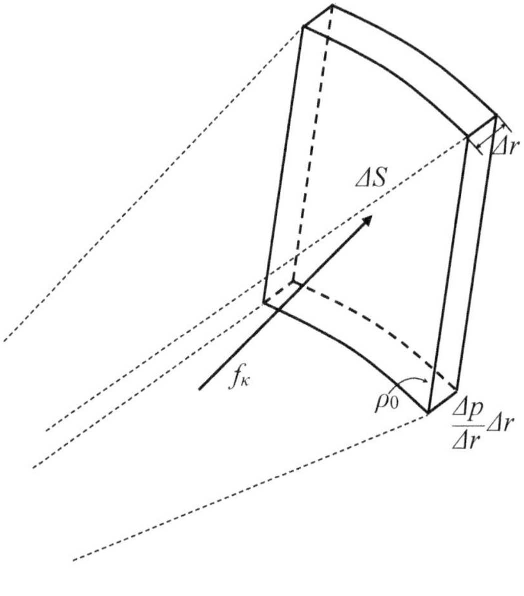

Fig. 8.7 Sound pressure in small portion of medium

$$f_K(r,t) = -\frac{\partial p(r,t)}{\partial r} dr 4\pi r^2 \quad \text{(N)} \tag{8.34}$$

applies to the total spherical shell including the small portion. Then the acceleration of the volume velocity of the total spherical shell can be written as

$$\frac{\partial v}{\partial t} 4\pi r^2 = \frac{\partial q(t-r/c)}{\partial t}, \quad (\text{m}^3/\text{s}^2) \tag{8.35}$$

and thus

$$\frac{\rho_0}{4\pi r^2} \frac{\partial q(t-r/c)}{\partial t} = -\frac{\partial p(r,t)}{\partial r} \quad (\text{Pa/m}) \tag{8.36}$$

holds well, following the equation of motion for the spherical shell. Therefore, the sound pressure becomes

$$p(r,t) = \frac{\rho_0}{4\pi r} \frac{\partial q(t-r/c)}{\partial t} \quad (\text{Pa}) \tag{8.37}$$

at the distance $r(\text{m})$ from the spherical source. The sound pressure is in proportion to the acceleration of the volume velocity of the source, and its magnitude decreases in inverse proportion to the distance from the source, while the phase delay is proportional to the distance as well as a plane wave.

Suppose that the volume velocity of a point source is $q(t) = Q_0 e^{i\omega t} (\text{m}^3/\text{s})$ in a complex function form. The sound pressure due to the spherical wave radiated from the source can be written as

$$p(r,t) = \frac{\rho_0}{4\pi r} \frac{\partial q(t-r/c)}{\partial t} = i\omega\rho_0 \frac{Q_0}{4\pi r} e^{i\omega t - ikr} \quad (\text{Pa}) \tag{8.38}$$

at the distance $r(\text{m})$ from the point source. The magnitude decreases and the phase is delayed as the distance increases.

As well as the sound pressure above, the particle velocity $v(r,t)$ is also formulated according to the equation of motion. The equation of motion can be rewritten as

$$\rho_0 \frac{\partial v(r,t)}{\partial t} = -\frac{\partial p(r,t)}{\partial r}, \quad (\text{Pa/m}) \tag{8.39}$$

where the acceleration is in proportion to the sound pressure gradient. Therefore, the particle velocity can be estimated as

$$v(r,t) = -\frac{1}{4\pi} \frac{\partial}{\partial r} \frac{q(t-r/c)}{r}. \quad (\text{m/s}) \tag{8.40}$$

The particle velocity is in proportion to the local slope of $q(t-r/c)/r$ instead of the acceleration of the volume velocity of the source, and thus the relationship between the sound pressure and particle velocity is more complicated than that for the plane waves. The particle velocity can be rewritten as

$$v(r,t) = \frac{1}{4\pi}\left(\frac{1}{cr}\frac{\partial q(t-r/c)}{\partial t} + \frac{q(t-r/c)}{r^2}\right) = \frac{p(t-r/c)}{\rho_0 c} + \frac{q(t-r/c)}{4\pi r^2}. \quad (\mathrm{m/s}) \quad (8.41)$$

The velocity is expressed as the sum of two types of effects that a sound source has on a distant point through wave propagation. Namely, in addition to the first term, which is in proportion to the volume velocity of the source and is of inversely proportional to the distance from the source, there is the second term, which linearly increases with the volume velocity but inversely decreases with the square of the distance from the source. However, the second term might be negligible, as the distance from the source increases and consequently the wavefront spreads over a wider area. In other words, the first term, independent of the distance r, represents the nature of a plane wave that is embedded evenly into a spherical wave, but it is noticeable as the distance increases.

The particle velocity due to the spherical wave can be written as

$$\begin{aligned} v(r,t) &= \frac{p(t-r/c)}{\rho_0 c} + \frac{q(t-r/c)}{4\pi r^2} \\ &= \mathrm{i}k\frac{Q_0}{4\pi r}\mathrm{e}^{\mathrm{i}\omega t - ikr} + \frac{Q_0}{4\pi r^2}\mathrm{e}^{\mathrm{i}\omega t - ikr}. \quad (\mathrm{m/s}) \end{aligned} \quad (8.42)$$

The first term is in phase with the sound pressure, while the second term is unique to a spherical wave and out-of-phase to the sound pressure. Moreover, the first term depends on the wavenumber of the sound, while the second one only depends on the distance from the source. By assuming a constant sound frequency, the magnitude of the first term becomes small as the speed of sound increases, while the magnitude of the second term is constant independent of the speed of sound.

Recall the definition of the specific impedance as the ratio of the sound pressure and particle velocity for a plane wave. Take the same type of ratio for a sinusoidal spherical wave. Then,

$$z_{00} = \frac{p(r,t)}{v(r,t)} = \frac{\mathrm{i}\omega\rho_0 r}{ikr+1} = \rho_0 c\frac{ikr}{ikr+1} \quad (\mathrm{Pa}\cdot\mathrm{s/m}) \quad (8.43)$$

depends on the distance $r(\mathrm{m})$ from the source and the wavenumber $k(\mathrm{rad/m})$, where

$$\rho_0\frac{\partial v(r,t)}{\partial t} = \rho_0 \mathrm{i}\omega v = -\frac{\partial p(r,t)}{\partial r} = p\frac{ikr+1}{r}. \quad (8.44)$$

The sound pressure is in phase with the particle velocity only when $kr \to$ large is taken as a limit such that

$$z_{00} \to \rho_0 c = z_0. \quad (8.45)$$

Here, z_0 indicates the specific impedance of the medium.

8.3.3 *Travelling Sound from Spherical Source*

The issues described in the previous subsection can be extended into a more practical case, such as a sound source with a finite radius length. Suppose that the spherical source uniformly vibrates following a sinusoidal function of the angular frequency $\omega = 2\pi f$. The sound pressure $p(r,t)$(Pa) conveyed by a spherical wave from the ideal spherical source can be written in a complex sinusoidal function form as

$$p(r,t) = \frac{A}{r}\mathrm{e}^{\mathrm{i}(\omega t - kr)} \tag{8.46}$$

$$v(r,t) = \frac{\mathrm{i}kr+1}{\mathrm{i}\rho_0 ckr}\cdot\frac{A}{r}\mathrm{e}^{\mathrm{i}(\omega t-kr)} = \frac{\mathrm{i}kr+1}{\mathrm{i}\rho_0 ckr}\cdot p(r,t), \tag{8.47}$$

where c(m/s) is the speed of sound in the medium, r(m) denotes the distance from the source. The constant A(Pa · m) is determined according to the strength of the source given by the volume velocity. Namely, the **strength of a source** or **volume velocity** $q(t)(\mathrm{m}^3/\mathrm{s})$ of a spherical source with a radius of a(m) is given by $q(t) = Q\mathrm{e}^{\mathrm{i}\omega t} = 4\pi a^2 v_X(t)(\mathrm{m}^3/\mathrm{s})$, where $v_X(t) = V_X\mathrm{e}^{\mathrm{i}\omega t}$(m/s) denotes the surface velocity of the spherical source.

Now consider the magnitude of sound pressure for a sinusoidal spherical wave radiated from a spherical source of radius a. Assume that the surface velocity of a spherical source must be equal to the particle velocity along the direction of propagation of the spherical wave. Then,

$$v(a,t) = V_X\mathrm{e}^{\mathrm{i}\omega t} = \frac{Q}{4\pi a^2}\mathrm{e}^{\mathrm{i}\omega t} = \frac{A}{a}\mathrm{e}^{\mathrm{i}(\omega t-ka)}\cdot\frac{\mathrm{i}ka+1}{\mathrm{i}\rho_0 cka} \tag{8.48}$$

must hold well where the volume velocity is expressed as a sinusoidal function of $Q\mathrm{e}^{\mathrm{i}\omega t}(\mathrm{m}^3/\mathrm{s})$. Consequently, the magnitude of sound pressure A can be written as

$$A = \frac{Q}{4\pi}\cdot\frac{\mathrm{i}\omega\rho_0}{\mathrm{i}ka+1}\mathrm{e}^{\mathrm{i}ka} \quad (\mathrm{Pa}\cdot\mathrm{m}) \tag{8.49}$$

as a complex magnitude form. The magnitude of the sound pressure wave approaches

$$A \cong \mathrm{i}\omega\rho_0\frac{Q}{4\pi} \qquad (\mathrm{Pa}\cdot\mathrm{m}) \tag{8.50}$$

as ka becomes small, such as for a point source, while it is approximated as

$$A \cong \frac{\rho_0 c}{a}\mathrm{e}^{\mathrm{i}ka}\frac{Q}{4\pi}. \quad (\mathrm{Pa}\cdot\mathrm{m}) \tag{8.51}$$

when ka is large. By taking the sound pressure at the surface of the source,

$$p(a,t) \cong \rho_0 a\dot{v}_X(t) \qquad ka \to \text{small} \tag{8.52}$$

$$p(a,t) \cong \rho_0 c v_X(t) \qquad ka \to \text{large} \qquad (\text{Pa}) \tag{8.53}$$

hold well. The sound pressure at the source position is in-phase with the acceleration of the source when the radius is as small as that for a point source, while it is in-phase with the velocity of the source as the radius increases.

The **acoustic radiation impedance of a source** is defined as the ratio of the sound pressure and volume velocity at the source position such that

$$Z_{A_{\text{rad}}} = \frac{p(a,t)}{q(t)} \cong \text{i}\frac{\omega\rho_0}{4\pi a} \qquad ka \to \text{small} \tag{8.54}$$

$$\cong \frac{\rho_0 c}{4\pi a^2}. \qquad (\text{Pa}\cdot\text{s/m}^3) \qquad ka \to \text{large} \tag{8.55}$$

The radiation impedance of a source might be purely imaginary as the size of a source decreases, while it can be purely real for a sufficiently large source. The radiation impedance is a significant issue in considering the sound power output of a source located in a room.

8.3.4 Incompressibility of Medium

The phase relationship between the sound pressure and velocity is an outcome of the wavefront being not a plane but a curved surface. The effect of a curved wavefront on the propagation of a wave can be interpreted as manifestation of the **incompressibility of a medium** in which the sound wave travels. Condensation or dilation might be necessary in a localized area of a medium surrounding a vibrating object, when sound is radiated into the medium from the vibrating object. Namely, if there is no condensation or dilation in the medium against the vibration of a object, no sound travels into the medium. The nature of a medium that disturbs the condensation or dilation is called incompressibility of the medium.

In general, a surrounding medium, in which an object oscillates very slowly, works as if it might be incompressible. Instantaneous non-uniformity of the medium density can be cancelled out by the movement or flow of the medium because the oscillation movement is slow enough for the medium to follow. Consequently, no sound is radiated into the medium from the object. However, there can be non-uniformity yielded in the medium as the frequency of oscillation increases. This is the outcome of the medium not being able to follow the motion any longer because the oscillation is fast. Thus, synchronized pressure variation is possible with the instantaneous non-uniformity in the medium density due to the vibration of the object. This type of pressure variation causes the propagation of sound into the medium. There are no effects of the incompressibility on propagation of a plane wave. Cancellation of the non-uniformity in the density of the medium is quite unlikely when the wavefront is infinitely extended or limited to a thin tube, such as that for a plane wave.

The particle velocity of a spherical wave is mostly due to the incompressibility, but only when the distance is very close to the source, however. In other words, a particle velocity that is out-of-phase to the sound pressure is in-phase with the vibration velocity of the source due to the incompressibility of the medium. Recall the particle velocity of the spherical wave given by Eq. 8.42. The first term representing the plane-wave-like velocity increases as the frequency of sound increases. This can be interpreted as the effect as incompressibility drops, because the surrounding medium is not able to synchronize with the vibration of the source of the high frequencies. On the other hand, the plane-wave-like velocity decreases with the sound speed. Namely, the incompressibility of a medium lessens as the sound speed decreases.

8.4 Interference of Waves

When a wave in space comes to a rigid wall, the wave returns into the space as a reflection wave. Consequently, there are two types of waves in the space; the approaching (incident) and returning (reflection) waves. Such a superposition of waves causes the distribution of sound pressure in the space. In other words, the sound pressure increases (decreases) as the two waves become of in-phase (anti-phase). **Wave interference** is when waves of the same frequency are "added" or "subtracted" according to the phase relation between the superposed waves.

8.4.1 *Sound Field Excited by Two Point Sources*

Suppose that there are two point sources which are closely located to each other, and take observation points far ($r_o >> d$) from the sources as shown in Fig. 8.8. The phase difference of the waves from the sources is determined by the difference

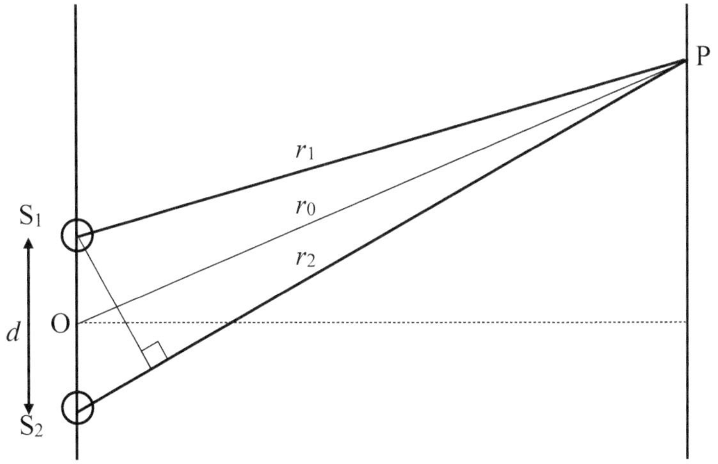

Fig. 8.8 Pair of closely located point sources

between the lengths of paths from the two sources to an observation point as shown in Fig. 8.8. Therefore, the additive effect of the two waves can be expected at positions where the path-length differences are $n\lambda$, while the two waves are subtractive at the positions where the differences are $\lambda(2n-1)/2$. Here, λ is the wavelength of sound and n is a positive integer.

Figure 8.9 shows an image of wave interference by the two point sources. The interference is identical at the positions where the path-length difference is the same. Therefore, by connecting such positions where the identical interference occurs, a family of hyperbolic curves is obtained as shown in the figure.

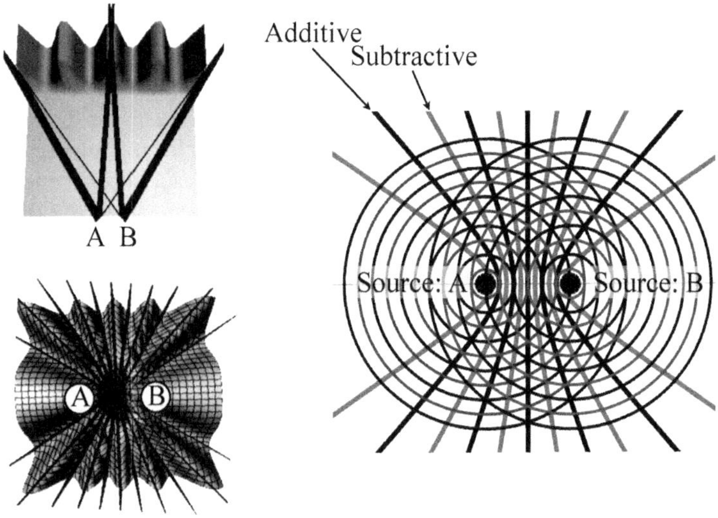

Fig. 8.9 Wave interference due to two closely located point sources from [9] (Fig. 7.13)

8.4.2 *Superposition of Two Sinusoidal Functions of Identical Frequency*

A **standing wave** is another example of superposition of waves. Suppose that there is a pair of progressive plane waves of a single frequency with equal magnitudes but opposite propagation directions. Taking the sum of the two waves such that

$$p(x,t) = A\sin(\omega t - kx) + A\sin(\omega t + kx) = 2A\sin\omega t\cos kx \tag{8.56}$$

shows that a **standing wave** is made in the space. In this wave, there are **loops** and **nodes**, while each progressive wave does not have any loops and nodes. This outcome is due to interference by the two progressive waves such that the two waves are always additive at the loops, while both are completely subtractive at the nodes. In other words, the loops are made at the locations where $kx = n\pi$; on the other hand, the nodes are at positions where $kx = (2n+1)\pi/2$ and n is a positive integer.

A standing wave is always made, independent of the frequency, by superposition of two progressive waves of a single frequency with the same magnitude but opposite propagation directions. The eigenfunction, for example, of an acoustic tube, is also a standing wave as described in the previous chapter, but the frequencies are limited to the eigenfrequencies. Therefore, a standing wave that is made in a free space, independent of the frequency, is not called the eigenfunction of the space.

Recall that there is no sound radiation from an ideal acoustical tube. There is no acoustic energy flow in space where there is a standing wave. However, there is no standing wave when there are two progressive waves with different magnitudes, such as incident and reflected waves. Suppose that there is a reflection wave with the magnitude of B for an incidental wave of the magnitude of unity such that

$$p(x,t) = \sin(\omega t - kx) + B\sin(\omega t + kx). \tag{8.57}$$

The sum of these two waves looks like a progressive wave to the right-hand side (positive direction) as the nodes are filled, when the magnitude B becomes smaller than unity.

8.4.3 Interference by Reflection Waves

There can be wave interference in a sound field where incident and reflected waves are travelling. Suppose a mirror image source by a rigid wall as shown in Fig. 8.3. The sound pressure observed at a position M can be written as the sum of waves radiated from the point source and the mirror image one:

$$p(M,t) = \mathrm{i}\omega\rho_0 \frac{Q}{4\pi}\mathrm{e}^{\mathrm{i}\omega t}\left(\frac{1}{r_1}\mathrm{e}^{-\mathrm{i}kr_1} + \frac{\mu}{r_2}\mathrm{e}^{-\mathrm{i}kr_2}\right) \quad \text{(Pa)} \tag{8.58}$$

in a complex function form, where μ represents the reflection coefficient of the wall.

The sound pressure at the observation point depends on the frequency of the source because of the interference by the reflection wave. The effect of the reflection wave can be expressed as the ratio of the observed pressure $p(M,t)$ and only for the direct wave from the source $p_d(M,t)$ such that

$$\hat{p}(M,t) = \frac{p(M,t)}{p_d(M,t)} = (1 + \frac{r_1}{r_2}\mu\mathrm{e}^{-\mathrm{i}\omega\tau_{dr}}), \tag{8.59}$$

where τ_{dr} denotes the time delay between the direct and reflection waves. By taking the magnitude for the ratio above, the equation above can be rewritten as

$$|\hat{p}(M,t)| \cong \sqrt{1 + \mu^2 + 2\mu\cos\omega\tau_{dr}} \tag{8.60}$$

where it is assumed that $r_1/r_2 \cong 1$.

Figure 8.10 illustrates an example of the sound pressure response at an observation point where $\mu = 1$. If the observation point (the position of a microphone) is located very close to the rigid wall, the magnitude of sound pressure is two times

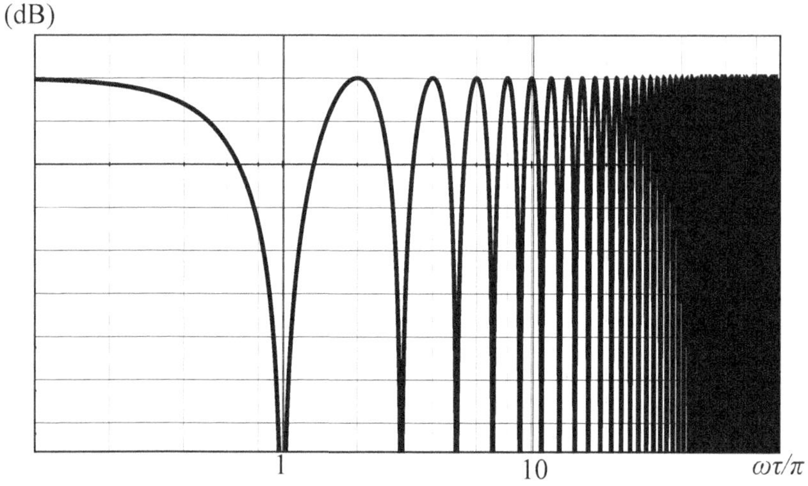

Fig. 8.10 Example of frequency dependency of sound pressure by interference with reflection from rigid wall

larger than the magnitude only for the direct sound (corresponding to the sound pressure without the rigid wall) independent of the frequency. This is because there is no significant time delay between the direct and reflected waves and thus the sum is always additive. However, when there is a considerable time delay between the two as the observation position is far from the wall, the interference depends on the sound frequency. That is, the interference is completely subtractive for the sound where $\omega\tau_{dr} = (2n-1)\pi$ for $n = 1, 2, 3, ...$, as shown by the periodic troughs in the figure. A recording position should be arranged so that such troughs might be out of the frequency range of interest.

Sound can be heard even if the sound source is not seen because of an obstacle between the source and listening positions. This type of sound transmission is called sound **diffraction** or **scattering**. Diffraction and scattering are important characteristics of waves, as are sound reflection and refraction. Therefore, there are many textbooks on these topics, such as[5][35].

Chapter 9
Wave Equations and Sound Radiation in Space

This chapter describes the wave equation in three-dimensional space and wave propagation in the space as well as in an acoustic horn. First, the derivation of the wave equation for spherical waves, similarly to the plane wave propagation due to an initial disturbance, is described. It might be interesting to see both compressive and dilated waves travel in the space, even if the initial disturbance of compression is given to only the local area of a medium. Next, sound power output from a source for the spherical waves is considered in terms of the radiation impedance. The sound power output is highly sensitive to the surroundings of the source. A coupled point sources is a good example of the dependency of sound power output on the surroundings. Finally, the wave propagation in an acoustic horn is briefly mentioned in terms of the radiation impedance. It can be interpreted that acoustic horns were developed so that sound waves might be well radiated from a source into the space.

9.1 Wave Equation of Spherical Waves

9.1.1 Wave Equation for Three-Dimensional Wave

The wave equation for plane waves of sound pressure $p(x,t)$ such that

$$\frac{\partial^2 p(x,t)}{\partial t^2} = c^2 \frac{\partial^2 p(x,t)}{\partial x^2} \tag{9.1}$$

is extended into three-dimensional space as

$$\frac{\partial^2 p(x,y,z,t)}{\partial t^2} = c^2 \nabla^2 p(x,y,z,t). \tag{9.2}$$

This is called the **wave equation** in general. By introducing Cartesian coordinates, the equation above can be rewritten as

$$\frac{\partial^2 p(x,y,z,t)}{\partial t^2} = c^2 \left(\frac{\partial^2 p(x,y,z,t)}{\partial x^2} + \frac{\partial^2 p(x,y,z,t)}{\partial y^2} + \frac{\partial^2 p(x,y,z,t)}{\partial z^2} \right), \tag{9.3}$$

M. Tohyama: *Sound and Signals*, Signals and Communication Technology, pp. 139–162.
springerlink.com

where

$$\nabla^2 = \left(\frac{\partial^2}{\partial x^2} + \frac{\partial^2}{\partial y^2} + \frac{\partial^2}{\partial z^2}\right) \tag{9.4}$$

is introduced.

9.1.2 Wave Equation for a Symmetric Spherical Wave

The wave equation is simplified for symmetric spherical waves, however. A symmetric spherical wave is represented by the distance from the source. Now consider the wave equation for the symmetric spherical waves in terms of the velocity potential. Suppose that the particle velocity $v(r,t)(\mathrm{m/s})$ is expressed as $v = -\frac{\partial \phi_v(r,t)}{\partial r}$ at the distance of $r(\mathrm{m})$ from the source. Here $\phi_v(r,t)$ denotes the **velocity potential**. Namely, by taking the derivative of the velocity potential with respect to a coordinate, the particle velocity along the coordinate is obtained as

$$v_r(r,t) = -\frac{\partial \phi_v(r,t)}{\partial r}, \qquad \phi_v = -\int v_r dr \tag{9.5}$$

with a negative sign. Thus, the volume velocity $q(r,t)(\mathrm{m^3/s})$ is expressed as

$$q(r,t) = -\frac{\partial \phi_v(r,t)}{\partial r} 4\pi r^2. \tag{9.6}$$

The small variation of the volume velocity Δq due to extension of the spherical wavefront between r and $r + \Delta r$ can be estimated as

$$\Delta q = \frac{\partial}{\partial r}\left(-\frac{\partial \phi_v}{\partial r} 4\pi r^2\right)\Delta r. \tag{9.7}$$

Such a difference of the volume in the unit time interval can be also rewritten as

$$\Delta q = -\frac{\partial s}{\partial t} 4\pi r^2 \Delta r, \tag{9.8}$$

where s denotes the condensation of a medium. In addition, by recalling the relationship between the sound pressure and particle velocity such that

$$\rho_0 \frac{\partial v_r(r,t)}{\partial t} = -\frac{\partial p(r,t)}{\partial r} \tag{9.9}$$

and

$$p = \kappa s \tag{9.10}$$

then

$$\frac{\partial \phi_v(r,t)}{\partial t} = c^2 s(r,t) \tag{9.11}$$

holds well. Consequently,

$$\frac{\partial^2 \phi_v(r,t)}{\partial t^2} = \frac{c^2}{r^2}\frac{\partial}{\partial r}\left(r^2 \frac{\partial \phi_v(r,t)}{\partial r}\right), \tag{9.12}$$

namely

$$\frac{\partial^2 (r\phi_v(r,t))}{\partial t^2} = c^2 \frac{\partial^2 (r\phi_v(r,t))}{\partial r^2}, \tag{9.13}$$

is derived. This is called the **wave equation for a symmetric spherical wave**.

9.1.3 General Solution for Symmetrical Spherical Wave

The wave equation for a symmetric spherical wave can be interpreted as an equation for the variable of $r\phi_v(r,t)$. By this interpretation, the equation is the same as that for a plane wave. Namely, the d'Alembert's solution can be applied to the variable $r\phi_v$ such that

$$r\phi_v(r,t) = f(ct-r) + g(ct+r), \tag{9.14}$$

where the first term shows an outgoing wave from a source and the second one represents a returning wave to the source. That is, the second one can be understood as a representation of a reflection wave from surroundings in general.

Suppose that there is an **ideal point source** (with a small radius of a) from which a symmetric spherical wave is radiated into a field, and that there can be both types of waves. Namely, if the outgoing and returning waves travel even after the source stops,

$$-\left.\frac{\partial \phi_v(r,t)}{\partial r}\right|_{r=a} 4\pi a^2 \to 0 \qquad (a \to 0) \tag{9.15}$$

must be imposed on the volume velocity at the source position. According to this condition, the relationship of

$$f(ct) + g(ct) = 0 \tag{9.16}$$

holds well between the outgoing wave $f(ct-r)$ and the returning wave $g(ct+r)$. Consequently, the general solution corresponding to d'Alembert's solution for plane waves can be formulated as

$$r\phi_v(r,t) = f(ct+r) - f(ct-r) \tag{9.17}$$

for the symmetric spherical waves.

9.1.4 Propagation of Initial Disturbance

Initial disturbance may propagate in three-dimensional space as a spherical wave as well as a plane wave traveling along a one-dimensional axis. Taking inspiration from a balloon filled with a gas, suppose that there is a portion of medium $B_R : r < r_B$ with condensation partitioned by a virtual wall as shown in Fig. 9.1. Consider

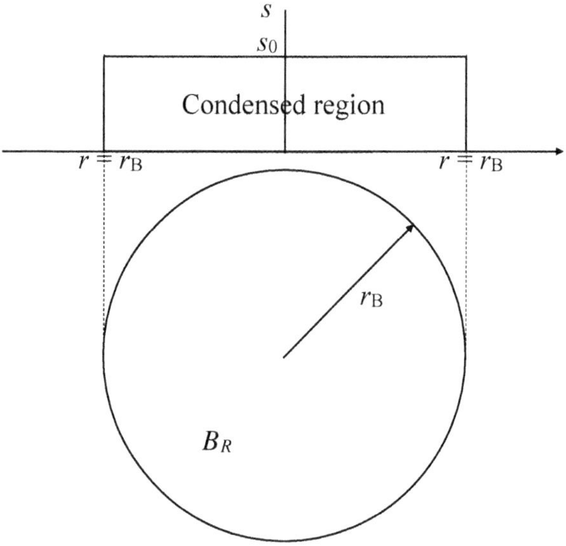

Fig. 9.1 Initial disturbance represented by condensation

the pressure wave travelling into the surrounding medium if the virtual wall were removed. Following Eq. 9.11, suppose that the pressure can be given in the small portion $(r < R_B)$ such that

$$p(r,0) = \kappa s(r,0) = \kappa s_0(r) = \rho_0 \left.\frac{\partial \phi_v(r,t)}{\partial t}\right|_{t=0} = \rho_0 \psi_0(r) \qquad \text{(Pa)} \tag{9.18}$$

is the initial $(t = 0)$ disturbance for $r < r_B$ where $\phi_v(r,t)|_{t=0} = 0$ and the sound pressure should be zero for $r > r_B$ at $t = 0$.

Recall the spherical waves expressed as Eq. 9.17. The next equations

$$r\phi_v(r,t)|_{t=0} = f(r) - f(-r) = 0 \tag{9.19}$$

$$r \left.\frac{\partial \phi_v(r,t)}{\partial t}\right|_{t=0} = cf'(r) - cf'(-r) = r\psi_0(r), \tag{9.20}$$

or

$$f(r) - f(-r) = 0 \tag{9.21}$$

$$f(r) + f(-r) = \frac{1}{c}\int_{\alpha}^{r} z\psi_0(z)dz \tag{9.22}$$

must hold well for the function of velocity potential $\phi_v(r,t)$. By following these equations, the velocity potential function can be obtained as

$$f(r) = \frac{1}{2c}\int_{\alpha}^{r} z\psi_0(z)dz. \tag{9.23}$$

Therefore, the spherical wave can be written as

$$r\phi_v = f(ct+r) - f(ct-r) = \frac{1}{2c}\int_{ct-r}^{ct+r} z\psi_0(z)dz \tag{9.24}$$

in terms of the velocity potential functions. This solution above corresponds to d'Alembert's solution for plane waves under the condition that initial velocity is given without displacement.

By introducing the primitive function $F_{\psi}^{+}(z)$ of $z\psi_0(z)$, Eq. 9.24 can be rewritten as

$$r\phi_v(r,t) = \frac{1}{2c}(F_{\psi}^{+}(ct+r) - F_{\psi}^{+}(ct-r)) = f(ct+r) - f(ct-r). \tag{9.25}$$

By taking the derivative with respect to the time variable,

$$\begin{aligned} r\frac{\partial \phi_v(r,t)}{\partial t} &= \frac{1}{2r}((ct+r)\psi_0(ct+r) - (ct-r)\psi_0(ct-r)) \\ &= \frac{1}{2r}((ct+r)\psi_0(ct+r) + (r-ct)\psi_0(r-ct)) \\ &= c^2 s(r,t) \end{aligned} \tag{9.26}$$

where

$$\psi_0(ct-r) = \psi_0(r-ct) \tag{9.27}$$

$$\tag{9.28}$$

is assumed.

Recall the initial condition given by Eq. 9.18 for the region $r < r_B$. Now consider the condensation observed at $r > r_B$ after the virtual partition was removed at $t = 0$. First, note that

$$\psi_0(ct_0 + r) = 0 \qquad ct_0 + r > r_B \tag{9.29}$$

always holds well at the time $t = t_0$. However, the condensation for $ct_0 < r$ can be observed as

$$s(r,t_0) = \frac{1}{2r}(r - ct_0)s_0 > 0 \tag{9.30}$$

only in the interval of $ct_0 < r < ct_0 + r_B$, because $r - ct_0 < r_B$. This means the condensation occurs in the area of r_B.

It is interesting to see that the condensation above is followed by the next wave of "dilation". Such a dilated wave can be seen for $r < ct_0$ such that

$$s(r,t_0) = \frac{1}{2r}(r - ct_0)s_0 < 0 \tag{9.31}$$

in the interval of $ct_0 - r_B < r < ct_0$ because $ct_0 - r < r_B$.

Consequently, the solutions can be summarized as follows:

$$s(r,t_0) = \frac{1}{2r}(r - ct_0)s_0 > 0 \qquad ct_0 < r < ct_0 + r_B \tag{9.32}$$

$$s(r,t_0) = 0 \qquad ct_0 + r_B < r \tag{9.33}$$

$$s(r,t_0) = \frac{1}{2r}(r - ct_0)s_0 < 0 \qquad ct_0 - r_B < r < ct_0 \tag{9.34}$$

$$s(r,t_0) = 0 \qquad r < ct_0 - r_B. \tag{9.35}$$

Note here again that no sound pressure is observed before the wave arrives and the sound pressure again disappears after the wave passes. In particular, the outcome of no traces after the spherical waves passes is different from that of the traveling of a plane wave. Therefore, the time interval when the sound wave is observed is only $r - r_B < ct < r + r_B$, and thus the spatial region where the sound wave can be seen is in a shell with the thickness of $2r_B$[4].

Figure 9.2 illustrates propagation of spherical waves due to the initial disturbance as shown in Fig. 9.1. The initial condensation, which is stored in the region $r < r_B$, travels as the pressure wave after the virtual partition is removed. The compressive wave travels followed by the dilated one. In other words, the compressive wave represented by the first half of the pressure waveform, which makes the local pressure increase, is observed until the wave comes from the center of the region of initial disturbance.

However, there is another wave even after the wave from the center passes. There can be propagation of waves inside the initial region, as well as of the outgoing waves to outside the region. The inside waves are due to dilation that is the outcome of the high-pressure medium exploding to the outside due to removal of the virtual partition. In other words, the dilations are "filled one after another" by the medium inside the region. Thus, when the inside waves to make the pressure lower reach the center, the dilation waves return to outside the initial region. Consequently, the dilated wave comes after the compressive wave in the later half of the pressure waveform.

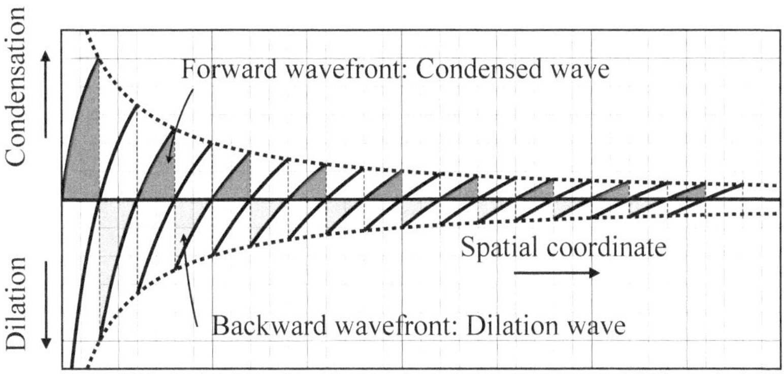

Fig. 9.2 Image of propagation of initial disturbance from [9] (Fig. 8.8)

9.2 Sound Power Radiation from Source

9.2.1 Sound Energy Conveyed by Spherical Wave

Suppose that there is a symmetric spherical wave in space, and let the velocity potential be $\frac{f(ct-r)}{r}$. The sound energy density per unit volume $E_0(\mathrm{J/m^3})$ is given by

$$E_0 = E_{P_0} + E_{K_0}, \tag{9.36}$$

where E_{P_0} $(\mathrm{J/m^3})$ denotes the potential energy density such that

$$E_{P_0} = \frac{1}{2}\kappa s(r,t)^2 = \frac{1}{2}\frac{\rho_0}{c^2}\left(\frac{\partial\phi_v(r,t)}{\partial t}\right)^2 = \frac{1}{2}\frac{\rho_0}{c^2 r^2}\left(\frac{\partial f(ct-r)}{\partial t}\right)^2, \tag{9.37}$$

and E_{K_0} shows the kinetic energy density such that

$$E_{K_0} = \frac{1}{2}\rho_0 v(r,t)^2 = \frac{1}{2}\rho_0\left(-\frac{\partial\phi_v(r,t)}{\partial r}\right) \cong \frac{1}{2}\frac{\rho_0}{c^2 r^2}\left(\frac{\partial f(ct-r)}{\partial t}\right)^2. \tag{9.38}$$

Both energies are the same as in a plane wave when the distance is far from the source.

Recall that the sound pressure $p(r,t)$ and velocity $v(r,t)$ of a sinusoidal spherical wave are written as

$$p(r,t) = \frac{-\omega\rho_0}{4\pi r}Q_0\sin(\omega t - kr) \tag{9.39}$$

$$v(r,t) = \frac{p(r,t)}{\rho_0 c} + \frac{Q_0}{4\pi r^2}\cos(\omega t - kr) \tag{9.40}$$

in the real function forms. By substituting the sound pressure and particle velocity for the total energy density defined in Eq.9.36 and taking the average for a single period,

$$E_{0_{Av}} = \frac{1}{2}\rho_0 \left(\frac{Q_0}{4\pi}\right)^2 \left(\frac{\omega^2}{c^2r^2} + \frac{1}{2r^4}\right)$$

$$\cong \frac{1}{2}\rho_0 k^2 \left(\frac{Q_0}{4\pi r}\right)^2 \qquad (\mathrm{J/m^3}) \tag{9.41}$$

is obtained. Here, only the first term represents the energy of sound to be consumed by wave traveling, while the second one corresponds to that for flow of the medium according to its incompressibility.

The first term related to the sound energy increases in proportion to the square of the frequency of the sound and decreases following the square of the distance from the source. In contrast, the second one is independent of the frequency but decreases according to the distance raised to the power of 4, and thus it is less than the first term as the distance is far from the source. Therefore, the effect of incompressibility on the sound power is remarkable for low frequencies and makes it difficult to radiate sound of low frequencies to long distances. In addition, the first term, namely the energy of sound, depends on the speed of sound in a medium, indicating that the sound energy of a spherical wave decreases according to the square of the sound speed. The wavelength becomes long as the speed of sound increases; this is equivalent to lowering the frequency and results in increase of the incompressibility.

9.2.2 *Sound Power Output of Point Source*

Let a closed surface surrounding a source but far from the source so that the effect of incompressibility might be negligibly small on the sound energy. Set the sound energy $E_{0_{Av}} 4\pi r^2 c\Delta t$ (J) in a thin shell between r and $r + c\Delta t$. The sound power output of the source is defined by the energy as Δt is set to be a unit time interval. Namely, the **sound power output of a source** indicates the total power in a unit time interval for sound radiated into the space where the sound source is located.

Suppose that there is a point source in ideally free space without a boundary, and take a virtual spherical surface centered at the point source of a sufficiently large radius r (m). The sound pressure $p(r,t)$ (Pa) and particle velocity $v(r,t)$ (m/s) can be expressed as

$$p(r,t) = \mathrm{i}\omega\rho_0 \frac{Q_0}{4\pi r} \mathrm{e}^{\mathrm{i}\omega t - \mathrm{i}kr} \tag{9.42}$$

$$v(r,t) = \frac{Q_0}{4\pi r} \mathrm{e}^{\mathrm{i}\omega t - \mathrm{i}kr} \cdot \frac{\mathrm{i}kr + 1}{r} \tag{9.43}$$

on the surface using the complex function forms, where $Q_0 e^{i\omega t}$ $(\mathrm{m}^3/\mathrm{s})$ denotes the volume velocity of the source. Recall that the **acoustic-energy flow density** $I(\mathrm{W/m}^2)$ across the closed surface in a unit time interval (**sound intensity**) is written as

$$I = \frac{1}{2}\mathrm{Re}[p(r,t)v(r,t)^*] = \frac{1}{2}\rho_0 k^2 \left(\frac{|Q_0|}{4\pi r}\right)^2 c \cong E_{0_{Av}} c \qquad (\mathrm{W/m}^2) \tag{9.44}$$

in the complex function form, where $\mathrm{Re}(a)$ and (a^*) indicate taking the real part and complex conjugate of a, respectively. Here the factor $1/2$ corresponds to a time average over the period of a squared sinusoidal function with a unit magnitude yielding $1/2$. The sound **power output of a point source** W_0 (W) is given by

$$W_0 = \int_S I dS = E_{0_{Av}} c \cdot 4\pi r^2 = \frac{1}{2}\rho_0 c k^2 \frac{|Q_0|^2}{4\pi}, \tag{9.45}$$

or the intensity can be rewritten as

$$I = \frac{W_0}{4\pi r^2} \qquad (\mathrm{W/m}^2) \tag{9.46}$$

on the surface at a distance of r from the source.

9.2.3 Phase Relationship for Sound Intensity between Sound Pressure and Velocity

Let the sound pressure and velocity be expressed as

$$p(t) = A\sin\omega t \tag{9.47}$$

$$v(t) = B\sin(\omega t + \phi). \tag{9.48}$$

By taking a time average for the product of $p(t)$ and $v(t)$ over a period, which corresponds to the acoustic intensity,

$$\overline{p(t)v(t)} = \frac{AB}{2}\cos\phi \tag{9.49}$$

is obtained. It can be seen that the average depends on the phase difference between the two functions.

Recall the sound pressure and particle velocity at a distance of r from a point source. There are in-phase and out-of-phase components for the velocity, such as

$$p(r,t) = \rho_0 c v_{in}(r,t) \qquad (\mathrm{Pa}) \tag{9.50}$$

$$v_{in}(r,t) = -\frac{Q_0}{4\pi}\frac{\omega\sin(\omega t - kr)}{rc} \qquad (\mathrm{m/s}) \tag{9.51}$$

$$v_{out}(r,t) = \frac{Q_0}{4\pi}\frac{\omega\cos(\omega t - kr)}{r^2}. \tag{9.52}$$

By taking the "in-phase" component of velocity, the sound power output written as Eq. 9.45 can be obtained, while it becomes zero for the "out-of-phase" component of the velocity. It can be reconfirmed that the out-of-phase component of the particle velocity with the sound pressure does not contribute sound power radiation at all. In other words, the in-phase component velocity produces the sound that travels into the space where the sound source is located.

9.3 Effects of Surroundings on Sound Power Radiation from Source

9.3.1 Sound Power Output and Radiation Impedance of Source

The effects of surroundings on the sound power output of a source can be represented by the **radiation impedance**. The **radiation acoustic impedance of a source** is given by the ratio of the sound pressure (Pa) on the surface of the source itself and the volume velocity (m^3/s) of the source. Suppose that there is a small spherical source with a radius of a. The sound pressure $p(a,t)$ yielded on the source surface is given by

$$p(a,t) = \frac{Q}{4\pi a} \cdot \frac{i\omega\rho_0}{ika+1} e^{i\omega t}, \tag{9.53}$$

where $Qe^{i\omega t}$ (m^3/s) is the volume velocity of the source. Thus, the radiation acoustic impedance $Z_{A_{\mathrm{rad}a}}$ ($\mathrm{Pa}\cdot\mathrm{s}/\mathrm{m}^3$) is given by

$$\begin{aligned} Z_{A_{\mathrm{rad}a}} &= \frac{p(a,t)}{Qe^{i\omega t}} = \frac{1}{4\pi a} \cdot \frac{i\omega\rho_0}{ika+1} \\ &= \frac{\rho_0 cka}{4\pi a^2}\left(\frac{ka}{1+k^2a^2} + i\frac{1}{1+k^2a^2}\right) \\ &= R_{A_{\mathrm{rad}a}} + iX_{A_{\mathrm{rad}a}}. \end{aligned} \tag{9.54}$$

Therefore the sound power output is given by

$$W_{X_a} = \frac{1}{2} R_{A_{\mathrm{rad}a}} |Q|^2 \qquad (\mathrm{W}) \tag{9.55}$$

is obtained in terms of the real part of the radiation impedance and the volume velocity.

The sound power output of a source can be expressed as

$$W_X = \frac{1}{2}\mathrm{Re}\int_S p(r,t) \cdot v(r,t)^* dS, \qquad (\mathrm{W}) \tag{9.56}$$

where dS is the area of the surface when the surface is taken ideally close to the sound source. If the sound pressure can be uniform on the closed surface, then the equation above can be rewritten as

$$W_X = \frac{1}{2}\mathrm{Re}\left[p_X(t)\int_S v^*(r,t)dS\right] = \frac{1}{2}\mathrm{Re}[p_X(t)\cdot q^*(t)], \tag{9.57}$$

where the sound power output is expressed using the volume velocity and sound pressure on the surface. In contrast, if the particle velocity is assumed to be uniform, then the power output is rewritten as

$$W_X = \frac{1}{2}\mathrm{Re}\left[v_X^*(t)\int_S p(r,t)dS\right] = \frac{1}{2}\mathrm{Re}[v_X^*(t)\cdot f_X(t)], \tag{9.58}$$

where the force f_X (N) and vibration velocity v_X (m/s) on the surface are used. Here, the ratio of the force and velocity is called the **radiation mechanical impedance of a source** $(\mathrm{N}\cdot\mathrm{s/m})$.

9.3.2 Effects of Mirror Image Source on Radiation Impedance

The sound power output of a source depends on the environment where the source is located. Such environmental effects on the radiation of sound from a source can be estimated using the radiation impedance. A mirror image source by a rigid wall provides a good example of the effects of surroundings on sound emanated from a source.

Suppose that there is a spherical source closely located to the rigid wall as shown in Fig. 9.3. The sound pressure $p(a,t)$ (Pa) can be expressed as the sum of the pressure from the "physical" source and the "mirror image" one:

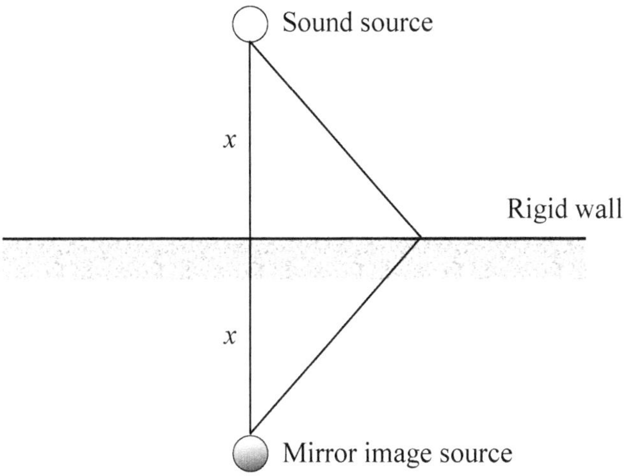

Fig. 9.3 Spherical source close to rigid wall

$$p(a,t) = \frac{Q}{4\pi} \cdot \frac{i\omega\rho_0}{ika+1} e^{ika} \left[\frac{1}{a} e^{-ika} + \frac{\mu}{2x} e^{-ik2x} \right] e^{i\omega t} \tag{9.59}$$

on the surface of the source, where a (m) shows the radius of the source, Q(m^3/s) is the volume-velocity magnitude of the source, and μ represents the reflection coefficient of the rigid wall.

The radiation acoustic impedance $Z_{A_{\mathrm{rad}}}$(Pa · s/m^3) becomes

$$Z_{A_{\mathrm{rad}}} = \frac{1}{4\pi a} \cdot \frac{i\omega\rho_0}{ika+1} \left[1 + \frac{a\mu}{2x} e^{ika-ik2x} \right]. \tag{9.60}$$

Here, assuming that $a \to 0$, the real part of radiation acoustic impedance can be written as

$$R_{A_{\mathrm{rad}}} = R_{A_{\mathrm{rad}_0}} \left[1 + \mu \frac{\sin(2kx)}{2kx} \right] \tag{9.61}$$

for the point source where $R_{A_{\mathrm{rad}_0}}$ is defined by

$$R_{A_{\mathrm{rad}_0}} = \frac{\rho_0 c k^2}{4\pi}. \tag{9.62}$$

Consequently, the sound power output W_X (W) of the point source is

$$W_X = \frac{|Q_0|^2}{2} R_{A_{\mathrm{rad}}} = W_0 \left[1 + \mu \frac{\sin(2kx)}{2kx} \right], \tag{9.63}$$

where W_0 denotes the sound power output in free space without the rigid wall.

The result indicates that the sound power output depends on the distance from the wall and the frequency of the source. Figure 9.4 illustrates the relationship between the sound power output and distance from the rigid wall, when $\mu = 1$. Here, the sound power output is normalized by that for free space, the distance is measured by kx, and k denotes the wavenumber. It can be found that the sound power output depends on the distance between the point source and the rigid wall. As a whole, the power increases as the source comes close to the wall.

In a limit case, namely when the point source is located exactly on the rigid wall, the radiated sound power is two times greater than that for free space. Note here that, even in such a limit case, it is never four times greater than the power obtained in free space. Although the sound pressure increases up to two times higher on the rigid wall, if it is compared with that for free space, the total sound power increases by only the power sum of two sound sources. This increase in the sound pressure can be interpreted as an outcome of the space into which the sound travels being limited to 1/2 the full space by the wall. The power increase effect is due to the mirror image in addition to the original source.

However, the power does not always increase, even if there is the additional virtual source. It decreases at some points differing from that for a single point source. This indicates that the radiated pressure that is in-phase to the velocity of the source can be negatively interfered with the sound from the virtual source. Any effects of

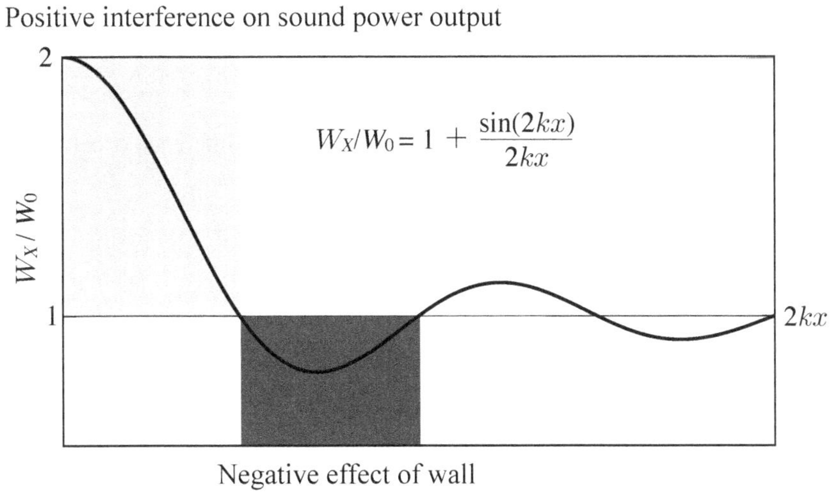

Fig. 9.4 Sound power output of point source closely located to rigid wall

the virtual source on the power output of a source fade, as the distance from the rigid wall increases.

9.3.3 *Sound Power Output of Pair of Anti-phase Sources*

The effects of the virtual source on the sound power output can also be achieved by using a pair of physical sources. It is possible to change the phase relationship between the two physical sources, which is difficult to do with a virtual source. A pair of point sources driven by anti-phased sinusoidal waves is a good example of sound power control from a source. Total power output for such a pair of sources can also be interpreted in terms of radiation impedance.

Suppose that there are two point sources in free space:

$$q_1(t) = Q_1 \mathrm{e}^{\mathrm{i}\omega t} \qquad (\mathrm{m}^3/\mathrm{s}) \tag{9.64}$$

$$q_2(t) = Q_2 \mathrm{e}^{\mathrm{i}(\omega t + \phi)}, \tag{9.65}$$

where $q_1(t)$ and $q_2(t)$ denote the volume velocities for the point sources and ϕ is the phase difference (rad) between the two. The sound power output W_X (W) of the pair of sources can be described as

$$W_X = \frac{1}{2} R_{A_{\mathrm{rad}_1}} |Q_1|^2 + \frac{1}{2} R_{A_{\mathrm{rad}_2}} |Q_2|^2 + R_{A_{\mathrm{rad}_{12}}} Q_1 Q_2 \cos\phi, \tag{9.66}$$

where $R_{A_{\mathrm{rad}_1}}$, $R_{A_{\mathrm{rad}_2}}$, and $R_{A_{\mathrm{rad}_{12}}}$ denote the real parts of the radiation impedance of the sources, and the "mutual" impedance between the two sources, respectively. In addition, the sound power output (W) of each source can be written as

$$W_{X_1} = \frac{1}{2}Q_1(R_{A_{\text{rad}_1}} Q_1 + R_{A_{\text{rad}_{12}}} Q_2 \cos\phi) \quad (9.67)$$

$$W_{X_2} = \frac{1}{2}Q_2(R_{A_{\text{rad}_2}} Q_2 + R_{A_{\text{rad}_{12}}} Q_1 \cos\phi). \quad (9.68)$$

The sound power output can be controlled by the phase difference ϕ between the two sources. Let the phase difference be π; namely, set

$$|Q_{2_{min}}| = (R_{12}/R_2)|Q_1|, \qquad \phi_{min} = \pi \quad (9.69)$$

where the radiation impedance terms are set

$$R_{A_{\text{rad}_1}} = R_1, \quad R_{A_{\text{rad}_{12}}} = R_{12}, \quad R_{A_{\text{rad}_2}} = R_2. \quad (9.70)$$

Then, the sound power output takes its minimum as

$$W_{X_{min}} = \frac{1}{2}|Q_1|^2 R_1 \left(1 - \frac{R_{12}^2}{R_1 R_2}\right), \quad \text{(W)} \quad (9.71)$$

and at the same time, the sound power output of source 2 is zero, i.e.,

$$W_{X_{2_{min}}} = \frac{1}{2}\frac{R_{12}}{R_2}|Q_1|(R_2\frac{R_{12}}{R_2}|Q_1| - R_{12}|Q_1|) = 0. \quad (9.72)$$

Namely, source 2 does not radiate sound at all; it works only to reduce the total power output.

Again suppose that there are two point sources of the volume velocity (m^3/s);

$$Q_1 = |Q_1|e^{i\omega t} \quad (9.73)$$

$$Q_2 = \mu|Q_1|e^{i(\omega t+\phi)}. \qquad (m^3/s) \quad (9.74)$$

By setting the distance between the two point sources as $2x$ (m) and assuming that $R_1 = R_2 = R$, then

$$R_{12}\cos\phi = R\frac{\sin(2kx+\phi)}{2kx} \quad (9.75)$$

according to Eq. 9.61. Therefore, the sound power output (W) from each source is formulated as

$$W_{X_1} = \frac{1}{2}R|Q_1|^2\left(1+\mu\frac{\sin(2kx+\phi)}{2kx}\right) \quad (9.76)$$

$$W_{X_2} = \frac{1}{2}R\mu^2|Q_1|^2\left(1+\frac{1}{\mu}\frac{\sin(2kx+\phi)}{2kx}\right) \quad (9.77)$$

$$W_{X_1}+W_{X_2} = \frac{1}{2}R|Q_1|^2\left(1+\mu^2+2\mu\frac{\sin(2kx+\phi)}{2kx}\right) \quad (9.78)$$

according to Eq. 9.68. Here, by setting $\mu = 1$ and $\phi = \pi$,

$$W_{X_{inv}} = W_{X_1} + W_{X_2} = R|Q_1|^2(1 - \frac{\sin 2kx}{2kx}) = 2W_0(1 - \frac{\sin 2kx}{2kx}) \quad \text{(W)} \qquad (9.79)$$

represents the total power output from the pair of sources, where W_0 denotes the power output for source 1 alone as given by Eq. 9.45.

Figure 9.5 illustrates an example of total power output measured for a pair of anti-phase sinusoidal point sources in a reverberation room. Here, the power output is normalized by W_0 as shown in the vertical axis of the figure. The sound power output decreases, as the distance between the pair of sources decreases. This is because the in-phase component of the sound pressure with the source velocity is faded at each source position due to anti-phase superposition of the sound waves. Consequently, there is no sound radiated from the sources at all at a limit when the distance approaches zero between the two. However, the total power output becomes two times greater than W_0 for the single source alone as the distance increases, even if the sources are excited by a pair of anti-phase sinusoids.

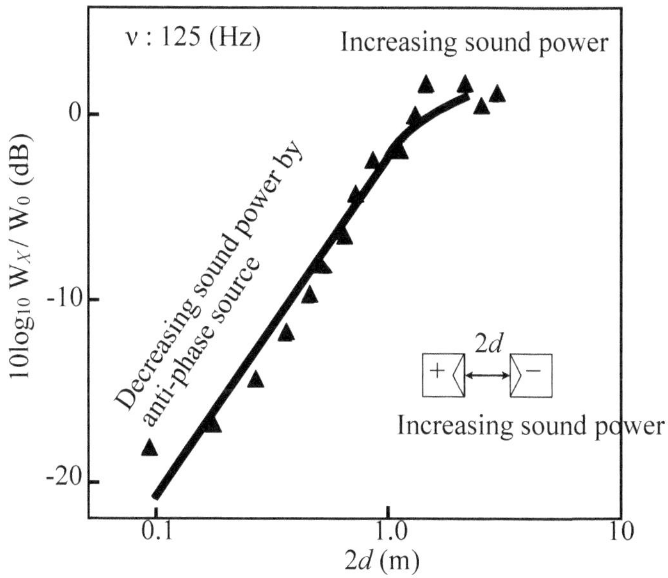

Fig. 9.5 Sound power output from pair of sources excited by sinusoidal signals in anti-phase from [37] (Fig.1)

9.4 Wave Propagation and Sound Radiation through Acoustic Horn

Sound radiation from the end of an acoustic pipe can also be understood according to the radiation impedance. Acoustic pipes are extended into **acoustic horns** where the cross sections are not always constant in this section. Acoustic horns were originally investigated in order to smoothly change the acoustic impedance from that inside the horn to that outside. Historical studies about wave propagation in an acoustic horn can be seen in reference[38]. Only introductory issues are very briefly summarized in this section.

9.4.1 Radiation Impedance for Circular Aperture in Rigid Wall

Suppose that there is an open aperture in an infinitely extending rigid wall as shown in Fig. 9.6. Now consider the radiation impedance of the open aperture, assuming

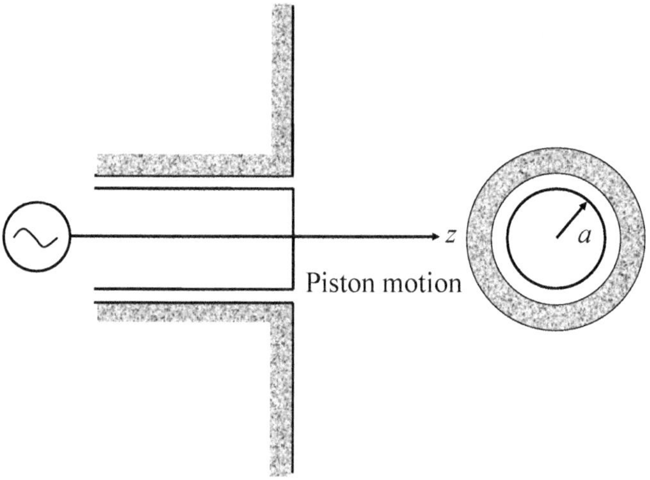

Fig. 9.6 Circular aperture of piston-like motion in infinitely extending rigid wall

that it moves along the z-axis in a piston motion with the velocity of v_X (m/s) such that

$$v_X = V_X e^{i\omega t}. \qquad (\mathrm{m/s}) \qquad (9.80)$$

The open aperture can be interpreted as an open end of an acoustic pipe embedded in the wall or a circular thin plate or membrane with piston motion on the wall.

According to Fig. 9.7, the sound pressure $p(da', r, t)$ (Pa) at a point on the aperture itself due to the vibration of a small area da' on the aperture can be described as[39]

$$p(da', r, t) = P(da', r)e^{i\omega t} = ick\rho_0 \frac{V_X da'}{2\pi r} e^{-ikr} e^{i\omega t}, \qquad (\mathrm{Pa}) \qquad (9.81)$$

where r (m) denotes the distance between the observation point and the location of small area da', $\rho_0(\mathrm{kg/m^3})$ gives the volume density of the surrounding medium, and

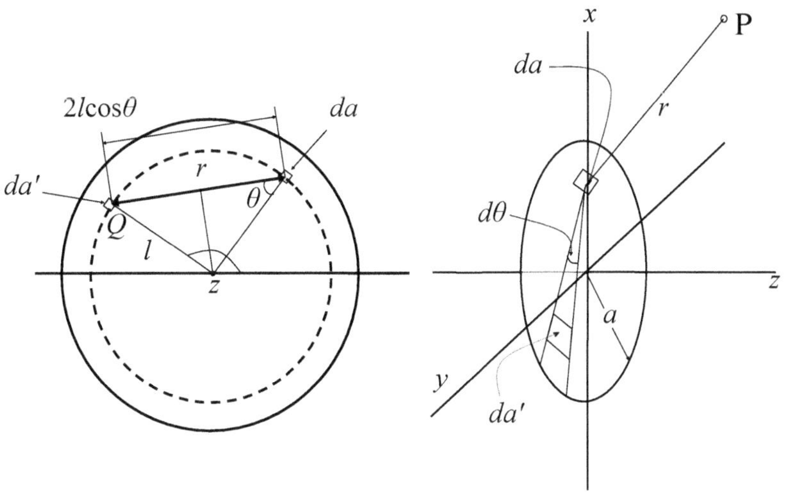

Fig. 9.7 Vibrating aperture in polar coordinate system

2π is introduced into the denominator instead of 4π according to the mirror image principle. Then the sound pressure $P(r)$ (Pa) can be written as

$$P(r) = \int_S P(da', r)da' \tag{9.82}$$

at a position P by summing the sound pressure radiated from all over the vibration surface. Here $\int_S da'$ denotes integration of the pressure over the surface.

Thus, the force, $f = Fe^{i\omega t}$ (N), applied to the vibration surface by the radiated sound itself can be formulated as

$$\begin{aligned} F &= \int_S P(r)da = ick\rho_0 \frac{V_X}{2\pi} \int_S da' \int_S \frac{e^{-ikr}}{r} da \\ &= ick\rho_0 \frac{V_X}{2\pi} \int_S I_r da' = ick\rho_0 \frac{V_X}{2\pi} I = AI, \end{aligned} \tag{9.83}$$

where da denotes a small area around an observation position on the vibrating surface.

The integral formula I can be evaluated as follows. Let the observation position be Q, following Fig. 9.7. The integration I_r can be divided into two parts,

$$I_r = \int_S \frac{e^{-ikr}}{r} da = I_{ra} + I_{rb}, \tag{9.84}$$

according to the observation position, where $I_{ra}(I_{rb})$ denotes the integration regarding the inside (outside) of the circle of radius l as shown in Fig. 9.7, and r shows the distance between the vibrational element and the observation position. Thus, the integration I can be rewritten as

$$I = \int_S I_r da = \int_S I_{ra} da + \int_S I_{rb} da = I_a + I_b. \tag{9.85}$$

By noting that the integrand is a function of r only, it can be found that the integration formula over the inside of the disc I_{ra} should be equal to I_{rb} over the outside of the disc, and thus

$$I_a = I_b = \int_S I_{ra} da = \int_S I_{rb} da = \frac{1}{2} I \tag{9.86}$$

holds well.

The integration I_{ra} can be performed such that

$$\begin{aligned} A \cdot I_{ra} &= \mathrm{i}\rho_0 c \frac{V_X}{2\pi} \int_{-\frac{\pi}{2}}^{\frac{\pi}{2}} d\theta \int_0^{2l\cos\theta} k\mathrm{e}^{-\mathrm{i}kr} dr \\ &= \frac{1}{2}\rho_0 c V_X \left[1 - \frac{2}{\pi} \int_0^{\frac{\pi}{2}} \mathrm{e}^{-\mathrm{i}kl\cos\theta} d\theta \right] \\ &= \frac{1}{2}\rho_0 c V_X \left[1 - \mathrm{J}_0(2kl) + \mathrm{i}\mathrm{S}_0(2kl) \right] \end{aligned} \tag{9.87}$$

according to the polar coordinate system as shown in Fig. 9.7, where $\mathrm{J}_0(*)$ represents the **Bessel function of 0-th order** and $\mathrm{S}_0(*)$ represents the **Struve function**[6][39]. Here, by noting that the equation is independent of the coordinate of θ,

$$\begin{aligned} F = AI &= 2A \int_S I_{ra} da = 2A \int_0^a I_{ra} 2\pi l dl \\ &= \rho_0 c V_X \int_0^a \left[1 - \mathrm{J}_0(2kl) + \mathrm{i}\mathrm{S}_0(2kl) \right] 2\pi l dl \\ &= \rho_0 c \pi a^2 V_X \left[1 - \frac{\mathrm{J}_1(2ka)}{ka} + \mathrm{i}\frac{\mathrm{S}_1(2ka)}{ka} \right] \end{aligned} \tag{9.88}$$

can be derived.

The **radiation mechanical impedance** can be applied when the vibration velocity is constant on the surface of a source. The mechanical radiation impedance $Z_{M_{\mathrm{rad}}}(\mathrm{N \cdot s/m})$ is derived as

$$\begin{aligned} Z_{M_{\mathrm{rad}}} &= R_{M_{\mathrm{rad}}} + \mathrm{i}X_{M_{\mathrm{rad}}} = \frac{F}{V_X} \\ &= \rho_0 c \pi a^2 \left[1 - \frac{\mathrm{J}_1(2ka)}{ka} + \mathrm{i}\frac{\mathrm{S}_1(2ka)}{ka} \right] \end{aligned} \tag{9.89}$$

and can be approximated as[40]

$$R_{M_{\mathrm{rad}}} \cong \frac{k^2 a^2}{2} \rho_0 c \pi a^2, \quad X_{M_{\mathrm{rad}}} \cong \frac{8ka}{3\pi} \rho_0 c \pi a^2 \tag{9.90}$$

for $ka < 1/2$, or

$$R_{M_{\rm rad}} \cong \rho_0 c \pi a^2, \quad X_{M_{\rm rad}} \cong \frac{2}{\pi ka} \rho_0 c \pi a^2 \tag{9.91}$$

for $ka > 2$.

Figure 9.8 illustrates the mechanical radiation impedance for (a) a circular vibration surface of piston motion and (b) a breathing sphere with uniform vibratory motion[41]. The impedance is normalized by $\rho_0 cS$ as shown in the vertical axis of the figure, where S is the surface of the vibration area. As shown in Fig. 9.8(a), the piston-like motion of the surface radiates mostly spherical waves in the low frequencies, while plane waves travel from the surface as the frequency increases. Figure 9.8(b) shows the radiation mechanical impedance for a breathing sphere with a radius of a. The radiation mechanical impedance is formulated as

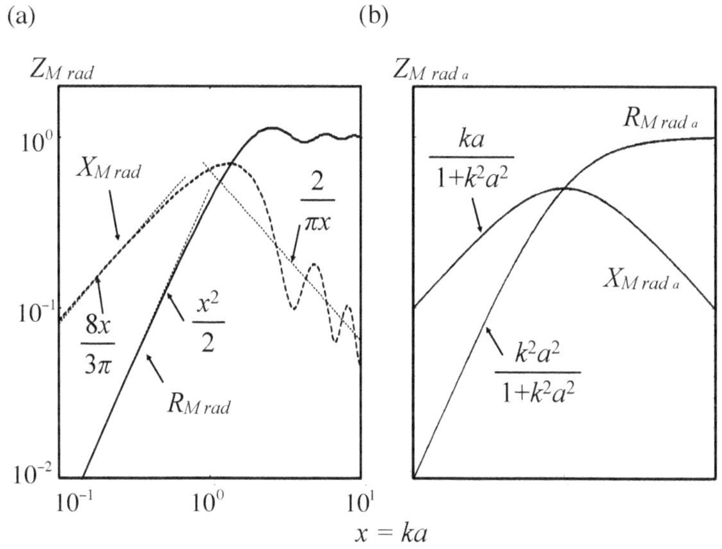

Fig. 9.8 Mechanical radiation impedance (a)for circular vibration surface with piston-like motion on rigid wall and (b) for small breathing sphere

$$Z_{M_{\mathrm{rad}_a}} = S\rho_0 c \left[\frac{k^2a^2}{1+k^2a^2} + \mathrm{i}\frac{ka}{1+k^2a^2} \right] \qquad (\mathrm{N \cdot s/m}) \tag{9.92}$$

instead of the radiation acoustic impedance defined by Eq. 9.54, where $S = 4\pi a^2 (\mathrm{m}^2)$. The radiation impedance of a circular surface under piston-like motion basically follows that for a breathing sphere[40]. As the frequency of sound increases up to

$ka \geq 2(\lambda \leq \pi a)$, where λ is the wavelength, the radiation impedance can be approximated by that for a plane wave, i.e., $Z_{M_{\text{rad}_a}} \cong S\rho_0 c$.

The real part of the driving point acoustical impedance for an acoustic tube $R_{A_{\text{in}}}$ was previously formulated as Eq. 7.37. By substituting the radiation acoustic impedance of a breathing sphere, such as $Z_{A_{\text{rad}_a}} = Z_{M_{\text{rad}_a}}/S^2 = R_{A_{\text{rad}_a}} + iX_{A_{\text{rad}_a}}$, for the impedance R_A and X_A, $R_{A_{\text{in}}}$ can be rewritten as

$$R_{A_{\text{in}}} = \frac{1 + \frac{1}{k^2a^2}}{[(1 + \frac{1}{k^2a^2})\cos kL - \frac{1}{ka}\sin kL]^2 + \sin^2 kL} Z_0. \qquad (\text{Pa} \cdot \text{s/m}^3) \qquad (9.93)$$

The real part of the driving point impedance can be approximated by that for a plane wave $Z_0 = \rho_0 c/S$ as ka increases.

Figure 9.9 illustrates an example of baffle effects on a point source that is located on a cylindrical baffle as shown in Fig. 9.10[42]. Here the vertical axis indicates the ratio of sound pressure at an observation point ($\phi = 0$ or $z = 0$) far from the source and that for a point source in free space. As shown in the figure, when the ratio approaches 2, the baffle effect can be equivalent to that for an infinitely extended rigid wall independent of the geometrical figure of a baffle. In other words, the baffle effects can be illustrated by the traces to 2.

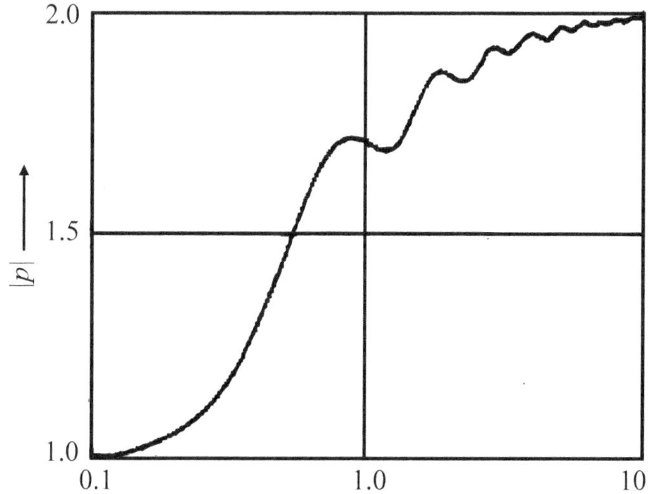

Fig. 9.9 Baffle effects of cylindrical baffle on point source from [42] (Fig.2)

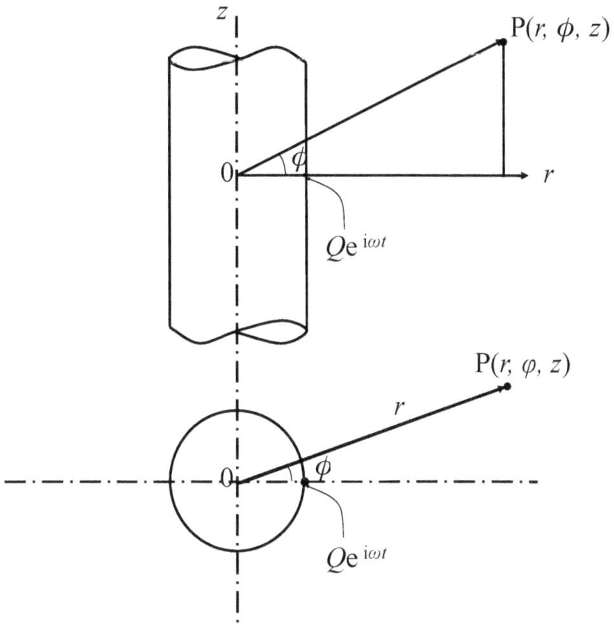

Fig. 9.10 Point source on cylindrical baffle and observation point of radiation sound

9.4.2 *Wave Equation for Traveling Waves in Acoustic Horn*

The frequency characteristics of radiation sound from a circular aperture on a rigid wall or a small breathing sphere can be interpreted as a process of extension for a spherical wavefront. If there is no extension of the wavefront, such as a plane wave, the **specific impedance** is independent of the frequency of sound such that $\rho_0 c$. Acoustic horns have been investigated so that the wavefront might be controlled by the extending process from that for a sound source to that of radiation sound in space.

Suppose that there is an acoustic horn where the area of cross section varies as shown in Fig. 9.11[43]. Assuming that a plane wave travels in an acoustic horn where the area of cross section varies,

$$\xi(x+dx) \cong \xi(x) + \frac{\partial \xi(x)}{\partial x}dx \tag{9.94}$$

$$S(x+dx) \cong S(x) + \frac{\partial S(x)}{\partial x}dx \tag{9.95}$$

might hold well for the sound displacement and the cross section at $x+dx$, subject to the displacement and cross section being given by $\xi\ (x)$ and $S(x)$ respectively at x. Following the relationship above, the dilation of the medium dV/V at the small interval dx due to the sound wave is expressed as

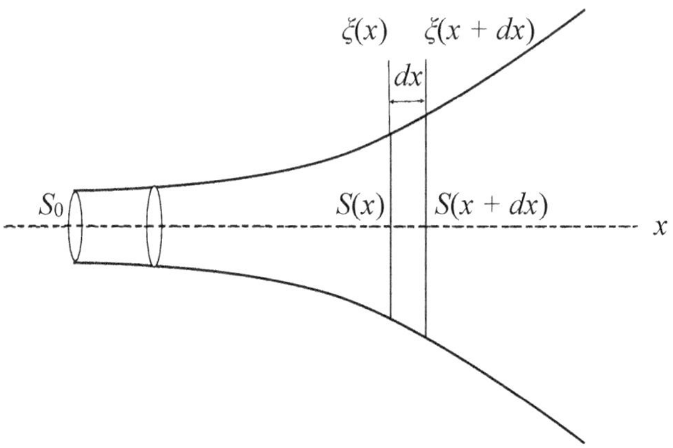

Fig. 9.11 Displacement of plane wave in acoustic horn and its area of cross section

$$\frac{dV}{V} \cong \frac{S(x+dx)\xi(x+dx) - S(x)\xi(x)}{S(x)dx} = \frac{1}{S(x)}\frac{\partial}{\partial x}[S(x)\xi(x)]. \quad (9.96)$$

Therefore, the sound pressure, $P(x)$ (Pa), due to dilation can be represented by

$$P(x) = -\kappa\frac{dV}{V} = -\kappa\frac{1}{S(x)}\frac{\partial}{\partial x}(S(x)\xi(x)) \quad (9.97)$$

where κ denotes the bulk modulus of the medium.

By rewriting the variable $\xi(x)$, $P(x)$ into $u(x,t)$, $p(x,t)$ by including a time variable, the equation of motion of the medium is formulated as

$$\rho_0 S(x)\frac{\partial^2 u(x,t)}{\partial t^2} = -\frac{\partial p(x,t)}{\partial x}S(x). \quad (9.98)$$

Consequently, the **wave equation**

$$S(x)\frac{\partial^2 p(x,t)}{\partial t^2} = c^2\frac{\partial}{\partial x}\left(S(x)\frac{\partial p(x,t)}{\partial x}\right) \quad (9.99)$$

can be derived. Assuming a sinusoidal stationary wave of

$$p(x,t) = P(x)\mathrm{e}^{\mathrm{i}\omega t}, \quad (9.100)$$

the **Helmholtz equation** can be written as

$$\frac{d^2P(x)}{dx^2} + \frac{1}{S(x)}\frac{dS(x)}{dx}\frac{dP(x)}{dx} + k^2P(x) = 0 \quad (9.101)$$

for a traveling wave in the horn.

9.4.3 Plane Wave Traveling in Exponential Horn

The second term of the wave equation, $(1/S(x))(dS(x)/dx)$ in Eq. 9.101, is an indicator for extension of the wavefront for a traveling wave in a horn. Assuming an exponential function for $S(x)$ such that

$$S(x) = S_0 e^{2m_h x}, \tag{9.102}$$

the second term could be constant:

$$\frac{1}{S(x)}\frac{dS(x)}{dx} = 2m_h \tag{9.103}$$

where S_0 denotes the area of the cross section at the entrance $x = 0$ into the horn.

The wave equation for an exponential horn can be written as

$$\frac{d^2}{dx^2}P(x) + 2m_h\frac{dP(x)}{dx} + k^2P(x) = 0. \tag{9.104}$$

If the horn is long enough that the reflection wave from the open end can be ignored, the wave can be represented as

$$P(x) = \mathrm{i}\rho_0 cke^{-m_h x}e^{-\mathrm{i}\sqrt{k^2-m_h^2}x} \tag{9.105}$$

$$V(x) = (m_h + \mathrm{i}\sqrt{k^2 - m_h^2})e^{-m_h x}e^{-\sqrt{k^2-m_h^2}x} \tag{9.106}$$

for the sound pressure $P(x)$ and particle velocity $V(x)$, respectively. Consequently, the driving point acoustic impedance $Z_{ha}(x,\hat{k})|_{x=0}$ at the entrance $(x = 0)$ of the horn is given by

$$Z_{A_{\mathrm{in}}}(0,\hat{k}) = \frac{P(0)}{S_0V(0)} = \frac{\rho_0 c}{S_0}\left(\frac{\sqrt{\hat{k}^2-1}}{\hat{k}} + \mathrm{i}\frac{1}{\hat{k}}\right) = R_{A_{\mathrm{in}}} + \mathrm{i}X_{A_{\mathrm{in}}}, \qquad \left(\frac{\mathrm{Pa}\cdot\mathrm{s}}{\mathrm{m}^3}\right) \tag{9.107}$$

where $\hat{k} = k/m_h$ and

$$R_{A_{\mathrm{in}}} = \frac{\rho_0 c}{S_0}\frac{\sqrt{k^2a^2 - m_h^2a^2}}{ka} \tag{9.108}$$

$$X_{A_{\mathrm{in}}} = \frac{\rho_0 c}{S_0}\frac{m_h a}{ka}. \tag{9.109}$$

The **cutoff frequency** in a horn is defined as that for making the real part of the driving point impedance be 0. This means that no sound traveling is possible into the horn below the cutoff frequency. The **cutoff frequency** is obtained as $\hat{k} = 1$ or $\omega_c = m_h c$ **for an exponential horn**. As described above the driving point impedance becomes pure imaginary when the frequency of sound is lower than the cutoff frequency. It is necessary to make m_h smaller in order to set the cutoff frequency lower, and thus a long horn is required to transmit a low frequency of sound into the horn.

However, note that the absolute of the driving point impedance could be constant such that

$$|Z_{A_{\text{in}}}| = \frac{\rho_0 c}{S_0}. \qquad (\frac{\text{Pa} \cdot \text{s}}{\text{m}^3}). \tag{9.110}$$

Recall the driving point impedance for an acoustic tube was described in Eq. 9.93, assuming that the radiation impedance of a breathing sphere is the same as that for the open end of the tube. Now instead suppose that an exponential horn is attached to the open end. By substituting the radiation acoustic impedance of an exponential horn for the impedance R_A and X_A, Eq. 9.93 can be rewritten as

$$R_{A_{\text{in}}} = \frac{\sqrt{1-(\frac{m_h a}{ka})^2}}{(\cos kL - \frac{m_h a}{ka}\sin kL)^2 + \frac{k^2a^2 - m_h^2 a^2}{k^2 a^2}\sin^2 kL}\frac{\rho_0 c}{S_0}. \tag{9.111}$$

Similarly, the imaginary part is obtained as

$$X_{A_{\text{in}}} = \frac{\frac{m_h a}{ka}\cos 2kL}{(\cos kL - \frac{m_h a}{ka}\sin kL)^2 + (1-(\frac{m_h a}{ka})^2)\sin^2 kL}\frac{\rho_0 c}{S_0}. \tag{9.112}$$

It can be reconfirmed that $R_{A_{\text{in}}}$ approaches $\rho_0 c/S_0$ just like a plane wave in free space for the frequencies higher than the cutoff frequency.

Chapter 10
Sound Waves in Rooms

Eigenfrequencies are found in sound fields in rooms by analyzing wave equations for a rectangular room surrounded by hard walls. The relationship between eigenfrequencies and eigenfunctions holds well for the wave equations as well as for linear equations. It might be quite rare to see a perfectly rectangular room in our daily life, but, room acoustics in rectangular rooms provide a good fundamental basis for better understanding of room acoustics in general. Representation of an impulse response and its frequency characteristics can be derived using the Green function in a rectangular room. Consequently, the wave theoretic representation of a sound field is derived from the expression in terms of geometric acoustics, such as the mirror image theory. Finally, a general solution form of the wave equation is presented using an integration formula in three-dimensional space.

10.1 Eigenfrequencies and Eigenfunctions for Rooms

10.1.1 Helmholtz Equation

Take a stationary wave for a symmetric spherical wave with an angular frequency of ω. Such a wave that follows the wave equation can be expressed in the form

$$p(\mathbf{r},t) = P(\mathbf{r})\mathrm{e}^{\mathrm{i}\omega t}, \tag{10.1}$$

where p denotes the sound pressure (Pa) and $\mathbf{r}$ represents the spatial coordinates at which the sound pressure is observed. Following the relationship between the sound pressure and velocity potential such that

$$p(\mathbf{r},t) = \rho_0 \frac{\partial \phi_v(\mathbf{r},t)}{\partial t}, \tag{10.2}$$

$$\nabla^2 P(\mathbf{r}) + k^2 P(\mathbf{r}) = 0 \tag{10.3}$$

M. Tohyama: *Sound and Signals*, Signals and Communication Technology, pp. 163–189.
springerlink.com

is obtained. Here, $k = \omega/c = (1/\mathrm{m})$ denotes the **wavenumber**. Note that the equation above has only the spatial variables represented by $\mathbf{r}$. That is, the sinusoidal time dependency of the sinusoidal wave is assumed here. This "spatial" equation for the "spatial function" $P(\mathbf{r})$ is called the **Helmholtz equation**.

10.1.2 Eigenfrequencies for Rooms

Recall a linear equation that can be represented by a matrix A and vectors $\mathbf{x}$ and $\mathbf{b}$ such that

$$A\mathbf{x} = \mathbf{b}. \tag{10.4}$$

Here, if a vector $\hat{\mathbf{x}}$ satisfies

$$A\hat{\mathbf{x}} = \lambda_u \hat{\mathbf{x}}, \tag{10.5}$$

$\hat{\mathbf{x}}$ is called the **eigenvector** of the matrix A with the **eigenvalue** λ_u.

Recall the Helmholtz equation such that

$$\nabla^2 P = -k^2 P. \tag{10.6}$$

The equation above states that if the operator ∇^2 is applied to the function P, then the function P is again obtained with a scaler multiplier of $-k^2$. This relationship is the same as that for the eigenfunction and eigenvalue for a linear operator such as a matrix. When P_i is a function that satisfies the Helmholtz equation such that

$$\nabla^2 P_i = -k_i^2 P_i, \tag{10.7}$$

the frequency of a sound wave with the wavenumber k_i is called the **eigenfrequency** in the sound field where the Helmholtz equation holds well following the boundary condition, and P_i is its **eigenfunction**.

Suppose that there is a rectangular room surrounded by hard walls whose lengths are L_x, L_y, and L_z (m). A perfectly rectangular room might be uncommon in our daily life, but a sound field in a rectangular room is a good example for qualitative and theoretical study of room acoustics from the view point of modal analysis. The Helmholtz equation can be rewritten as

$$\left(\frac{\partial^2}{\partial x^2} + \frac{\partial^2}{\partial y^2} + \frac{\partial^2}{\partial z^2}\right) P(x,y,z) + k^2 P(x,y,z) = 0 \tag{10.8}$$

in the Cartesian coordinate system. Here, by recalling that the boundary conditions are given such that vibration velocities are zero at all the walls because of perfectly rigid walls, one of the eigenfunctions that follows the equation can be found as

$$P(x,y,z) = \cos k_x x \cos k_y y \cos k_z z. \tag{10.9}$$

By substituting this function into the Helmholtz equation, the eigenvalue representing the wavenumber

$$k^2 = k_x^2 + k_y^2 + k_z^2 = k_{xyz}^2 \tag{10.10}$$

is given so that the Helmholtz equation might hold well. Consequently, the **eigen(angular)frequency** denoted by ω_{lmn} (rad/s) is expressed by

$$\omega_{lmn} = ck_{lmn} = c\sqrt{k_l^2 + k_m^2 + k_n^2} = c\sqrt{\left(\frac{l\pi}{L_x}\right)^2 + \left(\frac{m\pi}{L_y}\right)^2 + \left(\frac{n\pi}{L_z}\right)^2} \tag{10.11}$$

following the boundary conditions

$$\left.\frac{\partial P(x,y,z)}{\partial x}\right|_{x=0,L_x} = 0, \quad \left.\frac{\partial P(x,y,z)}{\partial y}\right|_{y=0,L_y} = 0, \quad \left.\frac{\partial P(x,y,z)}{\partial z}\right|_{z=0,L_z} = 0; \tag{10.12}$$

namely,

$$k_x = \frac{l\pi}{L_x} = k_l, \qquad k_y = \frac{m\pi}{L_y} = k_m, \qquad k_z = \frac{n\pi}{L_z} = k_n, \tag{10.13}$$

where l, m, and n are non-negative integers.

10.1.3 Number and Density of Eigenfrequencies

The eigenfrequencies for a rectangular room are located at "grids" of the orthogonal lattice in the wavenumber space as shown in Fig. 10.1. Following this schematic, the number of eigenfrequencies $N_V(k)$ lower than wavenumber k can be approximately estimated as

$$\begin{aligned} N_{V_{3D}}(k) &\cong \frac{1}{\frac{\pi^3}{L_xL_yL_z}}(N_{V_3}(k) + N_{V_2}(k) + N_{V_1}(k)) \\ &= \frac{Vk^3}{6\pi^2} + \frac{S_{3D}k^2}{16\pi} + \frac{L_{3D}k}{16\pi}, \end{aligned} \tag{10.14}$$

where

$$\begin{aligned} N_{V_3}(k) &= \frac{1}{8}\frac{4\pi k^3}{3} \\ N_{V_2}(k) &= \frac{\pi k^2}{4}\left(\frac{\pi}{2L_x} + \frac{\pi}{2L_y} + \frac{\pi}{2L_z}\right) \\ N_{V_1}(k) &= k\left(\frac{\pi}{2L_x}\frac{\pi}{2L_y} + \frac{\pi}{2L_y}\frac{\pi}{2L_z} + \frac{\pi}{2L_z}\frac{\pi}{2L_x}\right). \end{aligned} \tag{10.15}$$

and $V(\mathrm{m}^3), S_{3D}(\mathrm{m}^2)$, and L_{3D} (m) are denoted by

$$V = L_xL_yL_z$$

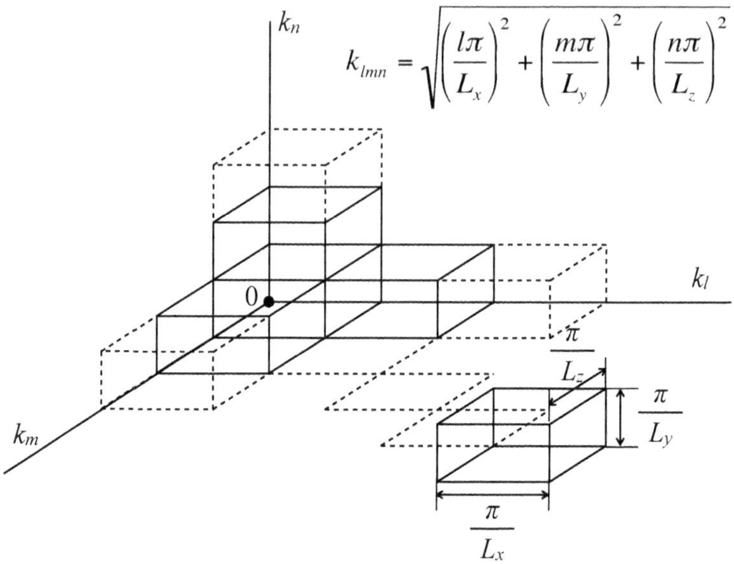

Fig. 10.1 Wavenumber space representing eigenfrequencies for rectangular room

$$S_{3D} = 2(L_xL_y + L_yL_z + L_zL_x)$$
$$L_{3D} = 4(L_x + L_y + L_z). \tag{10.16}$$

Thus, the number of eigenfrequencies in a unit interval of eigenfrequencies, namely, the **density of eigen(angular)frequencies**, can be written as

$$n_{V_{3D}}(\omega) = \frac{\partial N_{V_{3D}}(\omega)}{\partial \omega} \cong \frac{V\omega^2}{2\pi^2 c^3}. \tag{10.17}$$

That the density itself is a function of a frequency that increases in proportion to the square of the frequency for a rectangular room.

The density of eigenfrequencies was estimated for the sound field in a rectangular room; however, it can also be applied to estimate the density in general. A sound wave with the eigenfrequency ω_{lmn}, where none of l, m, and n are zero, is called an **oblique wave**. Similarly, if one of l, m, and n is zero, such a wave is called a **tangential wave**; the wave is named an **axial wave** if two of them are zero. The tangential or axial waves are typical for a sound field surrounded by parallel and perpendicular walls like those in a rectangular room. By removing these tangential and axial waves as shown in Fig. 10.2, the number of eigenfrequencies only for the oblique waves in a non-rectangular room can be estimated as

$$N_{V_{ob}}(k) \cong \frac{Vk^3}{6\pi^2} - \frac{S_{3D}k^2}{16\pi} + \frac{L_{3D}k}{16\pi}. \tag{10.18}$$

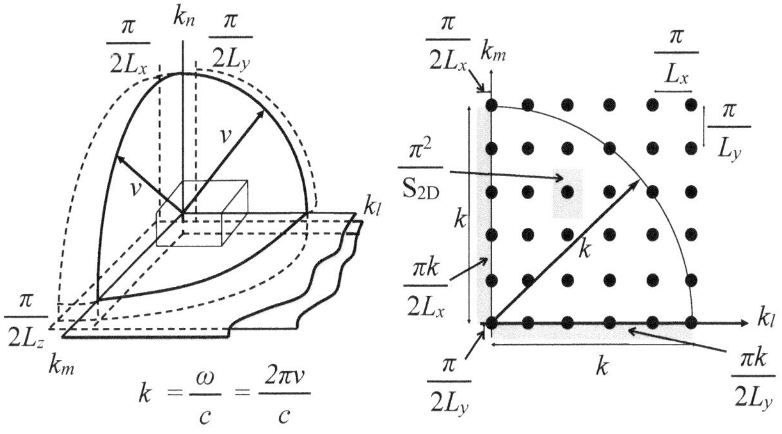

Fig. 10.2 Eigenfrequencies for oblique waves removing tangential and axial waves

Here, similarly, the number of tangential waves can be expressed as

$$N_{\nu_{xy}}(k) \cong \frac{L_x L_y}{4\pi} k^2 - \frac{L_x + L_y}{2\pi} k, \tag{10.19}$$

and

$$N_{\nu_x}(k) \cong \frac{L_x}{\pi} k \tag{10.20}$$

can also be derived only for the axial waves.

The number of eigenfrequencies can also be estimated for a two-dimensional field. The eigenfrequencies are expressed as

$$\omega_{lm} = c\sqrt{\left(\frac{l\pi}{L_x}\right)^2 + \left(\frac{m\pi}{L_y}\right)^2} \tag{10.21}$$

for a two-dimensional field. The number of those eigenfrequencies is approximately estimated by

$$N_{\nu_{2D}}(k) \cong \frac{S_{2D}k^2}{4\pi} + \frac{L_{2D}k}{4\pi} \tag{10.22}$$

for a two-dimensional rectangular field. Similarly

$$N_{\nu_{tan}}(k) \cong \frac{S_{2D}k^2}{4\pi} - \frac{L_{2D}k}{4\pi} \tag{10.23}$$

is obtained for a non-rectangular two-dimensional field, where $S_{2D}(\text{m}^2)$ and L_{2D} (m) are given by

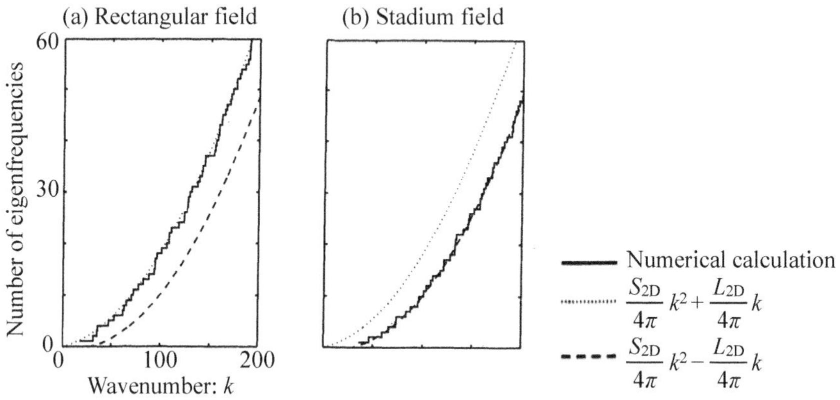

Fig. 10.3 Examples of distribution of eigenfrequencies for two-dimensional fields

$$S_{2D} = L_x L_y, \qquad L_{2D} = 2(L_x + L_y), \tag{10.24}$$

respectively. Figure 10.3 shows examples of the distribution of eigenfrequencies in two-dimensional fields. Figure 10.3(a) is for a rectangular field, and Fig. 10.3(b) illustrates a sample of a non-rectangular field like a stadium field. Approximation formulas for the number of eigenfrequencies give good estimate. No axial waves are included in a non-rectangular two-dimensional field like the stadium field.

There are three types of sound field; a three-dimensional field in a room, a two-dimensional one for vibration of a membrane, and a one-dimensional wave in an acoustic tube or vibration of a string. However, only one-dimensional waves consist of harmonic eigenfrequencies. It might be intuitively understood that almost all musical instruments have more or less a one-dimensional construction. A timpani, which sounds a part of the musical scale, is an exception[44].

The density of eigenfrequencies increases is proportional to the frequency itself in a two-dimensional field. Thus, the density is possibly higher in a three-dimensional field than in a two-dimensional one. However, note that the density also depends on the sound speed of the medium. By recalling flexural vibration of an elastic plate[21], the wave speed can be given by Eqs. 6.42 and 6.43 such that

$$c_b = \sqrt{\omega}\left(\frac{B_p}{\rho d_{th}}\right)^{1/4}, \quad \text{(m/s)} \tag{10.25}$$

where

$$B_p = \frac{Y_M}{1-\sigma^2}\frac{d_{th}^3}{12}, \tag{10.26}$$

Y_M is Young's modulus (N/m^2), d_{th} denotes thickness of the plate (m), σ is a Poisson's ratio, and ρ is volume density of the plate (kg/m^3). The sound speed of plate vibration becomes fast in proportion to the square root of the frequency of the sound.

The density of eigenfrequencies for a non-rectangular two-dimensional field can be written as

$$n_{\nu_{2D}}(\nu) = \frac{\partial N_{\nu_{2D}}(k)}{\partial \nu} \cong \frac{S_{2D} k}{2\pi} \frac{dk}{d\nu}, \tag{10.27}$$

where S_{2D} denotes the area of the plate (m^2). By introducing the surface density of the plate $M_{sf} = \rho d_{th}(\mathrm{kg/m^2})$ and substituting

$$k = \frac{2\pi\nu}{c_b} = \frac{2\pi\nu}{\sqrt{2\pi\nu}}\left(\frac{M_{sf}}{B}\right)^{1/4} \tag{10.28}$$

$$\frac{dk}{d\nu} = \frac{2\pi}{2\sqrt{2\pi\nu}}\left(\frac{M_{sf}}{B}\right)^{1/4} \tag{10.29}$$

for the density, the density of the eigenfrequencies becomes constant independent of the frequency such that

$$n(\nu) \cong \frac{S_{2D}}{2}\left(\frac{M_{sf}}{B_p}\right)^{1/2}. \tag{10.30}$$

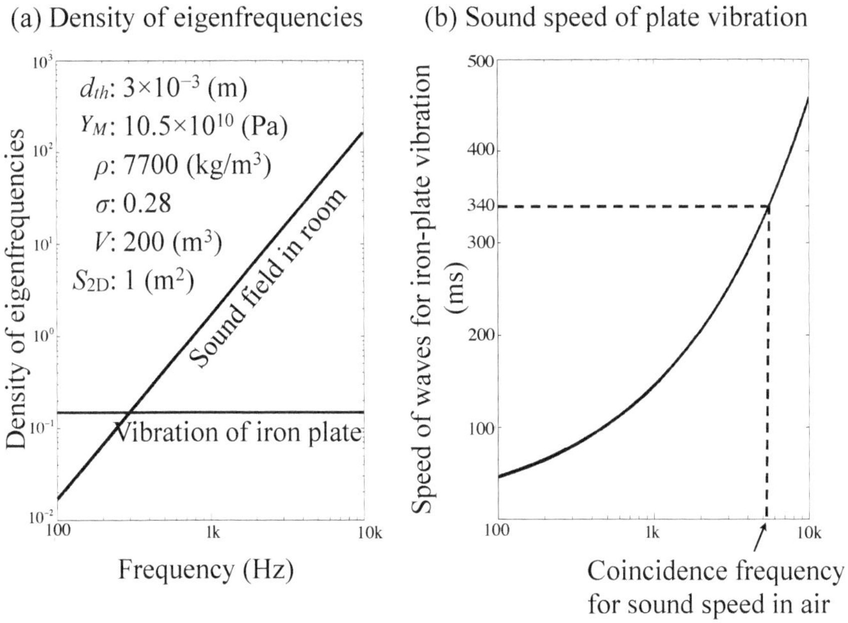

Fig. 10.4 Density of eigenfrequecies and speed of propagation for plate vibration

Figure 10.4 shows examples of the density of eigenfrequencies (a) and wave propagation speed for plate vibration (b). The parameters used in the calculation are thickness of the iron plate $d_{th} = 3 \cdot 10^{-3}$ (m), Young's modulus $Y_M = 10.5 \times 10^{10}$ (Pa), volume density $\rho = 7700(\text{kg/m}^3)$, Poisson's ratio $\sigma = 0.28$, and surface area of the iron plate $S_{2D} = 1(\text{m}^2)$. The density of the eigenfrequencies for a room, i.e.,

$$n_{3\text{D}}(\nu) \cong \frac{V\nu^2}{\pi c^3}, \tag{10.31}$$

is shown as a reference in the figure. It is interesting to see that the density of eigenfrequencies for plate vibration could be higher in low frequencies than that for the sound field in a room, despite the plate vibration being composed of two-dimensional waves. This is the outcome of the frequency dependency of wave propagation speed for plate vibration on the modal density[45].

10.1.4 Orthogonality of Eigenfunctions and Green Functions

The einenfunctions, which have spatially fixed loops and nodes, are also called **wave functions** or **eigenmodes**. The Helmholtz equation has solutions such that

$$\begin{aligned} p(x,y,z,t) &= \sum_{l,m,n} A_{lmn} \cos(k_l x)\cos(k_m y)\cos(k_n z)\mathrm{e}^{\mathrm{i}\omega_{lmn} t} \\ &= \sum_{l,m,n} A_{lmn} P_{lmn}(x,y,z)\mathrm{e}^{\mathrm{i}\omega_{lmn} t} \end{aligned} \tag{10.32}$$

for sound pressure waves propagating through the three-dimensional sound field in a rectangular room surrounded by rigid walls. Here, the eigen(angular)frequencies are given by Eq. 10.11 and eigenfunctions defined by

$$P_{lmn}(x,y,z) = \cos k_l x \cos(k_m y)\cos(k_n z) \tag{10.33}$$

and are **orthogonal**[46] such that

$$\int_V P_{lmn}(x,y,z)P_{l'm'n'}(x,y,z)dxdydz = 0 \quad \text{for} \quad (l,m,n) \neq (l',m',n'). \tag{10.34}$$

Following this orthogonality of eigenfunctions, the solution of the wave equation is obtained for the sound field in a room that is excited by a sound source.

Recall the **velocity potential** $\phi_v(\mathbf{r},t)$ that gives particle velocity v_x of a sound wave along an x-direction such that

$$v_x(\mathbf{r},t) = -\frac{\partial \phi_v(\mathbf{r},t)}{\partial x}, \quad \phi_v(\mathbf{r},t) = -\int v_x(\mathbf{r},t)dx, \tag{10.35}$$

And thus sound pressure of a wave can be expressed as

$$p(\mathbf{r},t) = \kappa s(\mathbf{r},t) = \rho \frac{\partial \phi_v(\mathbf{r},t)}{\partial t}, \tag{10.36}$$

where κ denotes bulk modulus of the medium (Pa), s is **condensation** defined by

$$s(\mathbf{r},t) = \frac{\rho(\mathbf{r},t) - \rho_0}{\rho_0}, \tag{10.37}$$

and ρ_0 is volume density of the medium when the sound wave is not excited. The wave equation represented can be rewritten as

$$\frac{1}{c^2}\frac{\partial^2 \phi_v(\mathbf{r},t)}{\partial t^2} = \nabla^2 \phi_v(\mathbf{r},t) \tag{10.38}$$

by using velocity potential.

Suppose that sound sources are distributed in a room and they can be defined as a density function of volume velocity in a unit volume of the room such that $Q_d(x,y,z)\mathrm{e}^{\mathrm{i}\omega t}$ (1/s). The sound field excited by the distributed sources follows the wave equation such that

$$\frac{1}{c^2}\frac{\partial^2 \phi_v(\mathbf{r},t)}{\partial t^2} = \nabla^2 \phi_v(\mathbf{r},t) + Q_d(\mathbf{r})\mathrm{e}^{\mathrm{i}\omega t}, \qquad (1/\mathrm{s}) \tag{10.39}$$

where $\mathbf{r}$ denotes the spatial coordinates of the field. Suppose that a stationary wave is expressed by

$$\phi_v(\mathbf{r},t) = \Phi_v(\mathbf{r})\mathrm{e}^{\mathrm{i}\omega t}. \qquad (\mathrm{m}^2/\mathrm{s}) \tag{10.40}$$

By introducing the stationary solution above, the so-called **non-homogeneous Helmholtz equation**

$$\nabla^2 \Phi_v(\mathbf{r}) + k^2 \Phi_v(\mathbf{r}) = -Q_d(\mathbf{r}) \qquad (1/\mathrm{s}) \tag{10.41}$$

is derived.

A solution for the non-homogeneous Helmholtz equation can be expressed as an orthogonal expansion series using the orthogonal eigenfunctions P_{lmn}, which satisfy

$$\nabla^2 \hat{P}_{lmn}(\mathbf{r}) + k_{lmn}^2 \hat{P}_{lmn}(\mathbf{r}) = 0 \tag{10.42}$$

$$k_l^2 + k_m^2 + k_n^2 = k_{lmn}^2 \tag{10.43}$$

When a function is represented by a series of orthogonal functions, the series expansion is called **orthogonal series representation** of a function. Suppose the spatial function for the velocity potential $\Phi_v(\mathbf{r})$ and the sound source distribution $Q_d(\mathbf{r})$ in the orthogonal series representation such that

$$\Phi_v(\mathbf{r}) = \sum_{l,m,n} A_{lmn}\hat{P}_{lmn}(\mathbf{r}) \qquad (\mathrm{m}^2/\mathrm{s}) \tag{10.44}$$

$$-Q_d(\mathbf{r}) = \sum_{l,m,n} B_{lmn}\hat{P}_{lmn}(\mathbf{r}), \qquad (1/\mathrm{s}) \tag{10.45}$$

where $\hat{P}_{lmn}$ denotes a set of **normalized orthogonal eigenfunctions** so that

$$\int_V \hat{P}^2_{lmn}(\mathbf{r})d\mathbf{r} = \frac{\Lambda_{lmn}}{V}\int_V P^2_{lmn}(\mathbf{r})d\mathbf{r} = 1. \tag{10.46}$$

Here, $\Lambda_{lmn}/V(=\Lambda_N/V)$, called the normalizing factor, is introduced into the eigenfunctions, and $V(\mathrm{m}^3)$ denotes the room volume. By introducing the orthogonal series representation into the non-homogeneous Helmholtz equation the velocity potential can be obtained:

$$\begin{aligned}\Phi_v(\mathbf{r}) &= \sum_{l,m,n} A_{lmn}\hat{P}_{lmn}(\mathbf{r}) \\ &= \sum_{l,m,n} \frac{\hat{P}_{lmn}(\mathbf{r})}{(k_l^2+k_m^2+k_n^2)-k^2}\int_V Q_d(\mathbf{r}')\hat{P}_{lmn}(\mathbf{r}')d\mathbf{r}' \\ &= \int_V G(\mathbf{r}',\mathbf{r})Q_d(\mathbf{r}')d\mathbf{r}'. \qquad (\mathrm{m}^2/\mathrm{s})\end{aligned} \tag{10.47}$$

Here,

$$G(\mathbf{r}',\mathbf{r}) = Q_0\sum_{l,m,n}\frac{\hat{P}_{lmn}(\mathbf{r})\hat{P}_{lmn}(\mathbf{r}')}{(k_l^2+k_m^2+k_n^2)-k^2} \qquad (\mathrm{m}^2/\mathrm{s}) \tag{10.48}$$

is called the **Green function** for sound field excited by a point source with $Q_0 = 1(\mathrm{m}^3/\mathrm{s})$.

If there is a point source at $\mathbf{r}'$ whose density is given by $Q_{0d}\delta(\mathbf{r}-\mathbf{r}')$ (1/s), a sound wave corresponding to the point source follows the equation such that

$$\nabla^2\Phi_{v_0}(\mathbf{r}',\mathbf{r}) + k^2\Phi v_{A0}(\mathbf{r}',\mathbf{r}) = -Q_{0d}\delta(\mathbf{r}-\mathbf{r}'). \qquad (1/\mathrm{s}) \tag{10.49}$$

Therefore, the solution $\Phi_{v_0}(\mathbf{r}',\mathbf{r})$ is given by

$$\Phi_{v_0}(\mathbf{r}',\mathbf{r}) = Q_0\sum_{l,m,n}\frac{\hat{P}_{lmn}(\mathbf{r}')\hat{P}_{lmn}(\mathbf{r})}{(k_l^2+k_m^2+k_n^2)-k^2} = G(\mathbf{r}',\mathbf{r}), \qquad (\mathrm{m}^2/\mathrm{s}) \tag{10.50}$$

where

$$\int_V Q_{0d}\delta(\mathbf{r}-\mathbf{r}')d\mathbf{r} = Q_0 = 1 \quad (\mathrm{m}^3/\mathrm{s}) \tag{10.51}$$

and

$$\int_V \delta(\mathbf{r}-\mathbf{r}')\hat{P}_{lmn}(\mathbf{r})d\mathbf{r} = \hat{P}_{lmn}(\mathbf{r}'). \tag{10.52}$$

That is, the Green function represents response of a sound field to a point source located at a single point whose strength $(\mathrm{m}^3/\mathrm{s})$ is unity.

10.1.5 Green Functions for One-Dimensional Sound Waves

Recall the traveling waves on a string or in an acoustic tube described in Chapters 5 and 7. Green functions for the vibration and waves in a one-dimensional system can be obtained in a closed form without using representation of the orthogonal series. Suppose the Helmholtz equation of $U(x)$ such that

$$\frac{d^2U(x)}{dx^2} + k^2U(x) = -F_X(x), \quad 0 \le x \le L \tag{10.53}$$

for the traveling waves in one-dimensional space. A solution of the equation is written as

$$U(x) = A(x)\cos kx + B(x)\sin kx \tag{10.54}$$

under the boundary conditions of

$$U(0) = U(L) = 0. \tag{10.55}$$

By substituting the solution form of $U(x)$ for the equation

$$-kA'(x)\sin kx + kB'(x)\cos kx = -F_X(x) \tag{10.56}$$

can be derived subject to

$$A'(x)\cos kx + B'(x)\sin kx = 0. \tag{10.57}$$

Consequently, $A'(x)$, and $B'(x)$ can be given by

$$A'(x) = \frac{F_X(x)\sin kx}{k} \tag{10.58}$$

$$B'(x) = \frac{-F_X(x)\cos kx}{k} \tag{10.59}$$

as solutions for the simultaneous equations. Therefore, by substituting the solutions for $U(x)$,

$$U(x) = \frac{\cos kx}{k}\int_{c_1}^{x} F_X(y)\sin ky dy - \frac{\sin kx}{k}\int_{c_2}^{x} F_X(y)\cos ky dy \tag{10.60}$$

is obtained, where c_1 and c_2 should be determined following the boundary conditions.

For meeting the boundary condition $U(0) = 0$,

$$U(0) = \frac{1}{k}\int_{c_1}^{0} F_X(y)\sin ky dy = 0 \tag{10.61}$$

must hold independent of the function $F_X(y)$. Therefore,

$$c_1 = 0 \tag{10.62}$$

is determined. Similarly, following $U(L) = 0$ and setting $c_1 = 0$ and $x = L$,

$$\begin{aligned} U(L) &= \frac{\cos kL}{k}\int_0^L F_X(y)\sin ky dy - \frac{\sin kL}{k}\left(\int_0^L F_X(y)\cos ky dy - \int_0^{c_2} F_X(y)\cos ky dy\right) \\ &= \frac{-1}{k}\int_0^L F_X(y)\sin k(L-y)dy + \frac{\sin kL}{k}\int_0^{c_2} F_X(y)\cos ky dy \\ &= 0 \end{aligned} \tag{10.63}$$

is obtained. Consequently, $U(x)$ can be rewritten as

$$\begin{aligned} U(x) &= \frac{\cos kx}{k}\int_0^x F_X(y)\sin ky dy - \frac{\sin kx}{k}\left(\int_0^x F_X(y)\cos ky dy - \int_0^{c_2} F_X(y)\cos ky dy\right) \\ &= -\frac{1}{k}\int_0^x F_X(y)\sin k(x-y)dy + \frac{\sin kx}{k}\int_0^{c_2} F_X(y)\cos ky dy \\ &= -\frac{1}{k}\int_0^x F_X(y)\sin k(x-y)dy + \frac{\sin kx}{k\sin kL}\left(\int_0^x F_X(y)\cos ky dy + \int_x^L F_X(y)\cos ky dy\right) \\ &= \int_0^x F_X(y)\frac{\sin ky\sin k(L-x)}{k\sin kL}dy + \int_x^L F_X(y)\frac{\sin kx\sin k(L-y)}{k\sin kL}dy \\ &= \int_0^L F_X(y)G(x,y)dy, \end{aligned} \tag{10.64}$$

where $G(x = x', y = x)$ represents the Green function defined as

$$G(x',x) = \begin{cases} \dfrac{\sin kx\sin k(L-x')}{k\sin kL} & 0 < x \le x' \\ \dfrac{\sin kx'\sin k(L-x)}{k\sin kL}. & x' \le x < L \end{cases} \tag{10.65}$$

The Green function for a one-dimensional wave field can be represented in a closed form without the orthogonal series of eigenfunctions.

10.1.6 Green Function for Three-Dimensional Wave Field According to Mirror Image Method

The Green function is defined according to the eigenfunctions of a sound field. However, the Green function for the sound field in a rectangular room can also be represented according to the mirror image theory without using eigenfunctions[46][47] [48]. The eigenfunction corresponding to an oblique wave is represented as

$$\begin{aligned} P_{lmn}(x,y,z) &= \cos k_l x\cos k_m y\cos k_n z \\ &= \left(\frac{e^{ik_l x}+e^{-ik_l x}}{2}\right)\left(\frac{e^{ik_m y}+e^{-ik_m y}}{2}\right)\left(\frac{e^{ik_n z}+e^{-ik_n z}}{2}\right). \end{aligned} \tag{10.66}$$

By using this expression of the eigenfunction by the complex function form, a product of two eigenfunctions can be rewritten as

$$P_{lmn}(x,y,z)P_{lmn}(x',y',z') = \sum_{P=1}^{8} \frac{\mathrm{e}^{\pm i\mathbf{k}_N \cdot \mathbf{R}_P}}{64} = \frac{\mathrm{e}^{\pm ik_l(x\pm x')\pm ik_m(y\pm y')\pm ik_n(z\pm z')}}{64}, \quad (10.67)$$

where the notation $\pm$ means eight combinations of the sign are taken. Define the vectors $\mathbf{k}_N$ and $\mathbf{R}_P$ as

$$\mathbf{k}_N = (k_l, k_m, k_n), \quad \mathbf{R}_P = (x \pm x', y \pm y', z \pm z'), \quad (10.68)$$

where $\mathbf{R}_P$ represents eight vectors composed of, $\mathbf{R}_1,...,\mathbf{R}_8$, according to the combinations of the positive or negative signs.

According to Eq. 10.50, the Green function was represented as a series of orthogonal eigenfunctions such that

$$G(\mathbf{x}',\mathbf{x}) = Q_0 \sum_N \frac{\Lambda_N}{V} \frac{P_N(\mathbf{x}')P_N(\mathbf{x})}{k_N^2 - k^2} \quad (\mathrm{m}^2/\mathrm{s}) \quad (10.69)$$

for velocity potential of the sound field in a rectangular room surrounded by rigid walls, where $\mathbf{x}'$ and $\mathbf{x}$ denote position vectors for the sound source and receiving points, $Q_0(\mathrm{m}^3/\mathrm{s})$ is the unit strength of the source, $V(\mathrm{m}^3)$ is room volume, and Λ_N is the normalized factor for the N-th eigenfunction. By substituting the vectors $\mathbf{k}_N$ and $\mathbf{R}_P$ for the Green function,

$$G(\mathbf{x}',\mathbf{x}) = Q_0 \sum_{N=-\infty}^{+\infty} \sum_{P=1}^{8} \frac{1}{\Lambda_N V} \frac{\mathrm{e}^{-i\mathbf{k}_N \cdot \mathbf{R}_P}}{k_N^2 - k^2} \quad (10.70)$$

is derived, where $k_N{}^2 = |\mathbf{k}_N|^2$, and the wavenumber constants k_l, k_m and k_n are extended into negative values for the complex sinusoidal functions. In addition, by rewriting the function in the right-hand side of the equation into an integral representation according to an integration formula such as

$$\frac{\mathrm{e}^{-i\mathbf{k}_N \cdot \mathbf{R}_P}}{k_N^2 - k^2} = \int_{-\infty}^{+\infty}\int_{-\infty}^{+\infty}\int_{-\infty}^{+\infty} \frac{\mathrm{e}^{-i\mathbf{k}\cdot\mathbf{R}_P}}{|\mathbf{k}|^2 - k^2} \delta(\mathbf{k}-\mathbf{k}_N)dk_x dk_y dk_z, \quad (10.71)$$

$$G(\mathbf{x}',\mathbf{x}) = \frac{Q_0}{V} \sum_{P=1}^{8} \int_{-\infty}^{+\infty}\int_{-\infty}^{+\infty}\int_{-\infty}^{+\infty} \frac{\mathrm{e}^{-i\mathbf{k}\cdot\mathbf{R}_P}}{|\mathbf{k}|^2 - k^2} \sum_{N=-\infty}^{+\infty} \frac{1}{\Lambda_N} \delta(\mathbf{k}-\mathbf{k}_N)dk_x dk_y dk_z \quad (10.72)$$

is obtained, where

$$|\mathbf{k}|^2 = k_x^2 + k_y^2 + k_z^2 \quad (10.73)$$
$$\delta(\mathbf{k}-\mathbf{k}_N) = \delta(k_x - k_l)\delta(k_y - k_m)\delta(k_z - k_n). \quad (10.74)$$

Note here that $\delta(k_x - k_l)$, $\delta(k_y - k_m)$, and $\delta(k_z - k_n)$ can be represented as a Fourier series with the period of π/L_x, π/L_y, and π/L_z, such that

$$\sum_{l=-\infty}^{+\infty} \delta(k_x - k_l) = \Sigma_{l=-\infty}^{+\infty} \delta\left(k_x - \tfrac{l\pi}{L_x}\right) = \Sigma_{l=-\infty}^{+\infty} \tfrac{L_x}{\pi} e^{-i2lL_x k_x} \quad (10.75)$$

$$\sum_{m=-\infty}^{+\infty} \delta(k_y - k_m) = \Sigma_{m=-\infty}^{+\infty} \delta\left(k_y - \tfrac{m\pi}{L_y}\right) = \Sigma_{m=-\infty}^{+\infty} \tfrac{L_y}{\pi} e^{-i2mL_y k_y} \quad (10.76)$$

$$\sum_{n=-\infty}^{+\infty} \delta(k_z - k_n) = \Sigma_{n=-\infty}^{+\infty} \delta\left(k_z - \tfrac{n\pi}{L_z}\right) = \Sigma_{n=-\infty}^{+\infty} \tfrac{L_z}{\pi} e^{-i2nL_z k_z}, \quad (10.77)$$

where L_x, L_y and L_z denote lengths of the sides. By introducing the Fourier series into the Green function given by Eq. 10.72,

$$G(\mathbf{x}', \mathbf{x}) = \sum_{P=1}^{8} \sum_{N=-\infty}^{+\infty} \frac{Q_0}{\Lambda_N \pi^3} \int_{-\infty}^{+\infty} \int_{-\infty}^{+\infty} \int_{-\infty}^{+\infty} \frac{e^{-i\mathbf{k}\cdot(\mathbf{R}_P + \mathbf{R}_N)}}{|\mathbf{k}|^2 - k^2} dk_x dk_y dk_z, \quad (10.78)$$

can be obtained, where $\mathbf{R}_N$ denotes a vector represented by

$$\mathbf{R}_N = 2(lL_x, mL_y, nL_z). \quad (10.79)$$

Recall that the velocity potential for a spherical wave radiated from a point source can be expressed as

$$\Phi_{v0}(r) = Q_0 \frac{e^{-ik|\mathbf{R}|}}{4\pi|\mathbf{R}|} = Q_0 \frac{e^{-ikr}}{4\pi r}, \qquad (\mathrm{m}^2/\mathrm{s}) \quad (10.80)$$

where length of the vector $\mathbf{R}$ shows the distance from the point source. A spherical wave can also be represented using superposition of plane waves such that

$$\Phi_{v0}(r) = \frac{Q_0}{8\pi^3} \int\int\int \Phi_{v0}^{+}(k) e^{-i\mathbf{k}\cdot\mathbf{R}} dk_x dk_y dk_z, \quad (10.81)$$

where the range of integration is extended from $-\infty \sim +\infty$. By substituting $\Phi_{v0}(r)$ for the Helmholtz equation where a point source is located at the origin such that

$$\nabla^2 \Phi_{v0} + k^2 \Phi_{v0} = -Q_{0d}\delta(\mathbf{x}), \qquad (1/\mathrm{s}) \quad (10.82)$$

$$\Phi_{v0}^{+}(k) = \frac{1}{|\mathbf{k}^2| - k^2} \qquad (\mathrm{m}^2) \quad (10.83)$$

is derived. Here, $|\mathbf{k}^2| = k_x^2 + k_y^2 + k_z^2$ and

$$\delta(\mathbf{x}) = \frac{1}{8\pi^3}\int\int\int \mathrm{e}^{-\mathrm{i}\mathbf{k}\cdot\mathbf{R}} dk_x dk_y dk_z \qquad (10.84)$$

$$|\mathbf{R}|^2 = x^2 + y^2 + z^2 = r^2 \qquad (10.85)$$

are introduced.

Consequently, the Green function can be formulated as

$$\begin{aligned} &G(\mathbf{x}',\mathbf{x}) \\ &= \sum_{P=1}^{8}\sum_{N=-\infty}^{+\infty} \frac{Q_0}{\Lambda_N \pi^3}\int\int\int \frac{\mathrm{e}^{-\mathrm{i}\mathbf{k}\cdot(\mathbf{R}_P+\mathbf{R}_N)}}{|\mathbf{k}|^2 - k^2} dk_x dk_y dk_z \\ &= \sum_{P=1}^{8}\sum_{N=-\infty}^{+\infty} \frac{8Q_0}{\Lambda_N}\frac{\mathrm{e}^{-\mathrm{i}k|\mathbf{R}_P+\mathbf{R}_N|}}{4\pi|\mathbf{R}_P+\mathbf{R}_N|}. \quad (\mathrm{m}^2/\mathrm{s}) \end{aligned} \qquad (10.86)$$

This formulation says that the Green function is interpreted as a solution according to the mirror image method written by superposition of all the spherical waves radiated from the mirror image sources, which are located as shown in Fig.10.5. The length of the vector $\mathbf{R}_P + \mathbf{R}_N$ denotes the distance from the source or mirror image sources, and Λ_N shows the normalized factor for oblique, tangential, and axial waves[47], while it is set to $\Lambda_N = 8$ in reference[48].

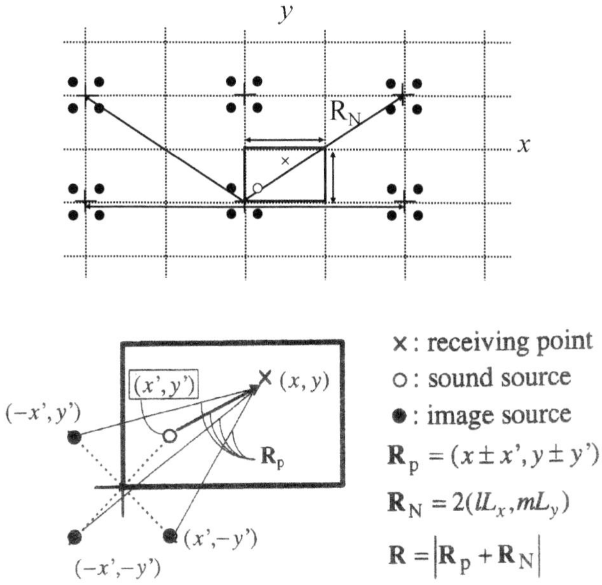

Fig. 10.5 Distribution of mirror image sources for rectangular room (two-dimensional sectional view) from [46](Fig. 6.5.1)

10.1.7 Impulse Response of Three-Dimensional Room

An **impulse response** of the sound field for a three-dimensional room can be represented by the image theory or modal-wave theory. The impulse response $h(\mathbf{x}',\mathbf{x},t)$ can be obtained by taking the inverse Fourier transform for the Green function such that

$$\begin{aligned}
&h(\mathbf{x}',\mathbf{x},t)\\
&= \frac{1}{2\pi\Delta\omega}\int_{-\infty}^{+\infty} G(\mathbf{x}',\mathbf{x},\omega)\mathrm{e}^{\mathrm{i}\omega t}d\omega\\
&= \hat{Q}_0\sum_{P=1}^{8}\sum_{N=-\infty}^{+\infty}\frac{8}{\Lambda_N}\frac{1}{4\pi|\mathbf{R}_P+\mathbf{R}_N|}\frac{1}{2\pi}\int_{-\infty}^{+\infty}\mathrm{e}^{\mathrm{i}\omega(t-(|\mathbf{R}_P+\mathbf{R}_N|/c))}d\omega\\
&= \hat{Q}_0\sum_{P=1}^{8}\sum_{N=-\infty}^{+\infty}\frac{8}{\Lambda_N}\frac{\delta(t-|\mathbf{R}_P+\mathbf{R}_N|/c)}{4\pi|\mathbf{R}_P+\mathbf{R}_N|},
\end{aligned}\tag{10.87}$$

where

$$\frac{1}{2\pi}\int_{-\infty}^{+\infty}\mathrm{e}^{\mathrm{i}\omega(t-\tau)}d\omega=\delta(t-\tau)\tag{10.88}$$

and $Q_0=\hat{Q}_0\Delta\omega$ are introduced. It can be found that the impulse response is represented by superposition of impulsive sound from the source and all the mirror image sources with time delays and magnitudes according to the distance between the receiving and source positions.

The representation of impulse response can be converted to a formulation based on the wave theory. Suppose that the Green function is written by Eq. 10.69 for the velocity potential according to the wave theory, where $\mathbf{x}'$ and $\mathbf{x}$ denote the source and receiving position vectors respectively, and $\hat{Q}_0\Delta\omega(\mathrm{m}^3/\mathrm{s})$ shows the strength of the impulsive source. The impulse response $h(\mathbf{x}',\mathbf{x},t)$ between the source and receiving positions can be derived as

$$\begin{aligned}
h(\mathbf{x}',\mathbf{x},t) &= \frac{1}{2\pi\Delta\omega}\int_{-\infty}^{+\infty} G(\mathbf{x}',\mathbf{x})\mathrm{e}^{\mathrm{i}\omega t}d\omega\\
&= \frac{\hat{Q}_0}{2\pi}\sum_N\frac{\Lambda_N}{V}\phi_N(\mathbf{x})\phi_N(\mathbf{x}')c^2\int_{-\infty}^{+\infty}\frac{\mathrm{e}^{\mathrm{i}\omega t}d\omega}{(\omega-\omega_{N_1})(\omega-\omega_{N_2})}\\
&= \frac{c^2\hat{Q}_0}{V}\sum_N\Lambda_N\phi_N(\mathbf{x})\phi_N(\mathbf{x}')\frac{\sin(\omega_{N_0}t)}{\omega_{N_0}}\mathrm{e}^{-\delta_N t}\quad (t>0)\quad (\mathrm{m}^2/\mathrm{s})
\end{aligned}\tag{10.89}$$

for the velocity potential where

$$\omega_{N_1}=\omega_{N_0}+\mathrm{i}\delta_N,\quad \omega_{N_2}=-\omega_{N_0}+\mathrm{i}\delta_N,\quad (\omega_{N_0}\gg\delta_N>0).\tag{10.90}$$

Thus, the impulse response for sound pressure is given by

$$p(\mathbf{x}',\mathbf{x},t) = \rho_0 \frac{\partial h(\mathbf{x}',\mathbf{x},t)}{\partial t}$$
$$\cong \frac{\rho_0 c^2 \hat{Q}_0}{V} \sum_N \Lambda_N \phi_N(\mathbf{x})\phi_N(\mathbf{x}') \sin(\omega_{N_0} t + \theta_N) \mathrm{e}^{-\delta_N t} \quad \text{(Pa)} \qquad (10.91)$$

where $\theta_N = \tan^{-1} \frac{\omega_{N_0}}{-\delta_N}$.

In the formulation above, decaying constants depending on the sound absorption coefficients of surrounding walls are denoted by δ_N. The impulse response corresponds to that based on the mirror image theory, taking account of the sound absorption coefficients of the walls. It can be seen that the impulse response can be written based on the wave theory as superposition of free oscillation with the eigenfrequencies instead of impulsive sound from all the mirror sources.

10.2 General Representation of Waves in Rooms

In the previous sections solutions were formulated in an orthogonal series expansion by taking a rectangular room as an example. This type of formulation can be theoretically extended into general cases according to the theory of eigenfrequency and eigenfunctions for the wave theory. In addition, there is also another way of representing sound waves as solutions of the wave equation instead of orthogonal series expansion. The wave solutions are theoretically formulated in this section.

10.2.1 Eigenfunctions and Eigenfrequencies for Wave Equations

The wave equation is written as

$$\frac{\partial^2 u(\mathbf{x},t)}{\partial t^2} = c^2 \nabla^2 u(\mathbf{x},t), \qquad (10.92)$$

where $\mathbf{x}$ represents the spatial coordinates of the sound field. Assuming a sinusoidal solution such as

$$u(\mathbf{x},t) = U(\mathbf{x})\mathrm{e}^{\mathrm{i}\omega t}, \qquad (10.93)$$

the wave equation can be rewritten as

$$\nabla^2 U(\mathbf{x}) + k^2 U(\mathbf{x}) = 0, \qquad (10.94)$$

where $k = \omega/c$ is the wavenumber.

By introducing a general notation indicating a linear operator L into the wave equation for a sinusoidal solution above, the equation above can be expressed in a symbolic form as (function U is simply written as u here)

$$Lu(\mathbf{x}) = \lambda_u u(\mathbf{x}), \tag{10.95}$$

where $u(\mathbf{x})$ and λ_u can be interpreted as the eigenfunction and its eigenvalue for the operator L, respectively. This is similar to the linear transform by a matrix,

$$A\mathbf{x} = \lambda_{\mathbf{x}}\mathbf{x}, \tag{10.96}$$

where A denotes the linear matrix, $\lambda_{\mathbf{x}}$ is the eigenvalue, and $\mathbf{x}$ denotes the eigenvector corresponding to the eigenvalue for the matrix A. In particular, a linear operator is called **Hermitian** if

$$\int_B u^*(\mathbf{x})Lv(\mathbf{x})d\mathbf{x} = \left[\int_B v^*(\mathbf{x})Lu(\mathbf{x})\right]^*, \tag{10.97}$$

where B is a specified region for the linear operator for which suitable boundary conditions are imposed. Both u and v are functions that obey the boundary conditions.

10.2.2 Eigenfunctions and Orthogonality

The eigenvalues of Hermitian operators are real, and the eigenfunctions corresponding to different eigenvalues are orthogonal in the specified region. By supposing that the linear operator L is Hermitian,

$$Lu_i(\mathbf{x}) = \lambda_i u_i(\mathbf{x}), \tag{10.98}$$

where λ_i denotes the i-th eigenvalue and $u_i(\mathbf{x})$ is the eigenfunction for the i-th eigenvalue. By taking another pair of the eigenfunction and its eigenvalue such that

$$Lu_j(\mathbf{x}) = \lambda_j u_j(\mathbf{x}), \tag{10.99}$$

the relations

$$\int_B u_j^*(\mathbf{x})Lu_i(\mathbf{x})d\mathbf{x} = \lambda_i \int_B u_j^*(\mathbf{x})u_i(\mathbf{x})d\mathbf{x} \tag{10.100}$$

and

$$\int_B u_i^*(\mathbf{x})Lu_j(\mathbf{x})d\mathbf{x} = \lambda_j \int_B u_i^*(\mathbf{x})u_j(\mathbf{x})d\mathbf{x} \tag{10.101}$$

hold well in the specified region B. According to L being Hermitian,

$$\int_B u_j^*(\mathbf{x})Lu_i(\mathbf{x})d\mathbf{x} = \lambda_j^* \int_B u_i(\mathbf{x})u_j^*(\mathbf{x})d\mathbf{x}, \tag{10.102}$$

and thus the orthogonality of the eigenfunctions are formulated as

$$(\lambda_i - \lambda_j^*) \int_B u_j^*(\mathbf{x})u_i(\mathbf{x})d\mathbf{x} = 0. \tag{10.103}$$

Consequently,

$$\int_B u_j^*(\mathbf{x})u_i(\mathbf{x})d\mathbf{x} = 0 \tag{10.104}$$

if $\lambda_i \neq \lambda_j^*$, and when $i = j$, $\lambda_i = \lambda_j^*$. Namely, the eigenvalues are real.

10.2.3 Orthogonal Expansion Using Eigenfunctions

The eigenfunctions for a Hermitian operator are orthogonal. Recall the wave equation

$$\frac{\partial^2 u(\mathbf{x},t)}{\partial t^2} = c^2 \nabla^2 u(\mathbf{x},t) + f_X(\mathbf{x},t) \tag{10.105}$$

under the condition that

$$f_X(\mathbf{x},t) = F_X(\mathbf{x})\mathrm{e}^{\mathrm{i}\omega t}. \tag{10.106}$$

Assuming a sinusoidal solution for the steady state i.e.,

$$u(\mathbf{x},t) = U(\mathbf{x})\mathrm{e}^{\mathrm{i}\omega t}, \tag{10.107}$$

the wave equation can be rewritten as

$$\nabla^2 U(\mathbf{x}) + k^2 U(\mathbf{x}) = -F_X(\mathbf{x}). \tag{10.108}$$

Namely, this introduces the notation for a liner operator L

$$LU(\mathbf{x}) - \lambda_u U(\mathbf{x}) = -F_X(\mathbf{x}), \tag{10.109}$$

where $k^2 = -\lambda_u$ suitable boundary conditions are imposed on the equation. This equation corresponds to the non-homogeneous Helmholtz equation.

Both of the functions $U(\mathbf{x})$ and $F_X(\mathbf{x})$ are represented by the orthogonal expansion using the eigenfunctions meeting the boundary conditions such that

$$U(\mathbf{x}) = \sum_n A_n u_n(\mathbf{x}) \tag{10.110}$$

$$-F_X(\mathbf{x}) = \sum_n B_n u_n(\mathbf{x}), \tag{10.111}$$

where the eigenfunctions are orthogonal and normalized so that

$$\int_B u_i^*(\mathbf{x})u_j(\mathbf{x})d\mathbf{x} = \delta_{ij}. \tag{10.112}$$

Those are the solutions for the homogeneous equation

$$Lu_n(\mathbf{x}) - \lambda_n u_n(\mathbf{x}) = 0 \tag{10.113}$$

under the imposed boundary conditions.

By substituting the eigenfunctions for the non-homogeneous Helmholtz equation, the coefficients can be derived as

$$A_n = \frac{B_n}{\lambda_n - \lambda_u} = \frac{-u_n \cdot F_X}{\lambda_n - \lambda_u}, \tag{10.114}$$

where $u_n \cdot F_X$ denotes

$$u_n \cdot F_X = \int_B u_n^*(\mathbf{x}')F_X(\mathbf{x}')d\mathbf{x}'. \tag{10.115}$$

Therefore, the solution can be given by

$$u(\mathbf{x}) = \int_B G(\mathbf{x}', \mathbf{x})F_X(\mathbf{x}')d\mathbf{x}', \tag{10.116}$$

where $G(\mathbf{x}', \mathbf{x})$ is called a **Green's function** that is given by

$$G(\mathbf{x}', \mathbf{x}) = \sum_n \frac{u_n(\mathbf{x}')u_n^*(\mathbf{x})}{\lambda_u - \lambda_n}. \tag{10.117}$$

As described in subsection 10.1.4, the Green function itself is a solution for the equation

$$LG(\mathbf{x}', \mathbf{x}) - \lambda_u G(\mathbf{x}', \mathbf{x}) = -\delta(\mathbf{x}' - \mathbf{x}). \tag{10.118}$$

A **reciprocal formula** can be seen for the Green's function as

$$G(\mathbf{x}', \mathbf{x}) = \left[\sum_n \frac{u_n(\mathbf{x})u_n^*(\mathbf{x}')}{\lambda_u - \lambda_n}\right]^* = [G(\mathbf{x},\mathbf{x}')]^* \tag{10.119}$$

with respect to the observation and source positions, subject to the eigenvalues being real. In other words, the operator is Hermitian. The Green function, eigenfunctions, and eigenvalues for the Hermitian operator are important and fundamental concepts representing wave propagation in acoustical systems such as room acoustics.

10.2.4 General Solution of Wave Equation by Integral Formula

Recall the wave equation and its Helmholtz equation such that

$$\frac{1}{c^2}\frac{\partial^2 \phi_v(x,y,z,t)}{\partial t^2} = \nabla^2 \phi_v(x,y,z,t) + Q_d(x,y,z)\mathrm{e}^{\mathrm{i}\omega t} \tag{10.120}$$

$$\nabla^2 \Phi_v(x,y,z) + k^2 \Phi_v(x,y,z) = -Q_d(x,y,z), \tag{10.121}$$

where $\phi_v(x,y,z,t)(\mathrm{m}^2/\mathrm{s})$ denotes the velocity potential of the sound field. By taking inspiration from the mirror image method, describe the solution as

$$\Phi_v(x,y,z) = \Phi_{v_D}(x,y,z) + \Phi_{v_R}(x,y,z), \tag{10.122}$$

which is a superposition of the direct and reflection waves from the boundary. Namely, the direct wave is given by an integration formula for approaching waves from the sources such that

$$\Phi_{v_D}(x,y,z) = \frac{1}{4\pi}\int_V Q_d(x',y',z')\frac{e^{-ikr_1}}{r_1}dx'dy'dz' \tag{10.123}$$

$$r_1 = \sqrt{(x-x')^2+(y-y')^2+(z-z')^2}, \tag{10.124}$$

which represent superposition of spherical waves approaching an observation point from the sources, and the other reflection waves are necessary to meet the boundary conditions. This type of formulation might be more intuitively understood than that by the Green's function method.

Now consider the second term which represents the reflection waves. For that purpose, first recall the Green's theorem between the volume and surface integrations [46]. Suppose that there are two vector functions of (x,y,z) such that

$$\mathbf{a} = \hat{\psi}\mathbf{b}, \tag{10.125}$$

where $\hat{\psi}$ is a scalar function of x,y,z. The **divergence** of vector **a** can be written as

$$\mathrm{div}\mathbf{a} = \hat{\psi}\cdot\mathrm{div}\mathbf{b} + \nabla\hat{\psi}\cdot\mathbf{b}, \tag{10.126}$$

where

$$\mathrm{div}\mathbf{b} = \nabla\cdot\mathbf{b} = \frac{\partial b_x}{\partial x} + \frac{\partial b_y}{\partial y} + \frac{\partial b_z}{\partial z} \tag{10.127}$$

$$\nabla\hat{\psi} = \mathrm{grad}\hat{\psi} = \frac{\partial\nabla\hat{\psi}}{\partial x}\mathbf{i} + \frac{\partial\nabla\hat{\psi}}{\partial y}\mathbf{j} + \frac{\partial\nabla\hat{\psi}}{\partial z}\mathbf{k} \tag{10.128}$$

and $(\mathbf{i},\mathbf{j},\mathbf{k})$ denotes the unit vector of the (x,y,z) space. The divergence of a vector function is a scalar function, while the **gradient** of a scalar function is a vector function.

By integrating the divergence of a vector over a volume,

$$\int_V (\mathrm{div}\mathbf{a})dv = \int_S \mathbf{a}\cdot\mathbf{dS} \tag{10.129}$$

shows that the divergence of a vector can be interpreted as an inner product between the vector and the small area vector. Here, S denotes a closed surface in the space where the vector function is defined, and V is the volume of the region surrounded by the surface S. Then, by introducing Eqs.10.125 and 10.126,

$$\int_S \hat{\psi}\mathbf{b}\cdot\mathbf{dS} = \int_V (\mathrm{div}\,\hat{\psi}\mathbf{b})dv = \int_V (\hat{\psi}\mathrm{div}\mathbf{b} + \nabla\hat{\psi}\cdot\mathbf{b})dv \tag{10.130}$$

is derived.

Let the vector **b** be

$$\mathbf{b} = \mathrm{grad}\hat{\phi} = \nabla\hat{\phi} \tag{10.131}$$

and thus div$\mathbf{b} = \nabla^2\hat{\phi}$. Then

$$\int_S \hat{\psi}\nabla\hat{\phi}\cdot \mathbf{dS} = \int_V (\hat{\psi}\nabla^2\hat{\phi} + \nabla\hat{\psi}\cdot\nabla\hat{\phi})dv \tag{10.132}$$

can be derived. By changing the variables in the equation above,

$$\int_S \hat{\phi}\nabla\hat{\psi}\cdot \mathbf{dS} = \int_V (\hat{\phi}\nabla^2\hat{\psi} + \nabla\hat{\phi}\cdot\nabla\hat{\psi})dv \tag{10.133}$$

can also be obtained. By subtracting these equations above,

$$\int_S (\hat{\psi}\nabla\hat{\phi} - \hat{\phi}\nabla\hat{\psi})\cdot \mathbf{dS} = \int_V (\hat{\psi}\nabla^2\hat{\phi} - \hat{\phi}\nabla^2\hat{\psi})dv \tag{10.134}$$

or

$$\int_S \left(\hat{\psi}\frac{\partial\hat{\phi}}{\partial n} - \hat{\phi}\frac{\partial\hat{\psi}}{\partial n}\right) dS = \int_V (\hat{\psi}\nabla^2\hat{\phi} - \hat{\phi}\nabla^2\hat{\psi})dv \tag{10.135}$$

is derived. This formula is called **Green's theorem**.

By applying the Green's theorem formulated above to the functions $f(x,y,z)$ and $h(x,y,z)$, which are the solutions of the Helmholtz equation,

$$\begin{aligned}\int_S \left(f\frac{\partial h}{\partial n} - h\frac{\partial f}{\partial n}\right) dS &= \int_V (h\nabla^2 f - f\nabla^2 h)dv \\ &= \int_V (-hk^2 f + fk^2 h)dv = 0;\end{aligned} \tag{10.136}$$

namely,

$$\int_S f\frac{\partial h}{\partial n}dS = \int_S h\frac{\partial f}{\partial n}dS \tag{10.137}$$

holds well where $\partial/\partial n$ represents the normal gradient to the inside of the region.

Now, newly define the functions such that[49]

$$\Phi_{V_R} = f(x,y,z) \tag{10.138}$$

$$h(x,y,z) = \frac{\mathrm{e}^{-ikr}}{r}. \tag{10.139}$$

Set the small sphere with a radius of r_s so that the singularity of the function h representing a spherical wave might be removed as shown in Fig. 10.6. and the function $h(x,y,z)$ follows the Helmholtz equation in the region $\widehat{V}$ outside the small region Σ surrounded by S.

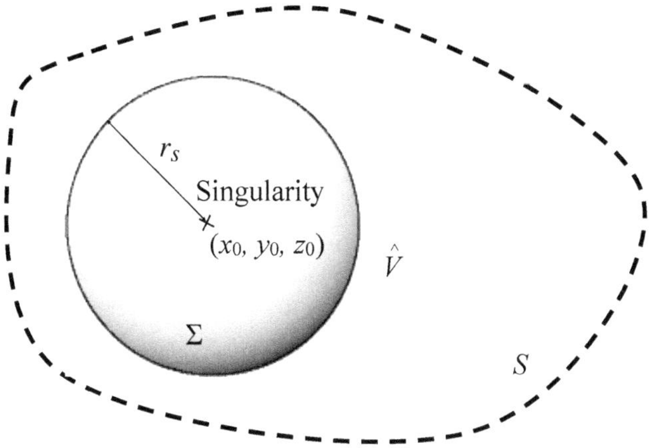

Fig. 10.6 Singularity in closed region

Recall

$$\int_{\hat{S}} \Phi_{\nu_R} \frac{\partial}{\partial n}\left(\frac{\mathrm{e}^{-ikr}}{r}\right) d\hat{S} = \int_{\hat{S}} \frac{\mathrm{e}^{-ikr}}{r} \frac{\partial \Phi_{\nu_R}}{\partial n} d\hat{S}, \tag{10.140}$$

where $\widehat{S}$ denotes $\Sigma + S$ according to the figure. It can be rewritten as

$$\int_{\Sigma} \Phi_{\nu_R} \frac{\partial}{\partial n}\left(\frac{\mathrm{e}^{-ikr}}{r}\right) d\Sigma + \int_{S} \Phi_{\nu_R} \frac{\partial}{\partial n}\left(\frac{\mathrm{e}^{-ikr}}{r}\right) dS$$
$$= \int_{\Sigma} \frac{\mathrm{e}^{-ikr}}{r} \frac{\partial \Phi_{\nu_R}}{\partial n} d\Sigma + \int_{S} \frac{\mathrm{e}^{-ikr}}{r} \frac{\partial \Phi_{\nu_R}}{\partial n} dS, \tag{10.141}$$

where the normal gradient can be set to

$$\frac{\partial}{\partial n}\left(\frac{\mathrm{e}^{-ikr}}{r}\right)\bigg|_{r=r_s} = \frac{\partial}{\partial r}\left(\frac{\mathrm{e}^{-ikr}}{r}\right)\bigg|_{r=r_s}. \tag{10.142}$$

Put

$$d\Sigma = r_s^2 d\Omega \tag{10.143}$$

at the small surface Σ where $d\Omega$ denotes the solid angle. Then, by taking a limit when r_s approaches 0, the integrals on the small surface Σ become

$$\int_{\Sigma} \Phi_{\nu_R} \frac{\partial}{\partial n}\left(\frac{\mathrm{e}^{-ikr}}{r}\right) d\Sigma \rightarrow -4\pi \Phi_{\nu_R}(x_0, y_0, z_0) \tag{10.144}$$

$$\int_{\Sigma} \frac{\mathrm{e}^{-ikr}}{r} \frac{\partial \Phi_{\nu_R}}{\partial n} d\Sigma \rightarrow 0. \tag{10.145}$$

Consequently,

$$\Phi_{v_R}(x_0,y_0,z_0) = -\frac{1}{4\pi}\int_S \frac{e^{-ikr}}{r}\frac{\partial \Phi_{v_R}}{\partial n}dS + \frac{1}{4\pi}\int_S \Phi_{v_R}\frac{\partial}{\partial n}\left(\frac{e^{-ikr}}{r}\right)dS \quad (10.146)$$

can be derived in the limit for $r_s \to 0$, where r is the distance (m) between the singular point (corresponding to the observation point) (x_0,y_0,z_0) in Fig. 10.6 and a point on the boundary S.

On the other hand, by taking the (singular observation) point outside the region, it is not necessary to divide the region in order to remove the singular point. Therefore,

$$\int_S \Phi_{v_R}\frac{\partial}{\partial n}\left(\frac{e^{-ikr}}{r}\right)dS = \int_S \frac{e^{-ikr}}{r}\frac{\partial \Phi_{v_R}}{\partial n}dS \quad (10.147)$$

holds well outside the region, and thus the solution formulated by Eq. 10.146 becomes

$$\Phi_{v_R}(x,y,z) = 0. \quad (10.148)$$

Consequently, by taking an observation point (x,y,z) inside the region, the sound field can be represented by superposition of the direct sound and reflection waves from the boundary such that

$$\begin{aligned}\Phi_v(x,y,z) &= \frac{1}{4\pi}\int_V Q_d(x'y'z')\frac{e^{-ikr_1}}{r_1}dx'dy'dz' \\ &\quad - \frac{1}{4\pi}\int_S \frac{e^{-ikr}}{r}\frac{\partial \Phi_{v_R}}{\partial n}dS + \frac{1}{4\pi}\int_S \Phi_{v_R}\frac{\partial}{\partial n}\left(\frac{e^{-ikr}}{r}\right)dS. \quad (10.149)\end{aligned}$$

10.2.5 Representation of Boundary Conditions

The expression in the previous subsection, represents a sound field in a region by using the values Φ_{v_R} (sound pressure) and $\partial\Phi_{v_R}/\partial n$ (particle velocity) on the boundary by which the region is surrounded. It is an intriguing question whether or not the sound pressure and particle velocity can be independent on the boundary. Take an observation point on the boundary. The next equation

$$\frac{1}{4\pi}\int_S \Phi_{v_R}\frac{\partial}{\partial n}\left(\frac{e^{-ikr}}{r}\right)dS - \Phi_{v_R}(\text{on}S) = \frac{1}{4\pi}\int_S \frac{e^{-ikr}}{r}\frac{\partial \Phi_{v_R}}{\partial n}dS \quad (10.150)$$

must be satisfied by the sound pressure and particle velocity on the boundary. Therefore, it can be found that the sound pressure and particle velocity cannot be independent on the boundary[50]. Here, r denotes the distance between the observation point and other positions on the boundary. It indicates that the sound field in the region might be represented by the sound pressure or particle velocity on the boundary.

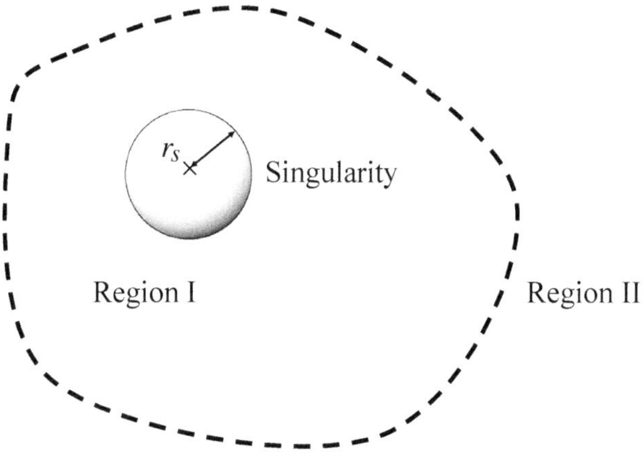

Fig. 10.7 Singularity in region I and region II outside region I

Set region I in which an observation point is located, and region II outside region I following Fig. 10.7. The sound field removing the direct sound in region I can be expressed as

$$\Phi_{vR_{\mathrm{I}}}(x,y,z) = -\frac{1}{4\pi}\int_S \frac{\mathrm{e}^{-ikr}}{r}\frac{\partial \Phi_{vR_{\mathrm{I}}}}{\partial n_{\mathrm{I}}} dS + \frac{1}{4\pi}\int_S \Phi_{vR_{\mathrm{I}}}\frac{\partial}{\partial n_{\mathrm{I}}}\left(\frac{\mathrm{e}^{-ikr}}{r}\right) dS \qquad (10.151)$$

according to the normal gradient toward the inside of region I, where r denotes the distance between the observation point and a position on the boundary. The sound field in region I can also be expressed by using the normal gradient with respect to region II (without including the observation point) as

$$0 = -\frac{1}{4\pi}\int_S \frac{\mathrm{e}^{-ikr}}{r}\frac{\partial \Phi_{vR_{\mathrm{II}}}}{\partial n_{\mathrm{II}}} dS + \frac{1}{4\pi}\int_S \Phi_{vR_{\mathrm{II}}}\frac{\partial}{\partial n_{\mathrm{II}}}\left(\frac{\mathrm{e}^{-ikr}}{r}\right) dS. \qquad (10.152)$$

Namely, by recalling the signs of the normal gradients, the sound field can be rewritten as

$$0 = \frac{1}{4\pi}\int_S \frac{\mathrm{e}^{-ikr}}{r}\frac{\partial \Phi_{vR_{\mathrm{II}}}}{\partial n_{\mathrm{I}}} dS - \frac{1}{4\pi}\int_S \Phi_{vR_{\mathrm{II}}}\frac{\partial}{\partial n_{\mathrm{I}}}\left(\frac{\mathrm{e}^{-ikr}}{r}\right) dS. \qquad (10.153)$$

Then, the sum of Eqs. 10.151 and 10.153 makes

$$\Phi_{vR_{\mathrm{I}}}(x,y,z) = -\frac{1}{4\pi}\int_S \frac{\mathrm{e}^{-ikr}}{r}\Phi_{v_1} dS + \frac{1}{4\pi}\int_S \Phi_{v_2}\frac{\partial}{\partial n_{\mathrm{I}}}\left(\frac{\mathrm{e}^{-ikr}}{r}\right) dS, \qquad (10.154)$$

where

$$\Phi_{v_1} = \frac{\partial \Phi_{vR_{\mathrm{I}}}}{\partial n_{\mathrm{I}}} - \frac{\Phi_{vR_{\mathrm{II}}}}{\partial n_{\mathrm{I}}} \qquad (10.155)$$

$$\Phi_{v_2} = \Phi_{v_{R_{\mathrm{I}}}} - \Phi_{v_{R_{\mathrm{II}}}}. \tag{10.156}$$

It can be interpreted that the sound field in region I is represented by controlling the boundry condition, i.e., only the difference of the particle velocity or sound pressure on the boundary.

Recall the sound field that is made by a point source is expressed as

$$\phi_v(r,t) = \frac{Q_0}{4\pi r} \mathrm{e}^{\mathrm{i}(\omega t - kr)} \tag{10.157}$$

for the velocity potential of the sound field, where r (m) denotes the distance from the point source and $Q_0(\mathrm{m}^3/\mathrm{s})$ shows the volume velocity of the source. On the other hand, suppose that there is a **dipole source** that is composed of a pair of closely located point sources with opposite phases as shown by Fig. 10.8. According to Fig. 10.8, the velocity potential is given by

$$\Phi_v(r) = \frac{-Q_0}{4\pi}\left(\frac{\mathrm{e}^{-ikr}}{r} - \frac{\mathrm{e}^{-ikr_1}}{r_1}\right) \to \frac{-Q_{ss}}{4\pi}\frac{\partial}{\partial z}\left(\frac{\mathrm{e}^{-ikr}}{r}\right), \qquad (\Delta z \to 0) \tag{10.158}$$

where $Q_{ss} = Q_0 \Delta z$ denotes the strength of the dipole source when $\Delta z \to 0$.

Look at the sound field given by integration on the boundary. The boundary S can be interpreted as representing discontinuity of the sound field. That is, assuming discontinuity in the particle velocity on the boundary between both sides, sound field can be expressed as

$$\Phi_{v_R}(x,y,z) = -\frac{1}{4\pi}\int_S \frac{\mathrm{e}^{-ikr}}{r}\Phi_{v_1} dS, \tag{10.159}$$

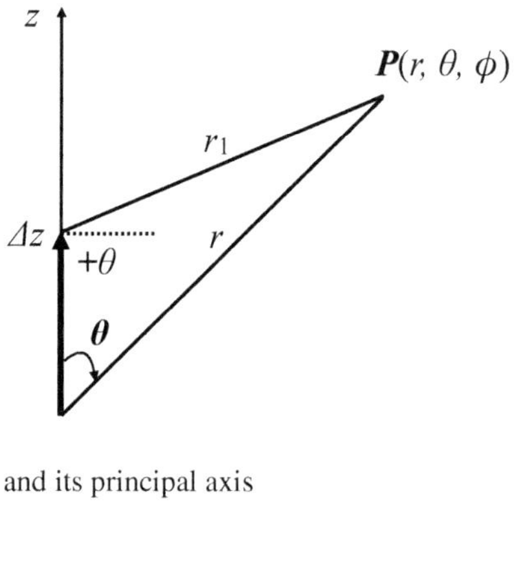

Fig. 10.8 Dipole source and its principal axis

where a distribution of virtual point sources is introduced on the boundary so that there might be continuity in the sound pressure between the two regions. In contrast, supposing a distribution of virtual dipole sources, the sound field can be given by

$$\Phi_{v_R}(x,y,z) = \frac{1}{4\pi}\int_S \Phi_{v_2}\frac{\partial}{\partial n_{\mathrm{I}}}\left(\frac{\mathrm{e}^{-ikr}}{r}\right)dS, \tag{10.160}$$

where the boundary represents the discontinuity of the velocity potential between both sides. The expression can be extended into a form including sinusoidal time factor such that

$$\phi_{v_{AR}}(x,y,z,t) = -\frac{1}{4\pi}\int_S \frac{\mathrm{e}^{\mathrm{i}\omega(t-r/c)}}{r}\Phi_{v_1}dS. \tag{10.161}$$

Chapter 11
Reverberation Sound in Rooms

Sound energy reaches the steady state in a room after a sound source starts to radiate sound in the room. The sound energy, however, decays after the source stops radiating sound. This sound-energy decaying process is called the **reverberation process in a room**. The reverberation process can be represented by an **impulse response** of the sound field in a room. It is possible to deal with the reverberation process on the basis of statistical analysis, assuming the sound reflection process with the boundary is random. Consequently, the correspondence of the geometrical properties to the wave theoretic nature of reverberation sound by a hybrid type of representation for the reverberation decay can be derived. Reverberation decay and its reverberation time for two-dimensional and almost-two-dimensional reverberation fields are formulated as well as the conventional formula for a three-dimensional field. Frequency characteristics due to geometric conditions and arrangement of absorption materials are also presented.

11.1 Sound Energy Density in Room

11.1.1 Sound Energy Balance in Room

When a listener listens to sound from a source, i.e., a musical instrument or a loudspeaker, in a room, sound normally travels from the source to the listener via many reflections at the boundary of the room. Sound energy is partly lost at every collision with the walls. Therefore, if no energy is supplied by the sound source in the room, the energy is consumed and the sound decays with the reverberation.

Suppose that there is a sound source that radiates the sound stationary in a room. Such a sound source can be characterized in terms of the sound power output W_X (W). An increase of the sound energy density $\Delta E_0(\mathrm{W/m^3})$ in a room for a unit time by the source can be expressed by

$$V\Delta E_0 = W_X - A_b, \qquad \text{(W)} \tag{11.1}$$

M. Tohyama: *Sound and Signals*, Signals and Communication Technology, pp. 191–218.
springerlink.com

where $V(\mathrm{m}^3)$ denotes the room volume, $V\Delta E_0$ (W) shows the energy increase in a unit time, $E_0(\mathrm{J/m}^3)$ shows the average energy density in the room, W_X is the sound power output of the source, and A_b (W) is the absorbed energy in a unit time in the room. The equation above shows the energy balance for the energy accumulation in the room, supplied energy from the source, and absorbed energy by the wall. Namely, the energy increase can be estimated by the difference between the energy coming from the source and the energy lost by absorption. It is the same as the relationship between the external source and vibration of a string described in subsection 5.3.3.

11.1.2 Sound Energy at Steady State

The sound energy increases in a room as a sound source radiates sound into the room. However, it reaches the steady state and stops increasing. According to the energy balance equation in the previous subsection, $W_X = A_b$ holds well at the steady state because $\Delta E_0 = 0$. There is a balance between the energy fed into the room by the source and that absorbed at the steady state.

Recall the relationship described in subsection 5.3.3 between the power injection into a vibrating string and the spatial average of a time-average for the squared displacement of the string. As described in the following sections, the absorbed energy for a unit time interval can be estimated by

$$A_b = \alpha \frac{E_{0_{st}} \cdot c}{4} S \qquad \text{(W)} \tag{11.2}$$

where α is the energy absorption coefficient of the wall per unit area, $E_{0_{st}}(\mathrm{J/m}^3)$ shows the energy density at the steady state, $c(\mathrm{m/s})$ denotes the sound speed, and $S(\mathrm{m}^2)$ denotes the area of the wall. This can be interpreted to mean that only a quarter of the intensity inside the room, just as denoted by $E_{0_{st}}c/4(\mathrm{W/m}^2)$, collides with the wall on average.

The energy density reaches

$$E_{0_{st}} = \frac{4W_X}{\alpha c S} \qquad (\mathrm{J/m}^3) \tag{11.3}$$

at the steady state in the room because $A_b = W_X$. The sound energy density increases in proportion to the sound power output of a sound source, while it decreases in inverse proportion to the sound absorption coefficient of the wall and its surface area, and the sound speed. Here, by multiplying the energy density by the volume of the room $V(\mathrm{m}^3)$,

$$VE_{0_{st}} = \frac{W_X}{\alpha \frac{c}{4V/S}} = \frac{W_X}{\alpha \frac{c}{M_{FP_3}}} \qquad \text{(J)} \tag{11.4}$$

or

$$\alpha V E_{0_{st}} \frac{c}{M_{FP_3}} = \alpha V E_{0_{st}} n_{c_3} = W_X \qquad \text{(W)} \tag{11.5}$$

is derived, where $M_{FP_3} = 4V/S$ is called the **mean free path** of a three-dimensional reverberation field, and $n_{c_3} = c/M_{FP_3}$ denotes the **average number of collisions with the walls in a unit time interval**. Namely, the sound energy is lost by $\alpha V E_{0_{st}}$ at every collision with the wall, and thus the consumed energy amounts to $\alpha V E_{0_{st}} n_{c_3}$ in a unit time interval because the number of collisions is given by n_{c_3}. Such consumption of energy must be compensated for by the source so that the sound energy might remain stationary at the steady state.

11.1.3 Energy of Decaying Sound at Reverberation Process

As stated above, the sound energy must be fed into the room by a source in order to keep the energy at the steady state. Suppose that the sound source stops after the sound field reaches the steady state. The energy must decay from that for the steady state because of sound absorption into the walls at every collision between the sound and the walls. This decay of sound after the source stops is called the **reverberation** process.

The energy-decaying equation indicating the reveberation process can be rewritten as

$$V\Delta E_0 = -A_b \qquad \text{(W)} \tag{11.6}$$

or

$$\frac{dE_0}{dt} = -\alpha M_{FP_3} E_0. \qquad \text{(W)} \tag{11.7}$$

Here αM_{FP_3} can be interpreted as the **eigenvalue of the reverberation process**. The energy density in the room becomes

$$E_{0_{Rev}}(t) = \frac{4W_X}{c\alpha S} e^{-\frac{c\alpha S}{4V}t} = \frac{4W_X}{c\alpha S} e^{-2\delta t} \qquad (\mathrm{J/m^3}) \tag{11.8}$$

at the reverberation process, where the decaying factor of the energy is given by

$$2\delta = \frac{c\alpha S}{4V} \qquad (1/\mathrm{s}). \tag{11.9}$$

The decay speed of reverberation becomes slower as the room volume increases, while it becomes faster as the ratio S/V becomes large. Such a decay speed can be characterized by the **reverberation time**, which shows the time interval until the sound energy decreases to $1/10^6$ of that for the steady state after the source stopped. The reverberation time can be expressed as

$$E_{0_{Rev}}(T_{R_3}) = \frac{4W_X}{c\alpha S} e^{-\frac{c\alpha S}{4V}T_{R_3}} = \frac{4W_X}{c\alpha S} 10^{-6}, \qquad (\mathrm{J/m^3}) \tag{11.10}$$

namely,

$$T_{R_3} = 6 \cdot \log_e 10 \frac{4V}{c\alpha S} \cong 13.8 \cdot \frac{4V}{c\alpha S} \cong 0.163 \frac{V}{\alpha S} \qquad (\mathrm{s}) \tag{11.11}$$

where $c = 340$ (m/s). The reverberation time is a historical parameter of room acoustics, but it is a fundamental one for the sound field in rooms.

11.1.4 Sound Field Representation Using Wavenumber Space

The energy balance equation that the sound field follows can be interpreted in terms of wavenumber space[51]. Suppose that the direction of sound propagation is random at every point, and set the random variables that represent the wavenumber components $k_x^2(t), k_y^2(t)$, and $k_z^2(t)$ as functions of time. Consequently, sound energy density E_{0_i} also becomes a random variable at position $i(\mathbf{r})$, and thus the time average of the density can be defined as

$$\overline{E}_{0_i}(k_{x_i}^2(t), k_{y_i}^2(t), k_{z_i}^2(t)) = \lim_{T \to +\infty} \frac{1}{2T} \int_{-T}^{+T} E_{0_i}(k_{x_i}^2(t), k_{y_i}^2(t), k_{z_i}^2(t)) dt, \quad (11.12)$$

where the constraint of the frequency, i.e.,

$$k_{x_i}^2(t) + k_{y_i}^2(t) + k_{z_i}^2(t) = k_0^2, \quad (11.13)$$

is assumed independent of the time and positions, $k_0 = \omega_0 / c$, ω_0 is the angular frequency of the sound waves, and c is the sound speed.

On the basis of the **ergodicity**, which says that the time average can be represented by the ensemble average,

$$\begin{aligned}
&\lim_{T \to +\infty} \frac{1}{2T} \int_{-T}^{+T} E_{0_i}(k_{x_i}^2(t), k_{y_i}^2(t), k_{z_i}^2(t)) dt = \lim_{N \to +\infty} \frac{1}{N} \sum_{i=1}^{N} E_{0_i}(k_{x_i}^2(t), k_{y_i}^2(t), k_{z_i}^2(t)) \\
&= \frac{1}{\int_{k_0^2 > k_x^2 + k_y^2} D(k_x^2, k_y^2) dk_x dk_y} \times \int_{k_0^2 > k_x^2 + k_y^2} E_0(k_x^2, k_y^2) D(k_x^2, k_y^2) dk_x dk_y \\
&= E_0 \quad (11.14)
\end{aligned}$$

holds well independent of the time and positions, where N denotes the number of observation samples and $D(k_x^2, k_y^2)$ shows the density of the samples on the disc, $k_0^2 > k_x^2 + k_y^2$.

Assuming that the density $D(k_x, k_y)$ should be uniform on the disc so that

$$\frac{1}{\int_{k_0^2 > k_x^2 + k_y^2} dk_x dk_y} = \frac{1}{\pi k_0^2} \quad (11.15)$$

holds well, the ensemble average can be rewritten as

$$E_0 =< E_0(k_x^2, k_y^2 >= \frac{1}{\pi k_0^2} \int_{k_0^2 > k_x^2 + k_y^2} E_0(k_x^2, k_y^2) dk_x dk_y. \quad (11.16)$$

This expression can be geometrically interpreted with respect to the sound propagating directions.

By introducing the transformations

$$k_x = k_0 \sin\theta \cos\phi \tag{11.17}$$
$$k_y = k_0 \sin\theta \sin\phi, \tag{11.18}$$

the vector product

$$dk_x dk_y = k_0^2 \cos\theta \sin\theta d\theta d\phi \tag{11.19}$$

is derived between the variables of

$$dk_x = k_0 \cos\theta \cos\phi d\theta - k_0 \sin\theta \sin\phi d\phi \tag{11.20}$$
$$dk_y = k_0 \cos\theta \sin\phi d\theta + k_0 \sin\theta \cos\phi d\phi. \tag{11.21}$$

Consequently, the ensemble average in the wavenumber space can be rewritten as

$$\frac{1}{\pi k_0^2}\int_{k_0^2>k_x^2+k_y^2} E_0(k_x^2,k_y^2)dk_x dk_y = \frac{1}{\pi}\int_0^{2\pi} d\phi \int_0^{\pi/2} E_0(\theta,\phi)\cos\theta\sin\theta d\theta$$
$$=< E_0(k_x^2,k_y^2) >= E_0 = \overline{E_0}_i(k_{x_i}^2(t),k_{y_i}^2(t),k_{z_i}^2(t)). \tag{11.22}$$

It shows the average of energy density for plane waves whose propagation is equally probable in all the directions of the space at any time in the steady state. Thus, the energy flow observed inside the space is given by $E_0 c(\mathrm{W/m^2})$, where c denotes the sound speed.

Now consider the portion of the energy flow of sound waves coming into the boundary from the inside of the space. For that purpose, the density D has to be changed from that stated above. The sound propagates in all the directions in inner space, so the sample points for k_x, k_y, and k_z in the wavenumber space can be equally probable on the surface of the sphere $k_0^2 = k_x^2 + k_y^2 + k_z^2$. Consequently, the portion of the energy colliding with the boundary can be expressed as

$$\frac{1}{\int_{k_0^2=k_x^2+k_y^2+k_z^2} dk_x dk_y dk_z} \times \int_{k_0^2>k_x^2+k_y^2} E_0(k_x^2,k_y^2)dk_x dk_y = \frac{1}{4\pi k_0^2}\int_{k_0^2>k_x^2+k_y^2} E_0(k_x^2,k_y^2)dk_x dk_y$$
$$= \frac{1}{4} < E_0(k_x^2,k_y^2) >= \frac{1}{4}E_0. \tag{11.23}$$

Namely, the energy flow into the boundary by one quarter of that for the inner space.

11.2 Sound Field in Room as Linear System

Sound fields in rooms such as space surrounded by hard walls can be assumed to be linear systems from a signal theoretic view point. According to the properties of linear systems, the response of the sound field to input signals can be formulated based on the **principle of superposition**. Namely, the response of a sound field to a sum of input signals can be estimated by the sum of the response to each input signal.

This type of principle can be formulated by the convolution between an input signal and the impulse response of the field. Here, the **impulse response** is defined as a response of the linear system to a signal defined by $\delta(n)$. Consequently, reverberation sound can be represented as responses to transient signals.

11.2.1 Transient and Steady State Response to Sinusoidal Input Signal

A sinusoidal signal is a basis of signals according to the principle of superposition on which any linear system is based. This is because any ordinary signal referred to as linear acoustics can be represented as superposition of sinusoidal signals following so-called Fourier analysis.

Suppose that a sinusoidal signal is fed to a sound field in a room at $n = 0$ from a sound source. Assuming the impulse response $h(n)$ with the length of N between the sound source and a receiving position, the observed response to the transient sinusoidal signal at the receiving position can be formulated as

$$\begin{aligned} y(n) &= \sum_{m=0}^{n} h(m)\mathrm{e}^{\mathrm{i}\Omega(n-m)} = \left(\sum_{m=0}^{n} h(m)\mathrm{e}^{-\mathrm{i}\Omega m}\right)\mathrm{e}^{\mathrm{i}\Omega n} \\ &= H(\mathrm{e}^{-\mathrm{i}\Omega}, n)\mathrm{e}^{\mathrm{i}\Omega n} \rightarrow H(\mathrm{e}^{-\mathrm{i}\Omega})\mathrm{e}^{\mathrm{i}\Omega n}, \qquad (\text{for} \quad n \geq N-1) \end{aligned} \tag{11.24}$$

where Ω denotes the (normalized) angular frequency of the input sinusoidal signal. This expression says that the response converges for $n \geq N-1$ to a sinusoidal signal with the same frequency as that for the input signal. Therefore, the response that can be observed after $n \geq N-1$ is called the response of the **steady state**. A longer time interval is required to reach the steady state, as the duration is long for the impulse response.

The magnitude and phase for the response to the input sinusoidal signal are different from the input ones. However the magnitude and phase of the corresponding sinusoidal one are given by the Fourier transform of the impulse response. Thus, the Fourier transform of the impulse response, namely,

$$H(\mathrm{e}^{-\mathrm{i}\Omega}, n) = \sum_{m=0}^{n} h(m)\mathrm{e}^{-\mathrm{i}\Omega m}, \tag{11.25}$$

is called the **steady state frequency response** of a linear system when $n \geq N-1$.

On the other hand, if the time n is earlier than $N-1$, then the magnitude and phase depend on the time. Therefore, such an interval before the steady state is called the **transient state**, and the response in the period is called the **transient response** to the input signal. Suppose that the linear system of interest is a single freedom of a resonant system that has a single frequency of free oscillation as stated in Chapter 2. Therefore, take an impulse response of the resonant system as a complex sinusoidal function form written as

$$h(m) = r^m \mathrm{e}^{\mathrm{i}\Omega_0 m}, \tag{11.26}$$

where Ω_0 denotes the (normalized) angular eigenfrequency. The response to a sinusoidal signal can be expressed as

$$y(n) = \sum_{m=0}^{n} r^m e^{i\Omega_0 m} e^{i\Omega(n-m)} = \left(\sum_{m=0}^{n} r^m e^{i(\Omega_0-\Omega)m} \right) e^{i\Omega n}$$

$$= \frac{1}{1-re^{i(\Omega_0-\Omega)}} \left(e^{i\Omega n} - r^{n+1} e^{i(\Omega_0-\Omega)} e^{i\Omega_0 n} \right). \tag{11.27}$$

Here, the first term of the right-hand side shows the response at the steady state, while the second one denotes the transient response.

The transient response in which there is free oscillation of the system is observed, while sinusoidal response to the input signal is obtained in the steady state. However, the transient response decays as the time passes after the input sinusoidal signal builds up, and thus the response reaches the steady state. The denominator that is common for both terms represents the resonant property of the system. When the frequency of the input signal comes close to the resonant frequency of the system, the magnitude of the response becomes its maximum, namely, the magnitude of the denominator takes its minimum.

11.2.2 Reverberant Response to Sinusoidal Signal

The decaying process after the sinusoidal input stops can also be formulated similarly to that in the previous subsection. Suppose that the sinusoidal signal stopped at $n = 0$. The transient response that is observed after $n = 0$ can be described as

$$y(n) = \sum_{m=n}^{N-1} r^m e^{i\Omega_0 m} e^{i\Omega(n-m)} = \left(\sum_{m=n}^{N-1} r^m e^{i(\Omega_0-\Omega)m} \right) e^{i\Omega n}$$

$$= \frac{1}{1-re^{i(\Omega_0-\Omega)}} \left(r^n e^{i\Omega_0 n} - r^N e^{i(\Omega_0-\Omega)N} e^{i\Omega n} \right) \tag{11.28}$$

where the impulse response of the system is assumed to be given by Eq.11.26 in the previous subsection. It is assumed that the response had reached the steady state before the sinusoidal source stopped, and thus the upper bound of summation in the equation above could be $N-1$. The free oscillation of the system can be seen in the first term independent of the frequency of the sinusoidal input, but the magnitude and phase depend on the input-signal frequency. On the other hand, the second term indicates the effect of the sinusoidal input after the signal stopped, but it might be only negligible, subject to the system being in steady state when the sinusoidal source stopped.

The free oscillation of a system can be seen independent of the input signal frequency. However, if the free oscillation is composed of several sinusoidal frequencies, the decay rates could be frequency dependent in general. Actual room reverberation is such a case. The free oscillation that has a frequency close to that for the sinusoidal input can be dominantly observed in the response. Therefore, room reverberation measurements are performed for a wide range of sinusoidal frequencies of input in general.

11.2.3 Reverberation Decay Curves in Rooms

The reverberation process could be interpreted based on the energy balance equation, and the sound decaying process could be formulated in terms of the **impulse response** of the sound field in a room. Suppose that there is a sound source that radiates a signal of $x(S,n)$ at position S, and another position R is taken as a sound receiving point in a room. The sound wave $y(R,n)$ that is observed at the receiving position can be expressed as[52]

$$y(R,n) = x(S,n) * h(S,R,n), \tag{11.29}$$

where $(*)$ represents convolution, and $h(S,R,n)$ denotes the impulse response from the positions S to R. The **impulse response** $h(S,R,n)$ specifies the sound waves received at the position R when a point source is located at S and an impulse signal $\delta(n)$ is radiated from the source. Here, the impulse signal $\delta(n)$ means that it takes unity for $n = 0$, but all the samples are zeros otherwise.

Again consider the sound field after the sound source stops, i.e., take a reverberation process in a room. Assuming that the sound source stopped at $n = 0$, then the receiving sound after $n = 0$ can be formulated as

$$y(R,n) = x(S,n) * h(S,R,n) = \sum_{m=n}^{N-1} x(S,n-m)h(S,R,m), \tag{11.30}$$

using **convolution**. The reverberation process is highly sensitive to the sound frequency of a source or locations of source and observation points in a room. Therefore, averaging procedures are necessary, in general, to obtain a stable parameter representing room acoustics from the reverberation process. One way to obtain such a stable and reproducible parameter is to take the ensemble average for the energy decay responses to random-noise signals.

Suppose that the source radiates **white noise**. By taking the ensemble average for the squared responses of $y(R,n)$ subject to the impulse response being time-invariant,

$$\begin{aligned} \mathrm{E}[y^2(R,n)] &= \sum_{l=n}^{N-1}\sum_{m=n}^{N-1} \mathrm{E}[x(S,n-m)x(S,n-l)]h(S,R,m)h(S,R,l) \qquad (11.31) \\ &= C\sum_{i=n}^{N-1} h^2(S,R,i) \end{aligned}$$

is obtained, where

$$\begin{aligned} \mathrm{E}[x(m)x(l)] &= 0 \qquad m \neq l \\ &= C \qquad m = l. \end{aligned} \tag{11.32}$$

Here, $E[x(m)x(l)]$ is called the **auto correlation** of a signal $x(n)$, **white noise** is a random noise whose auto-correlation sequence is given by $\delta(n)$, and $E[X]$ denotes the **expectation** of a random variable X.

The ensemble average of squared responses at a receiving position in a room after a white-noise source stops can be formulated by integration (or summation) of the squared impulse response between the source and receiving positions. This was already introduced for vibration of a string in subsection 5.5.5, and it was called the **reverberation decay curve**. In any case, the reverberation process or reverberation sound is quite significant for investigating perception of sound in rooms.

11.3 Reflection Sound from Mirror Image Sources

11.3.1 Mirror Image Sources in Rectangular Room

Figure 11.1 is an arrangement of the mirror image sources for a rectangular room surrounded by hard walls. Suppose that there is an impulsive source at the center of a rectangular room, and observe a history of reflection sound coming back to the source position after the impulsive signal explodes. The number of reflection waves coming back to the source position can be estimated by using the number of mirror images contained in a sphere with the radius of ct (m) as shown in Fig. 11.2, where $c(\mathrm{m/s})$ denotes the sound speed.

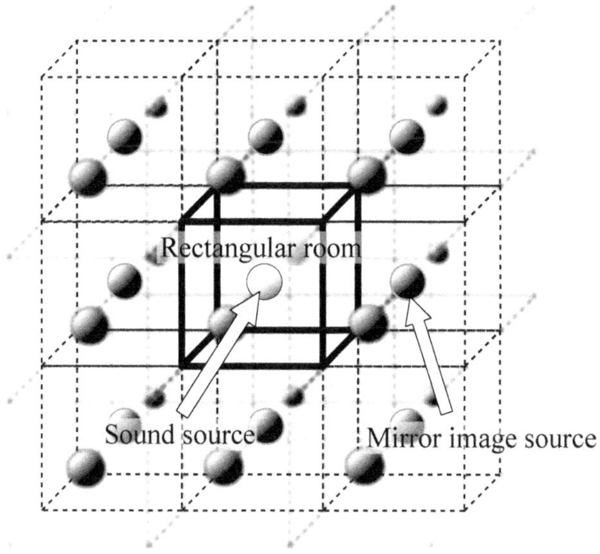

Fig. 11.1 Mirror image sources of rectangular room

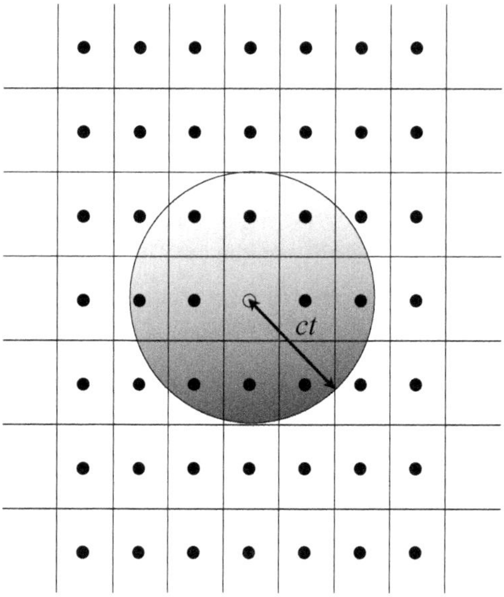

Fig. 11.2 Two-dimensional view of mirror image sources for rectangular room

The **number of reflection waves** can be estimated by recalling the arrangement of mirror image sources. Suppose that $V\,(\mathrm{m}^3)$ is the volume of the room. The number of mirror image sources from which reflection waves come back to the original sound source within t (s) after the direction sound is given by

$$N_{ims_3}(t) \cong \frac{4\pi(ct)^3}{3V}. \tag{11.33}$$

Therefore, if the reflection sound coming back in a unit time interval, namely, the density of reflection sound, is defined, then

$$n_{ims_3}(t) = \frac{dN_{ims}(t)}{dt} \cong \frac{4\pi c^3 t^2}{V} \tag{11.34}$$

is derived on the basis of the short-time average. The density of reflection sound in rooms decreases in inverse proportion to the room volume but is proportional to the square of the time.

11.3.2 Collision Frequency of Reflection Sound

The reverberation process can be interpreted as a history of collisions of sound with the walls surrounding a room. Now consider the number of collisions within t (s) after an impulsive sound starts at a sound source located at the center of the rectangular room whose sides are L_x, L_y, L_z (m) and volume is $V\,(\mathrm{m}^3)$. Suppose a rectangular room that is sketched as shown in Fig. 11.3 in the coordinate system of (r, θ, ϕ). The number of collisions at the x-walls perpendicular to the x-axis can be estimated by

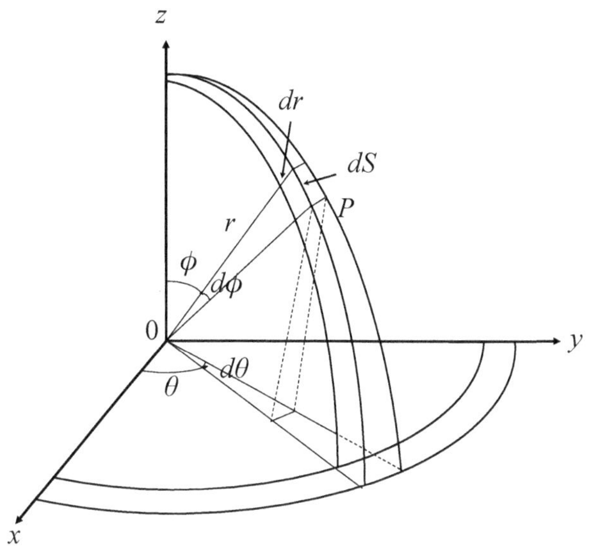

Fig. 11.3 Coordinate system for rectangular room with mirror image sources

$$N_{c_x}(\theta,\phi,t) = \frac{ct\sin\phi\cos\theta}{L_x} \tag{11.35}$$

for sound that comes back from the mirror image source located at $(r = ct, \theta, \phi)$. Similarly, those for the y- and z-walls can be obtained by

$$N_{c_y}(\theta,\phi,t) = \frac{ct\sin\phi\sin\theta}{L_y}, \qquad N_{c_z}(\theta,\phi,t) = \frac{ct\cos\phi}{L_z}, \tag{11.36}$$

respectively. The number of collisions depends on the locations of mirror image sources.

11.3.3 Mean Free Path for Reflection Sound

Take an average of the number of collisions for the reflection sound coming back to the original source position at the center of the room within t and $t+\Delta t$ (s). For that purpose, define the density for the distribution of mirror image sources as

$$w_{ims_3}(\theta,\phi) = \frac{dS}{4\pi(ct)^2} = \frac{\sin\phi d\phi d\theta}{\pi/2}, \tag{11.37}$$

assuming that the mirror image sources are densely distributed according to Fig. 11.3. The average of collisions at the walls is given by[53][54]

$$\begin{aligned}N_{c_3}(t)\\&=\frac{2}{\pi}\int_0^{\pi/2}d\theta\int_0^{\pi/2}[N_{c_x}(\theta,\phi,t)+N_{c_y}(\theta,\phi,t)+N_{c_z}(\theta,\phi,t)]\sin\phi d\phi\\&=\frac{ct}{M_{FP_3}},\end{aligned} \tag{11.38}$$

where

$$M_{FP_3}=\frac{4V}{S}=\frac{4L_xL_yL_z}{2(L_xL_y+L_yL_z+L_zL_x)} \tag{11.39}$$

is called the **mean free path (m) of reflection sound**. The mean free path, which can be defined by the room volume $V(\mathrm{m}^3)$ and the surface area of the room $S(\mathrm{m}^2)$, gives a length of a sound path for travelling without collisions with the walls. The formula of the mean free path that assumes a rectangular room can also be applied to a differently shaped room[54].

11.4 Reverberation Time Formulae

11.4.1 Three-Dimensional Reverberation Sound Field

The reverberation time formula was derived according to the energy balance equation in subsection 11.1.3. It can also be derived based on the mirror-image theory. Suppose that the sound power output of a source is W_0 (W), $N_{c_3}(t)$ denotes the number of collisions for the sound coming back when t (s) has passed after the impulse sound was radiated from the source, and α_3 is the averaged sound absorption coefficient of the walls. The energy flow density $I_3(t)(\mathrm{W/m^2})$ can be expressed as

$$I_3(t)=\frac{W_0(1-\alpha_3)^{N_{c3}(t)}}{4\pi(ct)^2}n_{ims_3}(t)\Delta t\cong\frac{W_0\Delta t(1-\alpha_3)^{N_{c3}(t)}c}{V}. \qquad (\mathrm{W/m^2}) \tag{11.40}$$

The sound energy coming from the mirror images decreases in inverse proportion to the square of the distance from the mirror images because of the nature of the spherical wave, but the increase in the density of mirror images always compensates for the decrease in the energy density. This is related to the so-called Albers' paradox[55]. If the universe could be infinitely extended, the light from the infinitely distributed stars in the universe might brightly shine on the earth even in the night. However, this is not the case for our universe.

In the acoustics in our real world, the sound energy decreases due to the collisions with the walls. By taking the 10-base-logarithm of the equation above,

$$\log I_3(t)\cong\log\left(W_0\Delta t\frac{c}{V}\right)+N_{c_3}(t)\log(1-\alpha_3) \tag{11.41}$$

is derived. By substituting the averaged number of collisions given by Eq.11.38 for $N_{c_3}(t)$,

$$\log I_3(t) \cong \log\left(W_0 \Delta t \frac{c}{V}\right) + \frac{ct}{M_{FP_3}} \log(1-\alpha_3) \tag{11.42}$$

is obtained. The reverberation time T_{R_3}(s),

$$T_{R_3} = \frac{6}{-\log(1-\alpha_3)}\frac{M_{FP_3}}{c} = \frac{\ln 10^6}{c}\frac{M_{FP_3}}{-\ln(1-\alpha_3)} \cong \frac{13.8}{c}\frac{M_{FP_3}}{-\ln(1-\alpha_3)} \cong 0.163\frac{V}{A_{b_3}}, \tag{11.43}$$

is derived where the sound speed c is assumed to be 340 (m/s), ln shows the natural logarithm (e-base-logarithm), and

$$A_{b_3} = -\ln(1-\alpha_3)S \cong \alpha_3 S \qquad (\mathrm{m}^2) \tag{11.44}$$

is called the **equivalent sound absorption area**.

The reverberation formula above is called **Eyring's reverberation formula**, while Eq. 11.11 is named **Sabine's reverberation formula**. These formulae can be applied to a room other than the rectangular room where Eyring's formula was derived in this section. The energy flow density can be estimated at the steady state as

$$I_{3_{st}} = \int_0^\infty \frac{W_0 c dt}{V}(1-\alpha_3)^{\frac{cSt}{4V}} = \frac{4W_0}{[-\ln(1-\alpha_3)]S}. \qquad (\mathrm{W/m}^2) \tag{11.45}$$

It increases in proportion to the sound-power output W_0 (W) of a source, but it decreases as the sound absorption α_3 increases.

11.4.2 *Initial Decay Rate of Reverberation Energy*

As described above, sound energy density in the reverberation process can be written as

$$E_{0_{Rev}}(t) = \frac{W_0}{2\delta V}\mathrm{e}^{-2\delta t}, \qquad (\mathrm{J/m}^3) \tag{11.46}$$

where

$$2\delta = \frac{cA_{b3}}{4V} \qquad (1/\mathrm{s}) \tag{11.47}$$

and the energy density is independent of the room volume $V(\mathrm{m}^3)$. Now take the first derivative of the decaying energy. Then the initial decay rate (speed of decay) is given by[56]

$$-V\frac{d}{dt}E_{0_{Rev}}(t)\bigg|_{t=0} = W_0 \qquad (\mathrm{W}). \tag{11.48}$$

The initial decay rate that is given by the speed of decay at the start of the decay curve shows the energy balance equation in the sound field. Although the energy density is independent of the room volume at the steady state, the initial decay rate depends on the room volume itself independent of the sound absorption of the room. Thus, the initial decay rate might be an important spatial factor for a sound field.

Similarly, take the second derivative of the decay curve, then the relationship between the speed and acceleration of the decay curve, i.e.,

$$-\frac{\frac{d^2}{dt^2}E_{0_{Rev}}(t)}{\frac{d}{dt}E_{0_{Rev}}(t)}=2\delta, \qquad (1/\mathrm{s}) \tag{11.49}$$

can be derived, where the decaying factor is given by the ratio of the second and first derivatives of the decay curve.

11.4.3 Energy Ratio of Direct and Reverberation Sound

Spatial impression by sound in space depends on several attributes of the reverberation of the space. In particular, the **ratio of the direct and reverberation sound** is a classical but important factor related to the subjective impression. Suppose that there is a point source whose sound power output is W_0 (W). **Energy flow density of the direct sound**, $I_D(\mathrm{W/m^2})$, is given by

$$I_D=\frac{W_0}{4\pi r^2}, \tag{11.50}$$

where r (m) denotes the distance between the source and receiving position. The energy ratio D_{R_0} can be written as

$$D_{R_0}=\frac{I_D}{I_R}=\frac{[-\ln(1-\alpha_3)]S}{16\pi r^2}. \tag{11.51}$$

It depends on the sound absorption coefficient of the walls, subject to the distance being constant. It is called the **critical distance of the sound field** where the energy ratio becomes unity. Namely, the critical distance r_c(m) is given by

$$r_c=\sqrt{\frac{[-\ln(1-\alpha_3)]S}{16\pi}}. \tag{11.52}$$

It can be approximately estimated as 80 cm in a room where reverberation time is 1 s and volume is 200 (m^3).

As stated before, the ratio between the direct and reverberation sound is an important factor for spatial impression of the sound field. However, from a closer to perceptual view point, a subjective ratio between the two kinds of energy was introduced to room acoustics in the past. It is called **definition**, **deutlichkeit**, or $D_{R_{50}}$ [57],[58]. This is based on the empirical results that reflective sound which arrives at a receiving position within around 50 (or 30) ms after the direct sound cannot be separately perceived but only reinforces the energy of the direct sound. According to these results, the energy density of subjective direct sound $I_D(\mathrm{J/m^3})$ can be expressed as[59].

$$E_{0_{50}}=\frac{W_0}{4\pi r^2}\frac{1}{c}+\frac{W_0(1-\alpha_3)}{V}\int_0^{0.05}\mathrm{e}^{-\frac{c\alpha_3 S}{4V}t}dt, \tag{11.53}$$

where W_0 (W) denotes sound power output of the point source, r (m) is the distance between the source and listening position, c (m/s) is the sound speed, ($\alpha_3 \cong -\ln(1-\alpha_3)$) is the sound absorption coefficient averaged over the walls, $V(\mathrm{m}^3)$ is the room volume, and $S(\mathrm{m}^2)$ is the surface area of the room. Here, the first term represents energy density of the direct sound that propagates to the receiving position on a spherical wave, while the second one gives the energy of the initial reflection sound, which does not include the direct sound, but arrives at the receiving point within 50 ms after the direct sound. No interference of direct and reflection sound is assumed and thus the total energy is only superposed.

Similarly, the total energy density of sound $E_0(\mathrm{J/m}^3)$, which is composed of the direct and all the reflection sound, is given by

$$E_0 = \frac{W_0}{4\pi r^2}\frac{1}{c} + \frac{W_0(1-\alpha_3)}{V}\int_0^{\infty} \mathrm{e}^{-\frac{c\alpha_3 S}{4V}t}dt. \tag{11.54}$$

Therefore, the subjective ratio $D_{R_{50}}$ becomes[59]

$$D_{R_{50}} = \frac{E_{0_{50}}}{E_0} = \frac{N}{D}, \tag{11.55}$$

where

$$N = 1 - \exp(-0.692/T_R) + \frac{\alpha_3 S}{16\pi r^2(1-\alpha_3)} \tag{11.56}$$

$$D = 1 + \frac{\alpha_3 S}{16\pi r^2(1-\alpha_3)}. \tag{11.57}$$

The ratio $D_{R_{50}}$ or $D_{R_{30}}$ is a function of the distance from the source, and it becomes less as the distance increases. It indicates that speech intelligibility, a representative parameter of speech perception in rooms, decreases as a listener moves away from the source. Figure 11.4 illustrates an example of $D_{R_{30}}$ in a reverberation room[16]. It can be seen that $D_{R_{30}}$ rapidly decreases as the distance exceeds the critical distance of the sound space. In contrast, free-field-like properties free from reverberation can be expected even in a reverberation room in the area very close to the sound source. The sound field within the critical distance is often called the **coherent field** of the sound field[60][61].

11.4.4 Two-Dimensional Reverberation Field

As the sound field in a rectangular room indicates, there is a **two-dimensional reverberation field** that is composed of "tangential" waves as well as a three-dimensional field composed of oblique waves. The reverberation field in a long tunnel is a good example of the two-dimensional reverberation field[47]. If the conventional reverberation formula is applied to the sound field in a tunnel, it is almost impossible to estimate a long reverberation in a tunnel. Such a reverberation sound field can be represented according to the two-dimensional reverberation theory.

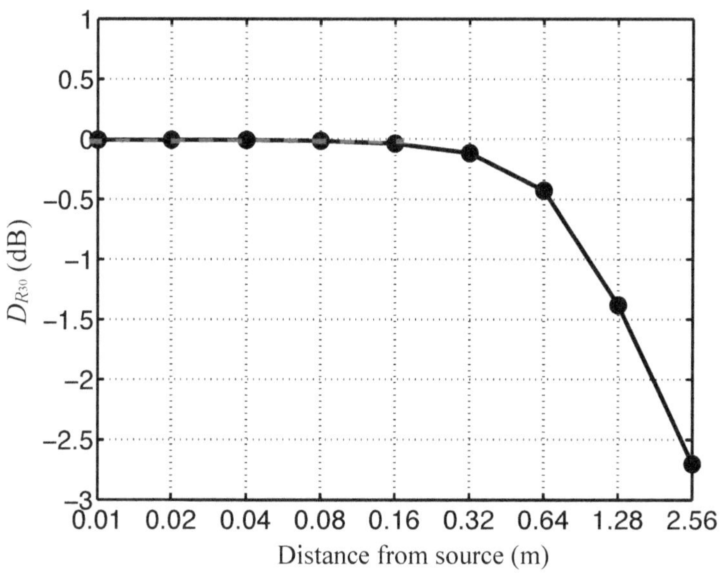

Fig. 11.4 Example of D in reverberation room as function of distance from source

The density of mirror image sources on a two-dimensional array can be written as

$$n_{ims_2}(t)dt \cong \frac{\pi c^2}{S_{2D}} 2t dt, \tag{11.58}$$

where $S_{2D}(\mathrm{m}^2)$ denotes the area inside the room, which increases in linear proportion to the time. Thus, the power output of the mirror sources becomes

$$W_{ims_2}(t) = W_0 \frac{2\pi c^2 t}{S_{2D}} dt, \qquad (\mathrm{W}) \tag{11.59}$$

where W_0 is the sound power output of a single mirror image source. Consequently, the energy flow density $I_2(t)(\mathrm{W/m^2})$ due to reflection sound from the mirror image sources in a time interval Δt is given by

$$I_2(t) = \frac{W_0}{2S_{2D}t}(1-\alpha_2)^{N_{c2}(t)}, \tag{11.60}$$

where $N_{c_2}(t)$ denotes an average of the number of collisions at the two-dimensional boundary and α_2 is the averaged absorption coefficient of the walls.

Similar to in subsection 11.3.2, consider the collision frequency of sound waves in a two-dimensional field. The number of collisions for the waves from the mirror images located at position (ct, θ) can be written as

$$N_{c_x} \cong \frac{ct\cos\theta}{L_x}, \qquad N_{c_y} \cong \frac{ct\sin\theta}{L_y} \tag{11.61}$$

for x-walls (perpendicular to the x-axis), and y-walls (perpendicular to the y-axis). By introducing the density of image sources as

$$w_{ims_2}(\theta) = \frac{d\theta}{2\pi}, \tag{11.62}$$

which corresponds to Eq.11.37 as defined for that in a three-dimensional field, an average for the collision frequency becomes

$$N_{c_2}(t) = \frac{2}{\pi}\int_0^{\pi/2}(N_{c_x}+N_{c_y})d\theta = \frac{ct}{M_{FP_2}} = \frac{ct}{\pi S_{2D}/L_{2D}}. \tag{11.63}$$

Here, M_{FP_2}(m), i.e.,

$$M_{FP_2} = \frac{\pi S_{2D}}{L_{2D}} = \frac{\pi L_x L_y}{2(L_x+L_y)}, \tag{11.64}$$

is called the **mean free path for a two-dimensional sound field**, where

$$S_{2D} = L_x L_y \quad (\mathrm{m}^2), \qquad L_{2D} = 2(L_x+L_y) \quad (\mathrm{m}). \tag{11.65}$$

According to the collision frequency, the energy flow can be rewritten for the two-dimensional field as

$$\log I_2(t) \cong \log\frac{W_0}{2S_{2D}} + \frac{L_{2D}ct}{\pi S_{2D}}\log(1-\alpha_2) - \log t, \tag{11.66}$$

which shows the steep slope of the initial energy decay for the two-dimensional field. The energy decay, however, approaches

$$\log I_2(t) \cong \frac{L_{2D}ct}{\pi S_{2D}}\log(1-\alpha_2) \tag{11.67}$$

as the reverberation process goes on, namely, ($t \rightarrow$ large). Consequently, the reverberation time T_{R_2}(s) can be estimated as

$$T_{R_2} \cong \frac{13.8}{c}\frac{M_{FP_2}}{-\ln(1-\alpha_2)} \cong \frac{0.128 S_{2D}}{-\ln(1-\alpha_2)L_{2D}} = 0.128\frac{S_{2D}}{A_{b_2}} \tag{11.68}$$

for the two-dimensional field, where $A_{b_2} = -\ln(1-\alpha_2)L_{2D}$. The formula above is called the **reverberation time for a two-dimensional field**.

11.4.5 Reverberation Time in Almost-Two-Dimensional Reverberation Field

Figure 11.5 shows examples of reverberation time measured in a special room rather than an ordinary rectangular room. This room has a lot of scattering obstacles and was designed as a reverberation variable room. In some configurations, the acoustic

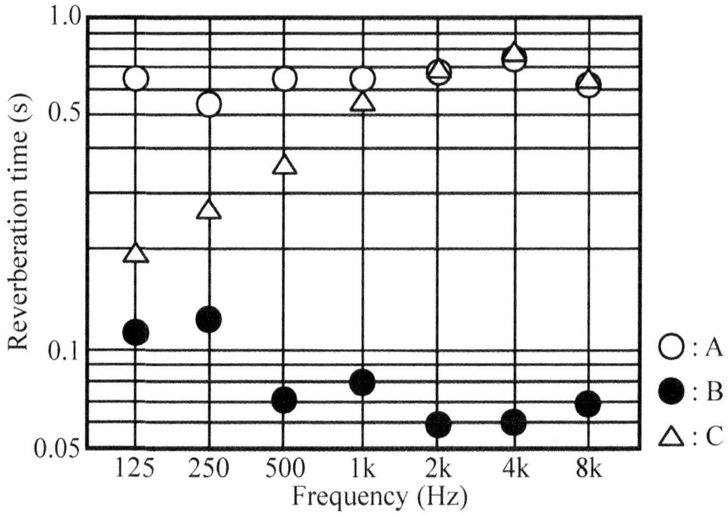

Fig. 11.5 Example of frequency characteristics of reverberation time in reverberation variable room under three room conditions, A(open circle): walls reflective, floor reflective (plastic tile), ceiling reflective (movable parts closed); B(solid circle): walls absorptive, floor absorptive (absorbing mat and carpet), ceiling absorptive (movable parts opened); C(triangle): walls reflective, floor absorptive (absorbing mat and carpet), ceiling absorptive (movable parts opened); from [62](Fig.2)

field in the room can be modeled by using **almost-two-dimensional reverberation** field theory[62].

Figure 11.6 shows the vertical (a) and horizontal (b) aspects of the room. The acoustic and geometric parameters are given in (c) and (d). The side walls are constructed of 32 rotatable cylinders. One half of the surface area of each cylinder is covered by sound-absorbing materials, and the other half is covered by reflective materials. The acoustic conditions in the room can be varied by removing or rotating the wall elements so that the reverberation time can be changed.

Figure 11.5 illustrates measurement results for the reverberation time under three different room conditions. In condition A, all the walls, the floor, and the ceiling are reflective. In condition B, all the walls, the floor, and the ceiling are absorptive. The reverberation time under condition C has a different frequency characteristics from those under condition A and B. In condition C, all the side walls are reflective as in condition A; however, the floor and ceiling are absorptive as in condition B. The reverberation time lengthens as the frequency increases, and above 1 kHz it reaches that for condition A. It is evident that the reverberation-time frequency characteristics under condition C cannot be predicted by using conventional reverberation theory.

The average number of reflections that a sound wave undergoes at the walls depends on the direction cosine of the sound wave. The average number of reflections of an oblique wave in a rectangular room N_{c_3} is given by

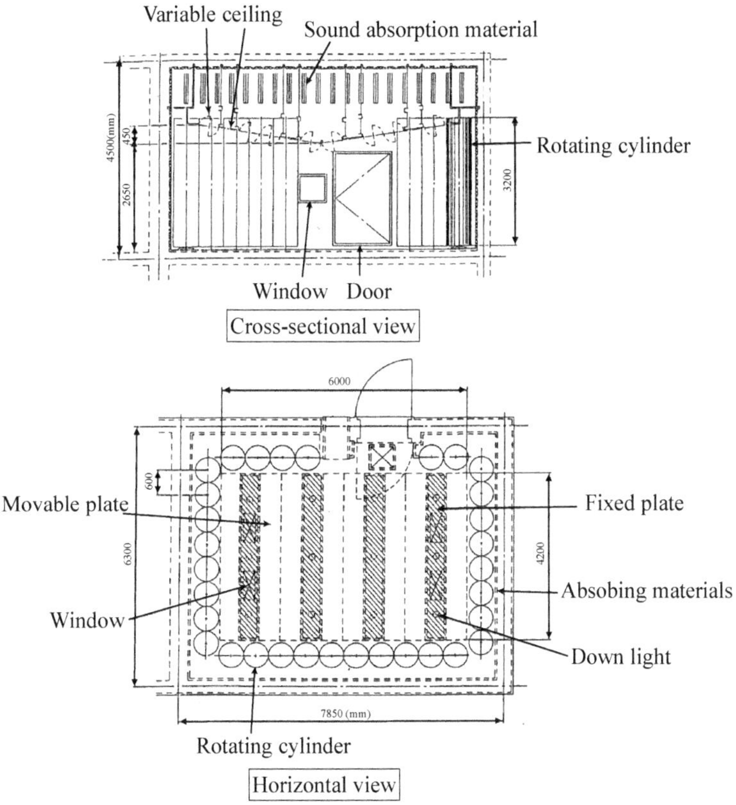

Averaged absorption coefficients for reflective walls

f (Hz)	Floor	$\bar{\alpha}$ (Ceiling)	$\bar{\alpha}$ (Side walls)
125	0.01	0.21	0.19
250	0.02	0.15	0.26
500	0.02	0.10	0.23
1000	0.03	0.09	0.23
2000	0.03	0.14	0.20
4000	0.03	0.12	0.18
8000	0.01	0.12	0.24

Geometical conditions

Floor	S_{xy}	25.2m^3 ($= L_x L_y$)
Circumference	L_{xy}	20.4m ($= 2(L_x+L_y)$)
Length	L_x	6.0m
Width	L_y	4.2m
Height	L_z	2.875m
Volume	V	72.45m^3 ($= L_x L_y L_z$)

Fig. 11.6 Cross-sectional(a) and horizontal(b) views, acoustical(c) and geometrical(d) properties for variable-reverberation room[62](Fig. 1 and Table 1)

$$N_{c_3} = \frac{c}{V}(L_xL_y|dcos_x| + L_yL_z|dcos_y| + L_zL_x|dcos_z|) = N_{c_x} + N_{c_y} + N_{c_z} \quad (1/\text{s})(11.69)$$

where L_x, L_y, and L_z are the respective lengths of the side walls of the rectangular room (m), V is the volume of the room (m^3), c is the sound speed in air (m/s), $dcos_x, dcos_y$, and $dcos_z$ are direction cosines, $dcos_x = k_x/k_0, docs_y = k_y/k_0$, and $dcos_z = k_z/k_0$, k_0 is the wave number of the oblique wave, ν_0 is the frequency of the oblique wave, $k_x^2 + k_y^2 + k_z^2 = k_0^2$, the spatial distribution of the oblique wave is expressed as $P(x,y,z) = \cos(k_x x)\cos(k_y y)\cos(k_z z)$, and N_{c_x}, N_{c_y}, and N_{c_z} are the respective average numbers of reflections that the sound wave undergoes at the x-, y-, and z-walls respectively.

The average number for z-walls N_{c_z} is given by

$$N_{c_z} = \left(\frac{c}{L_z}\right)\left(\frac{k_z}{k_0}\right). \qquad (1/\text{s}) \qquad (11.70)$$

Thus, when $N_{c_z} = n_c$, the z-component of the wave number becomes k_{n_c}, i.e.,

$$k_{n_c} = \frac{k_0 n_c L_z}{c}. \qquad (1/\text{m}) \qquad (11.71)$$

Then, by introducing

$$k_{n_c} = \frac{n\pi}{L_z}, \qquad (1/\text{m}) \qquad (11.72)$$

the equation

$$n = 2\left(\frac{L_z}{c}\right)^2 \nu_0 n_c \qquad (11.73)$$

can be obtained.

Consider the average number of reflections at the z-walls and that at the other side walls. The number of reflections at the z-walls is neglected in the conventional two-dimensional-reverberation theory. An **almost-two-dimensional reverberation theory** can be introduced in which reflections at the z-walls are taken into account. This field is assumed to be composed of both tangential and oblique waves that are close to tangential waves, called here **almost-tangential waves**. The ratio $Nc_{z/xy}$ of the average number of reflections at the z-walls to that at the other side walls for the almost-xy-two-dimensional reverberation field is needed to calculate the frequency characteristics of the reverberation time.

Suppose that the sound field contains the almost-xy-tangential waves whose z-components of the wavenumbers are positive or negative. The ratio defined above can be estimated by

$$Nc_{z/xy} \leq \frac{2n_c}{n \cdot c/M_{FP_{xy}}} = \frac{M_{FP_{xy}} c}{L_z^2 \nu_0}, \qquad (11.74)$$

where the number of reflections for the almost-xy-waves at the z-walls is $2n_c$ at most respectively, and $M_{FP_{xy}}$ denotes the mean free path for the xy-two-dimensional space. Namely, the number of reflections at the side walls is $c/M_{FP_{xy}}$ for a unit time interval respectively. This is a newly revised formula from the previous one in reference[62]. It is the same as that derived according to the geometrical acoustics in reference[47]. Consequently, by setting the ratio of the number of reflections at the z-walls and the total number of reflections,

$$N_{Ca2} \cong N_{C_{z/all}} = \frac{N_{c_z}}{N_{c_z} + N_{c_{xy}}} = \frac{N_{C_{z/xy}}}{1 + N_{C_{z/xy}}} \tag{11.75}$$

can be derived. Therefore, the averaged sound absorption coefficient for the almost-two-dimensional field can be estimated by

$$\alpha_{a2} = \alpha_{xy}(1 - N_{Ca2}) + \alpha_z N_{Ca2}, \tag{11.76}$$

where α_{xy} is the averaged absorption coefficient for the side walls and α_z shows that for the floor and ceiling. According to the absorption coefficient, the reverberation time for the almost-two-dimensional field can be estimated as

$$T_{R_{a2}} \cong \frac{13.8}{c} \frac{\pi S_{2D}}{-\ln(1 - \alpha_{a2}) L_{2D}} \quad \text{(s)} \tag{11.77}$$

following the reverberation formula for the two-dimensional reverberation field. Here, $S_{2D}(\mathrm{m}^2)$ represents the area of the floor, and L_{a2} (m) is the length of the circumference of the field.

The reverberation formula contains frequency as a parameter, and therefore the freguency characteristics of the reverberation time can be derived. The frequency characteristics described here are not those due to the freguency characteristics of the absorption coefficients of materials used in a room; that is, if there were no freguency dependency of the absorption coefficients of the materials, frequency characteristics of the reverberation time may exist and can be derived because of the almost-two-dimensional reverberation character of the field.

The ratio of the absorption coefficient of the z-walls to that of the other side walls in the total averaged absorption coefficient is determined by N_{Ca2}. As the frequency increases, N_{Ca2} decreases; conversely, when the frequency decreases, N_{Ca2} increases. For high frequency bands, the height of the side walls surrounding the side walls of a two-dimensional field is acoustically larger so the absorption of the z-walls is not significant. For low frequency bands, the height of the side walls is acoustically shorter, and the absorption of the z-walls becomes significant.

Figure 11.7 shows the reverberation time as calculated and measured under condition C. The results calculated from the almost-two-dimensional reverberation theory where $\alpha_z = 1$, displayed by the plots of inverse triangle, are for the most part in good agreement with the measured results. The data displayed here were revised from those in reference[62]. The frequency characteristics of reverberation time cannot be calculated by using conventional three-dimensional reverberation theory

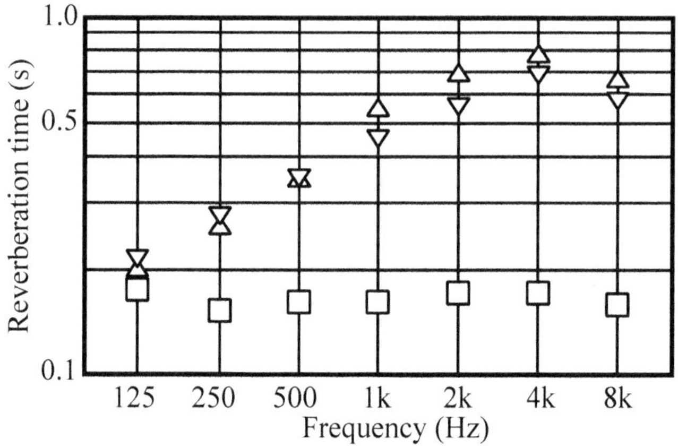

Fig. 11.7 Estimation of reverberation time in almost-two-dimensional reverberation field, measured: triangle, estimated by almost-two-dimensional reverberation theory (assuming $\alpha_z = 1$): inverse triangle, estimated by three-dimensional reverberation theory (assuming $\alpha_z = 1$): square from [62] (Fig.3)

plotted by square plots in the figure. It can be seen that the calculated frequency characteristics seem to be of the three-dimensional theory at the low frequency bands, and they reach that of the two-dimensional one as the frequency increases.

11.4.6 One-Dimensional Reverberation Field

Similar to the two-dimensional array of mirror image sources, which emanate spherical waves representing reflection sound, a **one-dimensional array of the image sources**, namely **one-dimensional reverberation field** is also possible for a sound field partitioned by parallel walls perpendicular to the floor. The reverberation decay curve can be written as

$$\log I_1(t) = \log\frac{W_0}{4\pi L_{1D}c} + \frac{ct}{L_{1D}}\log(1-\alpha_1) - \log t^2. \qquad (11.78)$$

Namely, the reverberation time that is estimated from the later stable part of the decay curve can be written as

$$T_{R_1} \cong \frac{13.8}{c}\frac{M_{FP_1}}{-\ln(1-\alpha_1)} \cong \frac{0.041L_{1D}}{-\ln(1-\alpha_1)} = 0.041\frac{L_{1D}}{A_{b_1}}, \qquad \text{(s)} \qquad (11.79)$$

where $M_{FP_1} = L_{1D}$ denotes the mean free path for the sound field, L_{1D} is the length between the two parallel walls, and α_1 is the averaged absorption coefficient of the walls. As well as in the two-dimensional field, a very steep slope can be seen in the initial decay portion independent of the absorption coefficient.

11.5 Modal Theoretic Analysis for Reverberation Process

An approximate formula of the spacc and ensemble averaged decay curve in a room is introduced in this section[63]. It is written as the summation of the decay curves in three types (oblique, tangential, and axial waves) of reverberation field. It was derived following modal theoretic analysis, but it is a kind of hybrid representation of both geometrical acoustic parameters such as the absorption coefficients with the mean free path and the wave theoretic parameters such as the modal density. Therefore, it can be applied to the sound field in rooms other than rectangular rooms, and it might be useful for estimating the averaged behavior of the sound field that is needed in acoustic measurements in rooms.

11.5.1 Superposition of Free Oscillation in Reverberation Decay

An impulse response for the sound field in a room can be expressed as a superposition of free oscillation of the sound field, as in the vibration of a string. Suppose that there is an impulsive sound source at $S(x',y',z')$ in a room. The sound pressure waveform at a receiving position $R(x,y,z)$ in the room is written as

$$p(S,R,t) \cong \sum_N A_N(S,R)\mathrm{e}^{-\delta_N t}\cos(\omega_N t + \phi_N), \tag{11.80}$$

where
A_N: modal coefficient,
ω_N: angular frequency for N-th free oscillation
ϕ_N: initial phase (rad)
δ_N: decaying constant (rad/s).
By integrating the squared pressure[52]

$$\frac{\Delta\omega}{\rho_0 c}\int_t^{\infty} p^2(S,R,t)dt \cong \frac{\Delta\omega}{\rho_0 c}\sum_N \frac{A_N^2(S,R)}{4\delta_N}\mathrm{e}^{-2\delta_N t} \qquad (\mathrm{W/m^2}) \tag{11.81}$$

is obtained. Thus, by taking the spatial average with respect to the positions of source and observation points all over the room,

$$I(t) \cong \frac{P^2\Delta\omega}{\rho_0 c}\sum_N \frac{1}{4\delta_N}\mathrm{e}^{-2\delta_N t} \qquad (\mathrm{W/m^2}) \tag{11.82}$$

is derived where the modal functions are normalized as

$$\langle A_N^2(S,R)\rangle_{S,R} = P^2 = 1. \qquad (\mathrm{Pa^2}) \tag{11.83}$$

The energy decay is described by superposition of the energy decay for free oscillation of the sound field. If the distribution of decay constants can be known, then the decay curve is rewritten by using simple parameters.

For that purpose, by introducing the averaged decaying constants

$$
\begin{aligned}
2\delta_{N_{ob}} &= c\alpha_{ob}/M_{FP_3} = cA_{b_{ob}}/4V \\
2\delta_{N_{tan}} &= c\alpha_{tan}/M_{FP_2} = cA_{b_{tan}}/\pi V \\
2\delta_{N_{ax}} &= c\alpha_{ax}/MFP_1 = cA_{b_{ax}}/2V, \quad (1/\mathrm{s})
\end{aligned}
\qquad (11.84)
$$

the energy decay curve becomes

$$
I(t) = \frac{P^2 \Delta\omega}{\rho_0 c} \sum_N \frac{\mathrm{e}^{-2\delta_N t}}{4\delta_N} = I_3(t) + I_2(t) + I_1(t) \qquad (\mathrm{W/m^2}) \qquad (11.85)
$$

where

α_{ob}: averaged absorption coefficient for oblique waves
α_{tan}: averaged absorption coefficient for tangential waves
α_{ax}: averaged absorption coefficient for axial waves
S: surface area of room $(\mathrm{m^2})$
S_{tan}: surface area of walls constructing tangential wave field $(\mathrm{m^2})$
S_{ax}: surface area of walls constructing axial wave field $(\mathrm{m^2})$
N_{ob}: number of oblique waves
N_{tan}: number of tangential waves
N_{ax}: number of axial waves
V: volume of room $(\mathrm{m^3})$
c: speed of sound in room (m/s)
$A_{b_{ob}} = \alpha_{ob} S_{ob}$ $(\mathrm{m^2})$
$A_{b_{tan}} = \alpha_{tan} S_{tan}$ $(\mathrm{m^2})$
$A_{b_{ax}} = \alpha_{ax} S_{ax}$ $(\mathrm{m^2})$

and

$$
I_3(t) \cong \frac{V}{c} \frac{2N_{ob}}{A_{b_{ob}}} \exp\left(\frac{-cA_{b_{ob}} t}{4V}\right) \qquad (11.86)
$$

$$
I_2(t) \cong \frac{V}{c} \frac{\pi N_{tan}}{2A_{b_{tan}}} \exp\left(\frac{-cA_{b_{tan}} t}{\pi V}\right) \qquad (11.87)
$$

$$
I_1(t) \cong \frac{V}{c} \frac{N_{ax}}{A_{b_{ax}}} \exp\left(\frac{-cA_{b_{ax}} t}{2V}\right). \qquad (11.88)
$$

The reverberation decay curve is expressed as a superposition of three types of modes that take different decay constants: oblique, tangential, and axial modes. The energy mixing ratio can be estimated by the densiy of those types of modes at the frequency of interest. Consequently, the decay curve in a room does not follow an exponential function with a single decay constant.

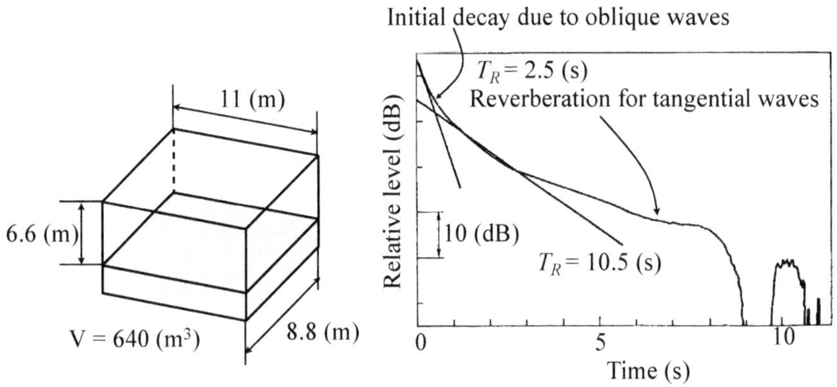

Fig. 11.8 Sample of reverberation decay curve for two-dimensional reverberation field constructed in rectangular reverberation room where only floor is covered by absorbing material[64] (Figs. 1 and 2)

Figure 11.8 is an example of the energy decay curve in a rectangular reverberation room where only the floor is covered by sound-absorbing materials. The energy decay curve that was calculated following the superposition formula shows shorter reverberation in the initial portion of the decay followed by longer reverberation in the later part of the decay. The initial decay part is due to the oblique waves that construct the three-dimensional reverberation field, while the later part is mainly due to the two-dimensional reverberation field that is composed of the tangential waves. It is well known that the initial decay with a short reverberation time is helpful in improving speech intelligibility even in highly reverberant space[64][65].

Vibration of strings in a piano is another example where the energy decay curve is not fitted by a simple exponential function[20]. The vibration of a string excited by a piano hammer decays as time passes, because the vibration energy is lost by energy transmission into the sound board in the piano. However, the vibration of the string is composed of two-types of patterns, namely, vertical and horizontal vibration patterns to the sound board. The vertical vibration patterns of vibration decay faster than the horizontal ones because the vibration energy can be more largely transmitted into the sound board compared with the horizontal ones.

Figure 11.9 is a photograph of the vibration measurements of an up-right piano in the author's home by an acceleration vibration pick up. Figure 11.9 shows a sample of energy decay curve for the string of tone A where a double decay characteristic can be seen. It might be interpreted as the vibration of vertical patterns rapidly decaying, as if it were the decay curve for oblique waves in a room, followed by slow decay due to the horizontal patterns of vibration, like those for the tangential or axial wave modes.

Fig. 11.9 Vibration measurement of piano string

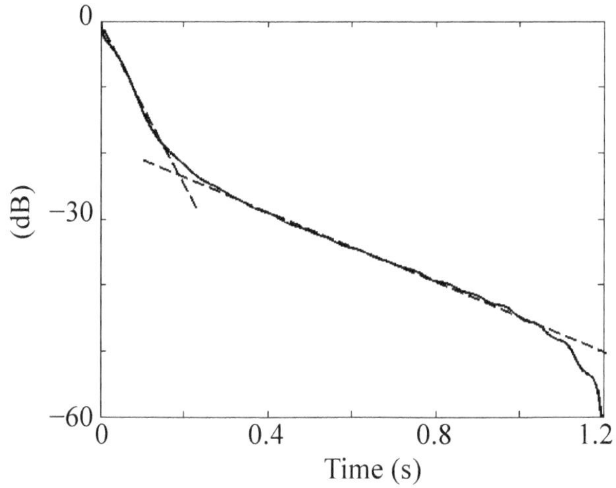

Fig. 11.10 Sample of vibration-energy decay curve of piano string

11.5.2 Frequency Characteristics of Reverberation Energy Decay Curve

By recalling the sound pressure impulse response in a room given by Eq. 11.80, the response can be rewritten as

$$p(\mathbf{x}',\mathbf{x},t) = \frac{\rho_0 c^2 \hat{Q}_0}{V} \sum_N \Lambda_N \phi_N(\mathbf{x})\phi_N(\mathbf{x}') \sin(\omega_N t + \theta_N) e^{-\delta_N t} \quad \text{(Pa)} \qquad (11.89)$$

where ω_N shows the angular frequency of the N-th free oscillation and $\hat{Q}_0$ shows the strength of the impulsive source (m^3). By integrating the squared sound pressure and taking the spatial average with respect to the sound source and receiving positions, the reverberation energy decay curve can be derived as[63]

$$I(t) = \frac{\rho_0 c^3 \hat{Q}_0^2/\Delta t}{V^2} \sum_N \frac{1}{4\delta_N} e^{-2\delta_N t}, \quad (\mathrm{W/m^2}) \tag{11.90}$$

where

$$\int_V \phi_N(\mathbf{x})\phi_M(\mathbf{x})dV = \begin{array}{ll} \dfrac{V}{\Lambda_N} & N = M \\ 0 & N \neq M. \end{array} \tag{11.91}$$

For a narrow-band response, the number of modes for the oblique, tangential, and axial modes that are taken into the superposition can be estimated by using the numbers contained in the **modal bandwidth** $B_M = \pi\delta_N = \pi\Delta\omega$[66]. Assuming the number of modes within half the modal bandwidth, namely $N(\omega) \cong n_{ob}(\omega)B_M/2 = M(\omega)/2$, the mean square sound pressure at the steady state that corresponds to the initial state of the reverberation decay becomes[61]

$$\begin{aligned} I(0) &\cong \frac{\rho_0 c^3(\hat{Q}_0^2/\Delta t)\pi\Delta\omega}{8V^2}\left(n_{ob}(\omega)\frac{1}{\delta_{N_{ob}}} + \sum_{xy} n_{xy}(\omega)\frac{1}{\delta_{N_{xy}}} + \sum_x n_x(\omega)\frac{1}{\delta_{N_x}}\right) \\ &\cong 4W_0\left[\frac{1}{A_{b_{ob}}}\left(1 - \frac{\pi Sc}{4\omega V} + \frac{\pi Lc^2}{8\omega^2 V}\right) + \sum_{xy}\frac{\pi c L_x L_y}{\omega V \tilde{A}_{b_{xy}}}\left(1 - \frac{c(L_x + L_y)}{\omega L_x L_y}\right) + \sum_x \frac{2\pi c^2 L_x}{\omega^2 V \tilde{A}_{b_x}}\right], (\mathrm{W/m^2}) \end{aligned} \tag{11.92}$$

where

$$\begin{aligned} n_{ob}(\omega) &\cong \frac{\omega^2 V}{2\pi^2 c^3} - \frac{\omega S}{8\pi c^2} + \frac{L}{16\pi c} \\ &= \frac{\omega^2 V}{2\pi^2 c^3}\left(1 - \frac{\pi Sc}{4\omega V} + \frac{\pi Lc^2}{8\omega^2 V}\right) \quad (\mathrm{s}) \end{aligned} \tag{11.93}$$

$$\begin{aligned} n_{xy}(\omega) &\cong \frac{\omega L_x L_y}{2\pi c^2} - \frac{L_x + L_y}{2\pi c} \\ &= \frac{\omega L_x L_y}{2\pi c^2}\left(1 - \frac{c(L_x + L_y)}{L_x L_y \omega}\right) \quad (\mathrm{s}) \end{aligned} \tag{11.94}$$

$$n_x(\omega) \cong \frac{L_x}{\pi c}, \quad (\mathrm{s}) \tag{11.95}$$

$$\tag{11.96}$$

and

$$L = 4(L_x + L_y + L_z) \qquad (\mathrm{m}) \tag{11.97}$$

$$S = 2(L_x L_y + L_y L_z + L_z L_x) \qquad (\mathrm{m^2}) \tag{11.98}$$

$$\hat{Q}_0^2/\Delta t \cdot \Delta\omega = Q_0^2 \quad (\mathrm{m^6/s^2}) \tag{11.99}$$

$$\frac{\omega^2 \rho_0 Q_0^2}{8\pi c} = W_0 \quad (\mathrm{W}) \tag{11.100}$$

$$\tilde{A}_{xy} = \frac{4}{\pi}A_{b_{xy}} \quad (\mathrm{m^2}) \tag{11.101}$$

$$\tilde{A}_x = 2A_{b_x}. \qquad (\mathrm{m}^2) \tag{11.102}$$

Here, $M(\omega)$ is called **modal overlap**, W_0 (W) denotes the sound power output of a point source whose magnitude of strength is given by $Q_0(\mathrm{m}^3/\mathrm{s})$, and $S(\mathrm{m}^2)$ is the surface area of the walls constructing the corresponding wave field. Similarly, α shows the corresponding average absorption coefficient such that $A_{b_{ob}} = \alpha_{ob}S_{ob}$, $A_{b_{xy}} = \alpha_{xy}S_{xy}$, and $A_{b_x} = \alpha_x S_x$.

Consequently, the energy decay curve is given by

$$\begin{aligned} I(t) \cong 4W_0 \Bigg[& \frac{1}{A_{b_{ob}}}\left(1 - \frac{\pi Sc}{4\omega V} + \frac{\pi Lc^2}{8\omega^2 V}\right) \mathrm{e}^{-\frac{cA_{b_{ob}}}{4V}t} \\ & + \sum_{xy} \frac{\pi c L_x L_y}{\omega V \tilde{A}_{b_{xy}}} \left(1 - \frac{c(L_x + L_y)}{\omega L_x L_y}\right) \mathrm{e}^{-\frac{cA_{b_{xy}}}{\pi V}t} \\ & + \sum_{x} \frac{2\pi c^2 L_x}{\omega^2 V \tilde{A}_{b_x}} \mathrm{e}^{-\frac{cA_{b_x}}{2V}t} \Bigg], \qquad (\mathrm{W/m}^2) \end{aligned} \tag{11.103}$$

which represents the frequency characteristics of the decay curve using the geometric acoustical parameters of the room of interest. Note that the frequency characteristics do not depend on the frequency characteristics of the sound absorption material but on the arrangement of the absorbing materials and the geometrical construction of the room[67]. If the sound absorption coefficients are assumed to be uniform all over the walls, the energy decay follows a simple exponential function as the frequency increases. This type of hybrid representation of the energy decay function is useful for acoustic measurements such as sound absorption measurements from the energy decay curves in a reverberation room. The same equation can also be derived according to the image theory based on the geometrical acoustics as described in reference[47].

The hybrid representation of a sound field can be interpreted as correspondence between the geometrical and wave-theoretic acoustics of the room reverberation. However, there is a big difference between the geometrical and modal acoustics in rooms from the viewpoint of chaotic properties of sound ray. Such the chaotic properties of sound propagation will be described in the next chapter.

Chapter 12
Spatial Distribution of Sound in Rooms

This chapter describes the nature of the spatial distribution of traveling waves in a reverberation field. It is greatly interesting to see the chaotic properties hidden in the ray tracing process of sound from the viewpoint of the relationship between geometrical and wave theoretical acoustics. In an irregularly shaped room where the sound ray trajectories might be chaotic, the distribution statistics for the eigenfrequency spacings following Rayleigh or Gamma distribution might represent the scar of the chaotic properties left in the linear acoustics. Spatial distribution properties for the squared sound pressure and two-point correlations are described from both the geometrical and wave theoretical viewpoints.

12.1 Chaotic Properties of Sound Ray Trajectories in Rooms

Sound ray tracing is a method for analyzing sound propagation in space by tracing the process of sound reflection at the boundary based on geometrical acoustics. A problem analyzed using the method is also called a **billiard problem**. Sound ray tracing has been a practical method for sound field analysis, but so-called **chaotic properties** are hidden in the trajectories of the sound rays.

12.1.1 Examples of Sound Ray Tracing

A **chaotic event** has an instable process such that even if only a slight error is produced in the initial stage or condition of the process, the effects of the error exponentially increase as the process goes on, and thus the outcome dramatically changes. Suppose a sound ray travels in a two-dimensional region. Figures 12.1 and 12.2 illustrate examples of sound ray trajectories in the space. In **billiard** problems, when the ray trajectories fill the space to a limit as the process goes on, such a system where the ray travels is called **ergodic**. A rectangular boundary as shown by Fig. 12.1 is an example of space that can be filled by sound ray trajectories. Figure 12.2 is another example of ray tracing in an ellipse[36]. Any ray that passes the focal point $F_1(F_2)$ comes back to $F_2(F_1)$, and thus the sound ray coming through F_1 or

M. Tohyama: *Sound and Signals*, Signals and Communication Technology, pp. 219–241.
springerlink.com

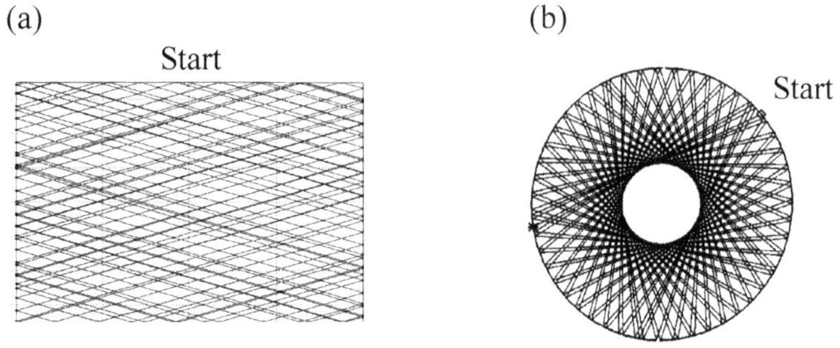

Fig. 12.1 Sound ray trajectory samples (a) rectangular boundary, (b) circular boundary from [68](Fig.3)

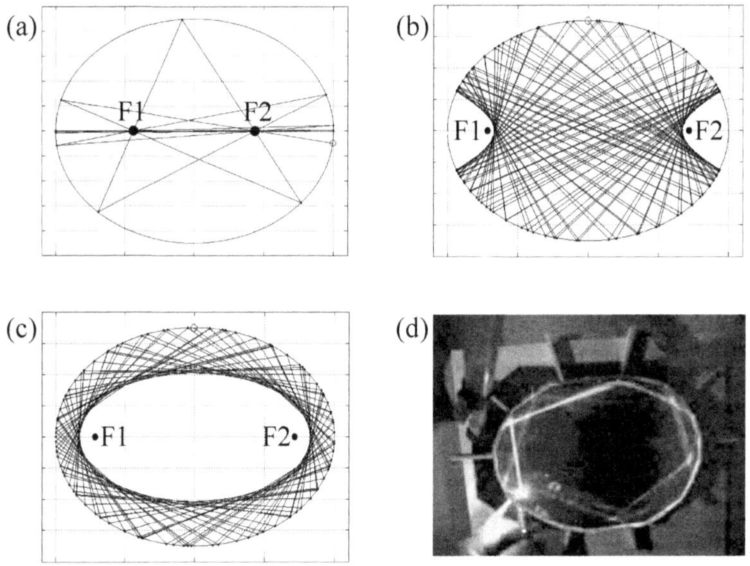

Fig. 12.2 Sound ray propagation in ellipse (a)-(c) numeric simulation, (d) photograph of scale model experiments using laser ray [69]

F_2 at the initial stage comes close to the axis that connects F_1 and F_2 as the process proceeds.

In contrast, if a sound ray cuts in at the initial stage between the two focal points, it goes through any point on the line between the focal points and is tangential to the hyperbolic. Instead, if a sound ray does not enter between the focal points at the initial stage, it travels along the inner surface of the ellipse without propagating across the space. This is a model for a **whispering gallery**. Figure 12.2 (d) shows an image of sound propagation along inside walls in a scale model experiment using a laser pointer[69].

Figure 12.3 presents another example of ray tracing for space such as the so-called **stadium type of field**, which is composed of a rectangle and hemi-circles. It looks very irregular compared with the other trajectories shown in Figs. 12.1 and 12.2. Suppose a pair of sound rays as shown in Fig. 12.4, which are closely located

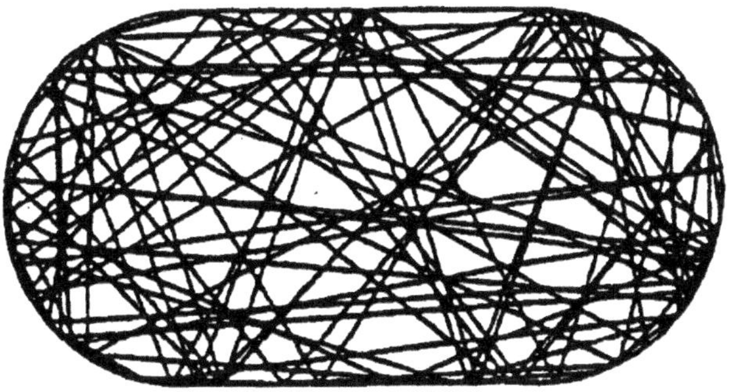

Fig. 12.3 Sound ray trajectory in stadium type of two-dimensional field from [34](Fig.5.13)

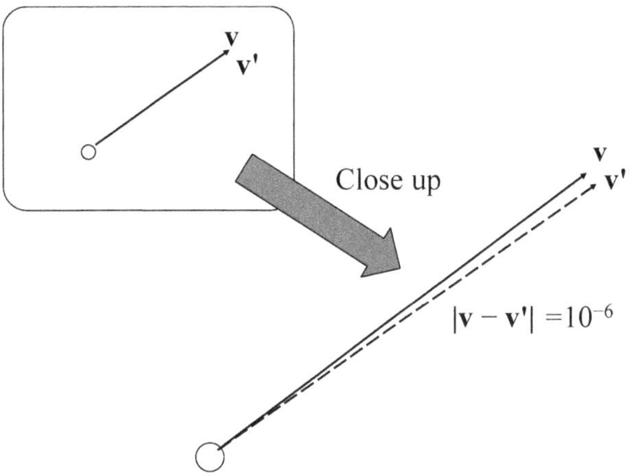

Fig. 12.4 Display of initial condition for closely located pair of sound rays from [70] (Fig.5)

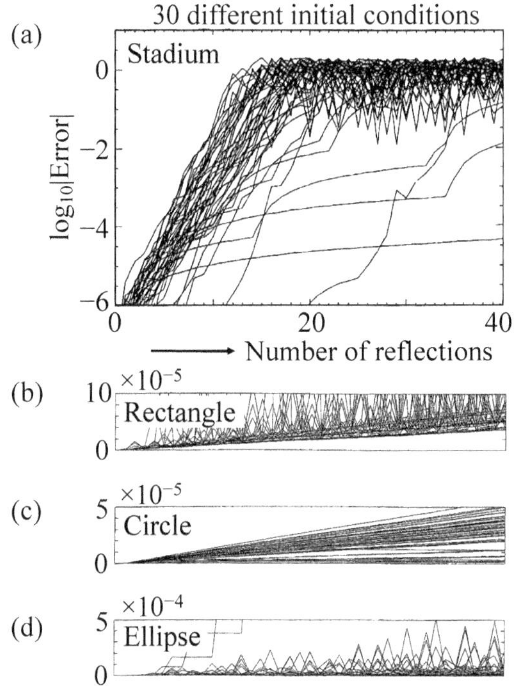

Fig. 12.5 Error propagation analysis for sound-ray tracing simulation from [70] (Fig.6)

with slightly different traveling directions at the initial stage. Measure the distance between the reflection points at the boundary for every reflection for the pair of rays. Such a distance is shown in Fig. 12.5, where other examples for rectangular, circular, and elliptic boundaries are also illustrated for comparison. The number of pairs used for the calculation was 30. A logarithmic scale was taken only for the results in the stadium field.

The distance of two reflection points due to the pair of rays increases as the process goes on. However, only the distance for the stadium field exponentially increases, while the other cases show linear increases. That is, if the initial distance of the pair of rays shows the error produced in the initial conditions, the effects of the error on the distance exponentially increases in the stadium type of field. Namely, the **chaotic properties** are hidden in the **ray tracing** process in the **stadium** field. The distance reaches the maximum of the field due to the size of the space when the number of reflections exceeds around 10. This outcome indicates that it is impossible to track the sound path according to the ray tracing method if the field of interest is surrounded by a boundary composed of curved surfaces and plane walls. On the other hand, the error increases linearly as the number of reflections increases in the area surrounded by rectangular, circular, and elliptic type of boundaries.

12.1.2 Distribution of Eigenfrequencies in Rooms

Correspondence between a billiard type of behavior of a particle in a surrounded area and its quantum wave-theoretic nature is called **quantum chaology**[71] in terms of quantum physics. It corresponds to the relationship between the geometrical acoustics represented by the sound ray and the wave-theoretic room acoustics characterized by the modal functions in the space in terms of acoustics, namely, **room acoustics chaos**[68]. It is well known that there is no room for chaotic properties in the linear wave-theoretic nature of room acoustics. However, as well as the quantum chaology, a **scar** of the chaotic properties can be seen in the distribution of eigenfrequencies for the space where the ray tracing might be chaotic.

The eigenfrequencies of a rectangular room of dimensions (L_x, L_y, L_z) with rigid boundaries are given by

$$\omega_{lmn} = \frac{c\pi}{L}\sqrt{(al)^2 + (bm)^2 + (cn)^2} \qquad \text{(rad/s)} \tag{12.1}$$

where l, m, and n are integer numbers, the lengths of the sides are $L = aL_x = b_{L_y} = cL_z$, and a, b, and c are the ratios of the lengths. The distribution of the eigenfrequencies appears complicated even for rectangular rooms. Therefore, the process by which the eigenfrequencies occur can be characterized by a **Poisson process**[72] that results in an **exponential distribution** of eigenfrequency spacing[73][74].

Figure 12.6 shows a numerical example of the **spacing distribution** for oblique modes in a rectangular room. The **oblique modes** denote eigenmodes whose eigenfrequencies have no zero entries for l, m, or n. The horizontal axis indicating the spacing distance between an adjacent pair of the eigenfrequencies is measured by the normalized variable for the averaged distance, and the vertical axis is the histogram indicating the statistical frequencies for the corresponding samples. It can be seen that pairs of eigenfrequencies with very narrow spacing are quite likely to occur following the exponential distribution.

The Poisson process might not be adequate for the rooms other than rectangular rooms, however. Groups of eigenmodes that have degenerate eigenfrequencies will suffer from the eigenfrequencies splitting due to **perturbations** in practical cases[73][74][75]. Figure 12.7 shows a series of semi-stadium fields where the field changes from a regular one ($i \to \infty$) to an irregular one ($i = 2$)[76][77]. The spacing statistics for eigenfrequencies similar to those in Fig. 12.6 for the rectangular area are shown in Fig. 12.8. The eigenfrequencies for the two-dimensional semi-stadium boundaries were numerically calculated by FEM instead of the theoretical one for the rectangular case. The solid lines in the figure represent **Gamma distributions** of X with the freedom of n such that

$$w_\Gamma(x, n) = \frac{n^n}{\Gamma(n)} x^{n-1} \mathrm{e}^{-nx}, \qquad x > 0 \tag{12.2}$$

where

$$\Gamma(r) = \int_0^\infty t^{r-1} \mathrm{e}^{-t} dt \qquad r > 0 \tag{12.3}$$

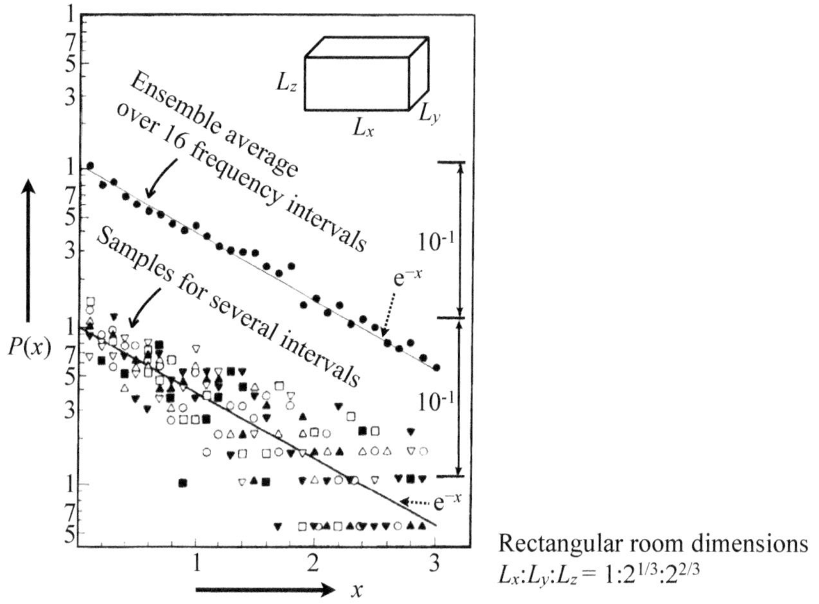

Fig. 12.6 Example of numerical analysis for modal spacing statistics of rectangular room with rigid walls where $P(x)$ denotes histogram of x and x is normalized distance between adjacent angular eigenfrequencies from [70] (Fig.2)

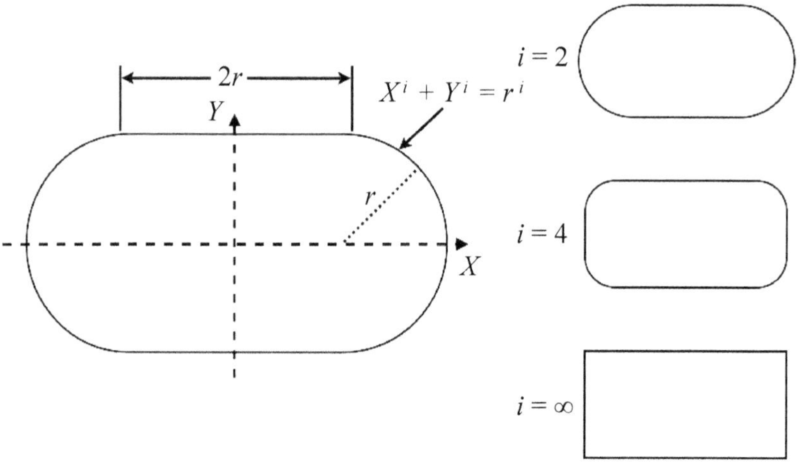

Fig. 12.7 Semi-stadium boundaries from [70] (Fig.4)

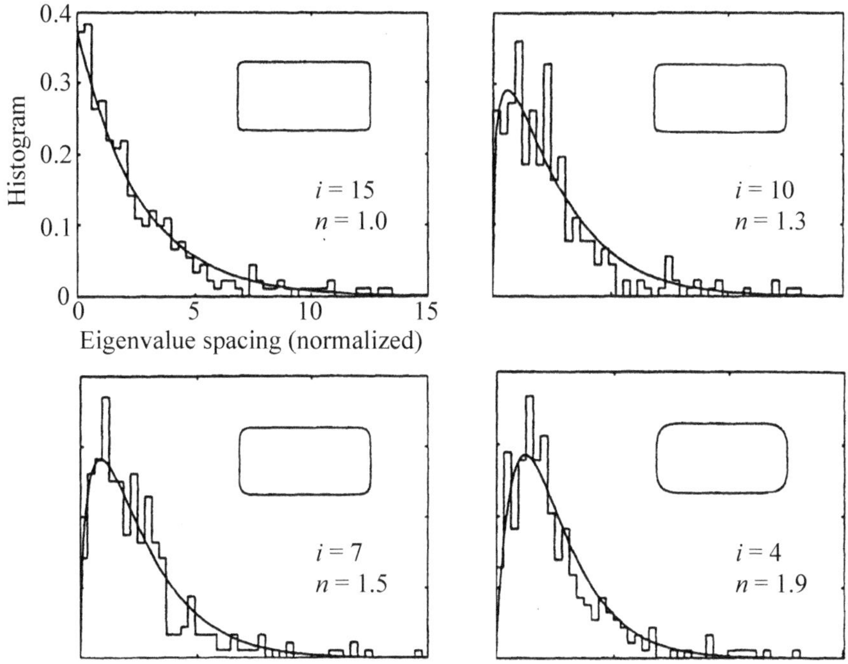

Fig. 12.8 Modal spacing statistics for semi-stadium fields from [70] (Fig.7)

and

$$\Gamma(n) = (n-1)! \tag{12.4}$$

for an integer $n > 1$.

It can be seen that the spacing distributions for eigenfrequencies when i =4,7,10, and 15 could be fitted to the distribution with a non-integer degree of freedom n =1.9,1.5,1.3, and 1.0 for the boundaries[76][77]. Here, x is the spacing normalized by the mean spacing, and the exponential distribution for the regular case corresponds to the freedom $n = 1$, while for the freedom of two that indicates the irregular case, the distribution follows[74]

$$w_{Wig}(x) = 4xe^{-2x}. \tag{12.5}$$

This transition in the degree of freedom might be interpreted as indicating the breakdown of regularity[75][76][77]. It can also be interpreted as the process in which the degeneration of eigenfrequencies becomes unlikely[73][74].

12.1.3 Eigenfrequencies and Random Matrices

The transition due to the effects of perturbation can also be illustrated by **coupled mechanical oscillations**. Suppose a coupled oscillator as shown in Fig. 12.9 where the masses of the two oscillators are M_1, and M_2, their spring constants are K_1, and K_2, and the coupling constant is K_{12}. The eigenfrequencies for the coupled system can be derived following the equation of motion;

$$M_1\frac{d^2x_1(t)}{dt^2} = -K_1x_1(t) - K_{12}(x_1(t) - x_2(t)) \quad (12.6)$$

$$M_2\frac{d^2x_2(t)}{dt^2} = -K_2x_2(t) - K_{12}(x_2(t) - x_1(t)). \quad \text{(N)} \quad (12.7)$$

Assuming free oscillation, $x_1(t) = A_1e^{i\omega t}$, and $x_2(t) = A_2e^{i\omega t}$, the equation of motion can be rewritten as

$$\left(\frac{K_1 + K_{12}}{M_1} - \omega^2\right)A_1 - \frac{K_{12}}{M_1}A_2 = 0 \quad (12.8)$$

$$-\frac{K_{12}}{M_2}A_1 + \left(\frac{K_2 + K_{12}}{M_2} - \omega^2\right)A_2 = 0. \quad (12.9)$$

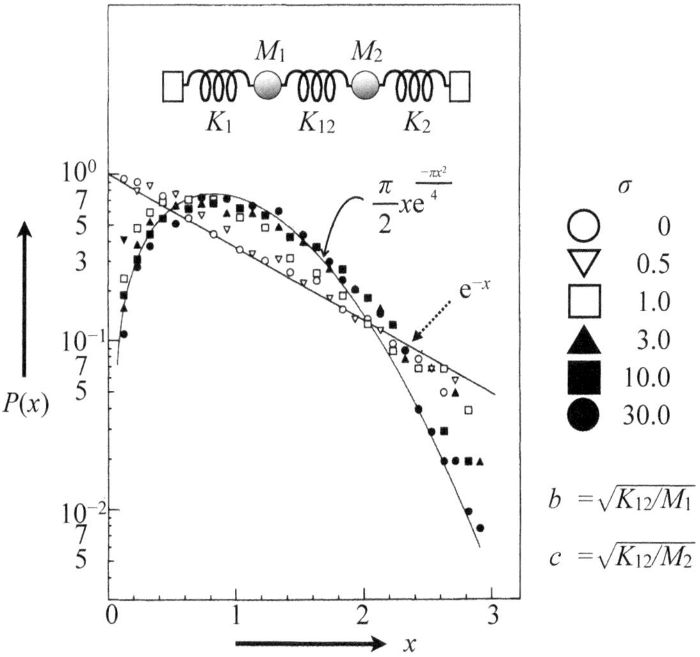

Fig. 12.9 Samples of eigenfrequency spacing for coupled oscillator from [70] (Fig.3)

Namely, the eigenfrequencies are obtained as the eigenvalues of the matrix such that

$$\begin{aligned} \Delta_\lambda &= \Delta_{\lambda_o} + \Delta_{\lambda_p} \\ &= \begin{pmatrix} \omega_1^2 & 0 \\ 0 & \omega_2^2 \end{pmatrix} + \begin{pmatrix} \Delta\omega_{12}^2 & -\Delta\omega_{12}^2 \\ -\Delta\omega_{21}^2 & \Delta\omega_{21}^2 \end{pmatrix}, \end{aligned} \tag{12.10}$$

where

$$\omega_1^2 = \frac{K_1}{M_1} \quad \omega_2^2 = \frac{K_2}{M_2} \qquad \Delta\omega_{12}^2 = \frac{K_{12}}{M_1} \quad \Delta\omega_{21}^2 = \frac{K_{12}}{M_2}. \tag{12.11}$$

Matrix Δ_{λ_o} in the equation above has the eigenvalues of two independent oscillators, and matrix Δ_{λ_p} represents the **coupling effect** on the eigenvalues.

The **perturbation** caused in actual acoustic space such as in an irregularly shaped room results in random coupling between adjacent modes[78]. Assume that the coupling parameters $\Delta\omega_{12}^2$ and $\Delta\omega_{21}^2$ of matrix Δ_{λ_p} are mutually independent **Gaussian variables**[79] with a zero mean value and a standard deviation of σ. The spacing of the two eigenfrequencies $\Delta\omega$ can thus be expressed by

$$\begin{aligned} (\Delta\omega)^2 &\cong \Delta\omega_{12}^2 + \Delta\omega_{21}^2 \qquad && \Delta\omega_{12}^2, \Delta\omega_{21}^2 \quad \rightarrow \quad \text{large} \\ &\cong (\omega_2 - \omega_1)^2 && \Delta\omega_{12}^2, \Delta\omega_{21}^2 \quad \rightarrow \quad \text{small}, \end{aligned} \tag{12.12}$$

where

$$\Delta\omega^2 = \Delta\omega_o^2 + \Delta\omega_p^2 \tag{12.13}$$

$$\Delta\omega_o^2 = \omega_1^2 + \omega_2^2 - 2\sqrt{N^2} \tag{12.14}$$

$$\Delta\omega_p^2 = \Delta\omega_{12}^2 + \Delta\omega_{21}^2 \tag{12.15}$$

$$N^2 = (\omega_1^2 + \Delta\omega_{12}^2)(\omega_2^2 + \Delta\omega_{21}^2) - \Delta\omega_{12}^2\Delta\omega_{21}^2. \tag{12.16}$$

The random variable $\Delta\omega$ can be the positive square root of the squared sum of the two independent Gaussian variables as standard deviation σ (perturbation) increases. The spacing for the two eigenfrequencies therefore follows a **Rayleigh distribution**[79][80] as the coupling effect increases.

Figure 12.9 illustrates the transition in a spacing histogram from a Poisson (exponential) to a Rayleigh distribution. Five thousand pairs of eigenfrequencies with a spacing that followed a Poisson distribution with a zero coupling effect were used for this calculation. It can be surmised that the Rayleigh distribution of the eigenfrequency spacing is produced by randomly coupled wave modes[78]. This outcome can be generalized by using studies about the eigenvalues for an ensemble of **real symmetric random matrices**[81] such that the spacing of successive eigenvalues has a probability density, $w_{Ray}(x)$, well approximated by the Rayleigh distribution

$$w_{Ray}(x) = \frac{\pi x}{2} \mathrm{e}^{-\frac{\pi x^2}{4}} \tag{12.17}$$

when spacing x is normalized by the mean spacing[79].

However, it is also known that some experimental data indicate that the spacing distribution could be well fitted by Eq. 12.5 [74][75] that is easier to handle analytically than that for Rayleigh distribution. Note that Eq. 12.5 is a function of the family of Gamma distributions that were introduced previously. It could be found that the **transition from regular to irregular systems** can be illustrated by the family of functions. In particular, Eq. 12.5 corresponds to that for the freedom of two in the Gamma family. If the transition is regarded as a process of perturbation, the **chaotic properties of the sound field** can be interpreted as a **diffuse field** in terms of classical acoustical theory, which states that the diffuse field is an outcome at a limit when the perturbation becomes strong[78].

12.2 Sound Energy Distribution in Reverberation Field

12.2.1 Superposition of Random Plane Waves and Squared Sound Pressure Distribution

Energy distribution in a random sound field can be represented by **superposition of plane waves with random amplitudes and phases**. This is because a sound field composed of many reflections, such as the sound field in a room surrounded by rigid walls, is highly sensitive to sound receiving and source positions. The composition of reflection sound, such as the magnitude and phase, varies almost randomly according to the sound source and observing positions, and most of those reflection sounds are made of the plane waves coming from the image sources far from the receiving positions.

Suppose that sound pressure due to the superposition of reflection waves simply with equal magnitude and random phases is written as

$$p(t) = A_0 \sum_{i=1}^{N} \cos(\omega t + \phi_i), \qquad \text{(Pa)} \tag{12.18}$$

where A_0 denotes the uniform magnitude, ϕ_i shows the phase angle for the i-th component of the plane waves, and N is the number of components to be superposed. The mean square sound pressure becomes

$$\overline{p^2(t)} = \frac{1}{T}\int_0^T p^2(t)dt = \frac{1}{2}A_0^2(X^2+Y^2), \qquad (\text{Pa}^2) \tag{12.19}$$

where T denotes the period of the waves and

$$X = \sum_{i=1}^{N} \cos\phi_i, \quad Y = \sum_{i=1}^{N} \sin\phi_i. \tag{12.20}$$

The phase angle can be assumed to be a random variable with respect to the source and receiving positions, and thus it can be interpreted to be $\phi_i = \omega\tau_i$, where τ_i denotes the delay time for the i-th reflection sound arriving at the receiving position with reference to the direction sound.

Suppose that the phase angle follows a uniform distribution from 0 to 2π, and introduce another random variable

$$Z^2 = X^2 + Y^2 = U_z \qquad (Z \geq 0). \tag{12.21}$$

Then, as the number N increases, the variables X and Y follow a **normal (Gaussian) distribution**, and U_z becomes a random variable from an **exponential distribution**[79][80]. That is, X and Y are mutually independent random variable following a normal distribution with zero mean and the variance of σ^2 such that

$$w_{Norm}(x) = \frac{1}{\sqrt{2\pi\sigma^2}} e^{-\frac{x^2}{2\sigma^2}} \tag{12.22}$$

$$w_{Norm}(y) = \frac{1}{\sqrt{2\pi\sigma^2}} e^{-\frac{y^2}{2\sigma^2}}. \tag{12.23}$$

Introduce transformation of the variables such that

$$X = Z\cos\theta \qquad Y = Z\sin\theta, \tag{12.24}$$

the probability density function of X and Y is thus

$$\begin{aligned} w(x,y)dxdy &= w(z\cos\theta, z\sin\theta)zdzd\theta \\ &= \frac{1}{2\pi\sigma^2} e^{-\frac{z^2}{2\sigma^2}} zdzd\theta. \end{aligned} \tag{12.25}$$

Consequently, by following the expression

$$w(z)dz = \frac{1}{2\pi\sigma^2}\int_0^{2\pi} e^{-\frac{z^2}{2\sigma^2}} zdzd\theta = \frac{z}{\sigma^2} e^{-\frac{z^2}{2\sigma^2}} dz, \tag{12.26}$$

the probability density function of Z can be written as

$$w_{Ray}(z) = \frac{z}{\sigma^2} e^{-\frac{z^2}{2\sigma^2}}, \tag{12.27}$$

which is called the **Rayleigh distribution** (already introduced in the previous section). Therefore, if the density function is rewritten as

$$\frac{z}{\sigma^2} e^{-\frac{z^2}{2\sigma^2}} = \frac{\sqrt{u_z}}{\sigma^2} e^{-\frac{u_z}{2\sigma^2}} \frac{du_z}{2\sqrt{u_z}} = w_{Exp}(u_z)du_z, \tag{12.28}$$

the **probability density function for the mean square sound pressure** is given by

$$w_{Exp}(u_z) = \frac{1}{2\sigma^2} e^{-\frac{u_z}{2\sigma^2}}, \tag{12.29}$$

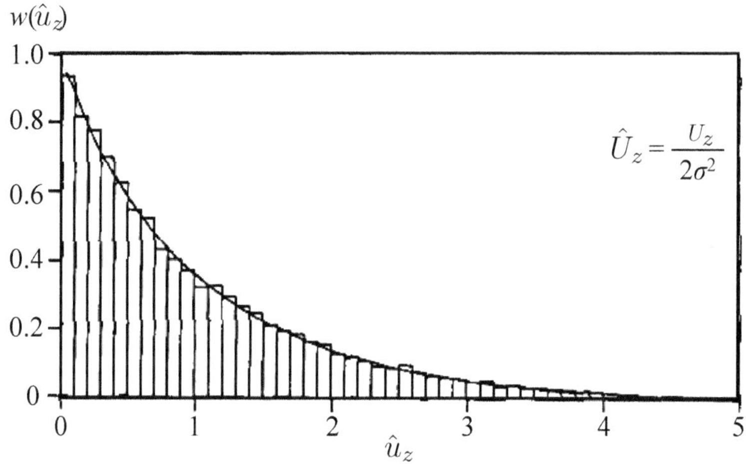

Fig. 12.10 Distribution for mean square sound pressure for reverberation room from [80] (Fig.3.4.1)

which is called an **exponential distribution**. Figure 12.10 illustrates an example of the distribution of mean square sound pressure in a reverberation room[80]. The sound field is not uniform any longer in a reverberation room where a sinusoidal wave travels.

Given a random variable to mean square sound pressure such that

$$\frac{\overline{p^2(t)}}{\frac{1}{2}A_0^2} = X^2 + Y^2 = Z^2 = U_z. \tag{12.30}$$

Recall that the random variables X and Y follow the normal distribution with zero mean and variance of $\sigma^2 = N/2$. The random variable U_z representing the mean square sound pressure follows the exponential function such that

$$w_{Exp}(u_z) = \frac{1}{N}\mathrm{e}^{-\frac{u_z}{N}}. \tag{12.31}$$

The expectation of the mean square sound pressure is given by

$$\mathrm{E}[U_z] = \int_0^{\infty} u_z w_{Exp}(u_z) du_z = N. \tag{12.32}$$

Similarly, the variance becomes

$$\mathrm{Var}[U_z] = \int_0^{\infty} u_z^2 w_{Exp}(u_z) du_z - N^2 = N^2. \tag{12.33}$$

Therefore, the normalized standard deviation Σ is expressed by

$$\Sigma = \frac{\sqrt{\mathrm{Var}[W]}}{\mathrm{E}[W]} = \frac{\sqrt{N^2}}{N} = 1, \tag{12.34}$$

which indicates that almost 100% error might be produced if the mean square sound pressure observed at a single receiving position were to provide an estimate of the space average for the mean square sound pressure observed in the whole room. This is why a broad band noise signal is usually used for measurements of room acoustics.

12.2.2 Distribution of Sound Pressure Level in Random Sound Field

By taking 10-base logarithm of the mean square sound pressure, the **sound pressure level** can be given as

$$L_p = 10\log\frac{\overline{p^2}}{P_M^2}, \qquad \text{(dB)} \tag{12.35}$$

where $P_M = 2\times 10^{-5}$ (Pa) indicates the **minimum audible sound** of a listener. Write the random variable of U_z by taking the natural logarithm again as

$$S = \ln R = \ln\frac{U_z}{N}, \tag{12.36}$$

whose probability density function can be written as

$$w(s) = \mathrm{e}^{s-\mathrm{e}^s}. \tag{12.37}$$

Therefore, by taking the expectation of the random variable S,

$$\mathrm{E}[S] = \int_{-\infty}^{\infty} s\mathrm{e}^{s-\mathrm{e}^s}ds = \int_0^{\infty} \mathrm{e}^{-r}\ln r dr = \Gamma^{(1)}(1) = -\mathrm{C} \tag{12.38}$$

can be derived, where $\Gamma^{(1)}(*)$ denotes the first derivative of the Gamma function, and C is Euler's constant[73]. Similarly, the variance becomes

$$\mathrm{Var}[S] = \int_{-\infty}^{\infty} s^2\mathrm{e}^{s-\mathrm{e}^s}ds - \mathrm{C}^2 = \int_0^{\infty} \mathrm{e}^{-r}(\ln r)^2 dr - \mathrm{C}^2 = \Gamma^{(2)}(1) = \frac{\pi^2}{6}, \tag{12.39}$$

where $\Gamma^{(2)}(*)$ denotes the second derivative of the Gamma function. Generally, the n-th derivative of the Gamma function is given by[79][82]

$$\Gamma^{(m)}(t) = \int_0^{\infty} x^{t-1}(\ln x)^m \mathrm{e}^{-x}dx. \tag{12.40}$$

Consequently, by converting the natural logarithm to the 10-base one, which gives the decibel unit, the **standard deviation of the mean square sound pressure** becomes about 5.57 (dB) in a decibel unit. If the nature of Gaussian variables were applied to the variable S, almost 70% of observation samples in a room would be distributed within about 10 dB around the mean level[73].

12.3 Spatial Correlation of Random Sound Field

Suppose a pair of observation points for sound pressure in a random sound field. A cross-correlation function can be defined for the sound pressure signals between the pair of points. If the pair of two points is the right and left ears of a listener, such a function is called the **inter-aural cross-correlation function**[83].

12.3.1 Cross-Correlation Functions for Sound Pressure in Sinusoidal Waves in Room

Take a pair of sound pressure signals in a sound field. The **cross-correlation function** of the sound pressure for the pair can be defined as

$$C_F(m) = \overline{\mathrm{E}\left[c_f(n,m)\right]} \quad (12.41)$$

$$c_f(n,m) = p_1(n)p_2(n-m), \quad (12.42)$$

where $\overline{(*)}$ denotes taking the average for a single period or a long term of the waves, and E represents the ensemble average that can be estimated by taking the spatial average in the room.

Suppose that the sound field is made of sinusoidal waves of a single frequency. A pair of sound pressure signals can be expressed as

$$p_1(n) = A\cos\Omega n \quad (12.43)$$

$$p_2(n) = B\cos(\Omega n - \phi), \quad (12.44)$$

where Ω is the normalized angular frequency in discrete signals. According to Fig. 12.11, the phase difference ϕ is given by

$$\phi = kr\cos\theta \quad (12.45)$$

for a pair of two signals, where k is the wavenumber (l/m) and θ represents an incident angle for a plane sinusoidal wave coming into the paired positions of interest[84]. Assuming that the angle of the incidence wave is equally probable in the three-dimensional reverberation sound field, the ensemble average for the pairs, with the equal distance of r (m), can be written as

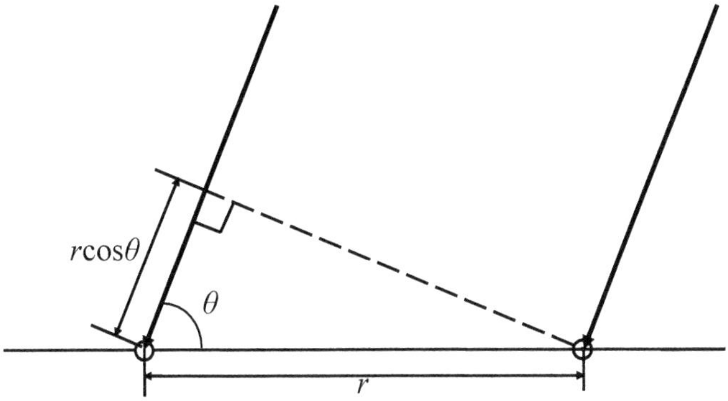

Fig. 12.11 Coupling plane wave to pair of observation points in random sound field

$$\begin{aligned}
&\mathrm{E}\left[p_1(n)p_2(n-m)\right] \\
&= \frac{AB}{2}\int_0^{\pi/2}\cos(kr\cos\theta)\sin\theta d\theta[\cos\Omega m+\cos(2\Omega n-\Omega m)] \\
&= \frac{AB}{2}\frac{1}{k}\int_0^{k}\cos(xr)dx[\cos\Omega m+\cos(2\Omega n-\Omega m)] \\
&= \frac{AB}{2}\frac{\sin(kr)}{kr}[\cos\Omega m+\cos(2\Omega n-\Omega m)]
\end{aligned} \tag{12.46}$$

where

$$x = k\cos\theta \tag{12.47}$$

and

$$\frac{AB}{4}\int_0^{\pi}\sin(kr\cos\theta)\sin\theta d\theta = 0. \tag{12.48}$$

Following the result above, by taking the time average for the single period,

$$C_F(kr,m) = \frac{AB}{2}\frac{\sin(kr)}{kr}\cos\Omega m \tag{12.49}$$

can be obtained. Here, the function

$$C_{F_3}(kr) = \frac{\sin kr}{kr} \tag{12.50}$$

is called the **cross-correlation coefficient of sound pressure for a three-dimensional reverberation field**[84].

12.3.2 Angular Distribution of Waves in Rectangular Reverberation Room

Figure 12.12 is an example of the cross-correlation coefficients measured in a reverberation room. In this example, a narrow-band noise (1/3 octave band) was used

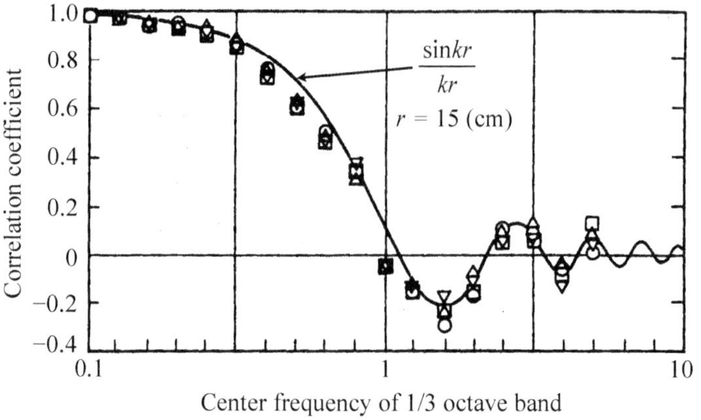

Fig. 12.12 Example of two-point cross-correlation coefficient for sound pressure in rectangular reverberation room where 1/3 octave-band noise source is located from [85] (Fig.1)

instead of a sinusoidal signal so that the ensemble average might be replaced by taking a longterm average even for a fixed measuring pair in a reverberation field. It can be confirmed that the cross-correlation coefficients follow the function given by the equation above in the reverberation room[85].

The distribution for angles of incidental waves into the observation points are called **angular distribution** of the sound field[86]. The distribution depends also on the location of a sound source in a reverberation room, such as where the measurements were performed as shown by Fig. 12.13. In a rectangular reverberation room, such the effects of the source position on the sound field might be strong. Figure 12.14 illustrates an example of mean square sound pressure distribution in a rectangular room where a sound source is located on a symmetric central line as shown by Fig. 12.13. The sound source radiates a narrow-band noise such as 1/3-octave band noise. The mean square sound pressure of $p(\mathbf{r},t)$ along the y-direction (x_0, y, z_0) is given by[87][88]

$$\begin{aligned}\overline{p^2} &= \sum_{m=0}^{m_{\mathrm{Max}}} D_{pm} \cos^2 \frac{m\pi y}{L_y} \\ &= \frac{1}{N} \sum_{\omega_a < \omega_{lmn} < \omega_b} A_{lmn}^2 \cos^2 \frac{l\pi x_0}{L_x} \cos^2 \frac{m\pi y}{L_y} \cos^2 \frac{n\pi z_0}{L_z},\end{aligned} \tag{12.51}$$

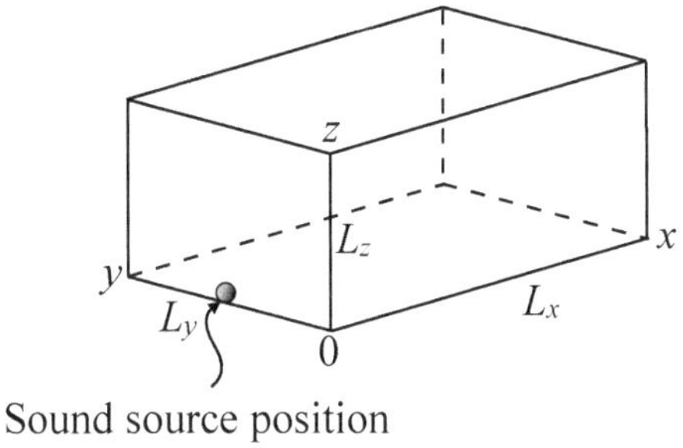

Fig. 12.13 Room dimensions of rectangular reverberant room. $L_x = 11$ (m), $L_y = 8.8$ (m), $L_z = 6.6$ (m), and source location on center line on floor from [87] (Fig.1)

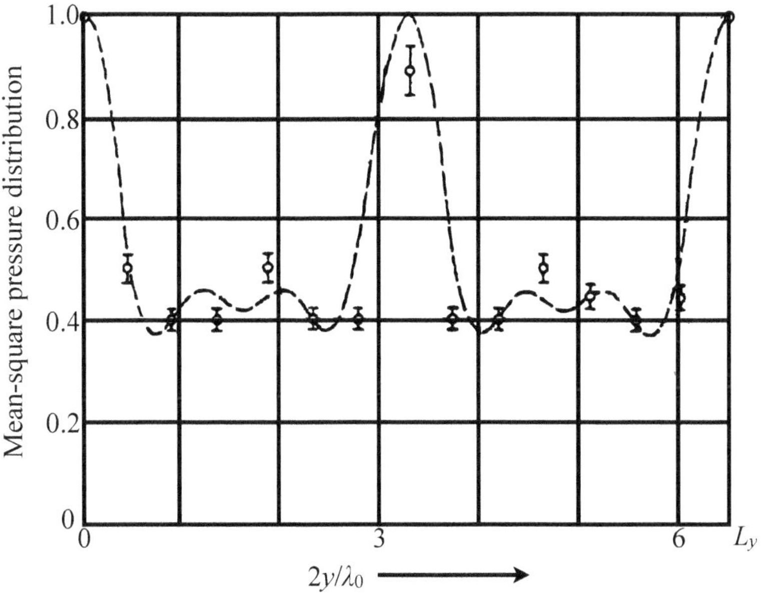

Fig. 12.14 Mean squared sound pressure distribution $(\overline{p^2})$ at $(L_x/2, y, L_z/2)$. ν=125 Hz (center frequency of 1/3 octave band) and λ_0 shows its wavelength, dotted line: calculated by Eq. 12.51, all data are shown with observational error range, from [88] (Fig.3)

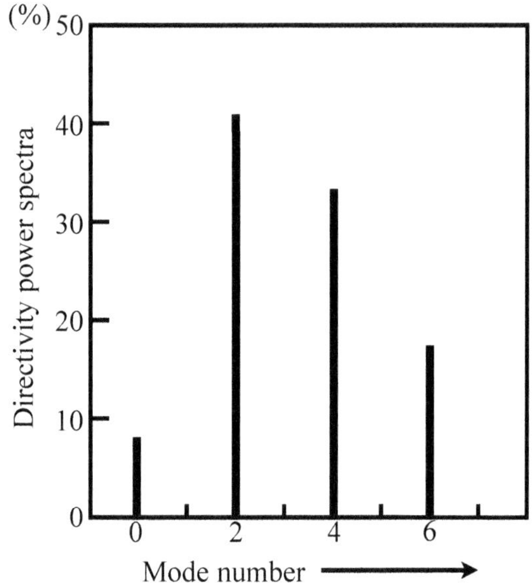

Fig. 12.15 Directivity power spectra D_{pm} along line $(L_x/2, y, L_z/2)$ in Eq. 12.51. ν=125 Hz (center frequency of 1/3 octave band) from [88] (Fig.4)

where D_{pm} is the **directivity power spectrum** representing the angular spectrum (distribution) along the y-direction (x_0, y, z_0)[86], ω_{lmn} is the angular eigenfrequency, and superposition is taken only for the modes whose eigenfrequencies are within the band of the noise source. Namely, $m_{Max}\pi/L_y \leq \omega_b/c < (m_{Max}+1)\pi/L_y$ when ω_a and ω_b give the frequency range of the band noise. The dotted line in Fig. 12.14 represents the numerical examples according to the equation above, and the directivity power spectra are shown in Fig. 12.15, where only the even-numbered modal components are included because of the symmetry along the y-direction[88]. The sound pressure distribution seems symmetric, which represents the nature of even-numbered modal response.

Such a symmetric property of the squared sound pressure might be possible in a sound field where there are odd-numbered modal responses. The symmetric nature including the phase can be confirmed by observing the cross correlation of the sound pressure. The cross-correlation coefficients can be expressed by modal functions:

$$c_f\left(y_A, y_A+\Delta y\right)\big|_{x_0,z_0} = \frac{N}{D}$$
$$N = \sum_{\omega_a<\omega_{lmn}<\omega_b} D_{pm}\cos\frac{m\pi y_A}{L_y}\cos\frac{m\pi(y_A+\Delta y)}{L_y}$$
$$D = \sqrt{\sum_{\omega_a<\omega_{lmn}<\omega_b} D_{pm}\cos^2\frac{m\pi y_A}{L_y}\sum_{\omega_a<\omega_{lmn}<\omega_b} D_{pm}\cos^2\frac{m\pi(y_A+\Delta y)}{L_y}}. \tag{12.52}$$

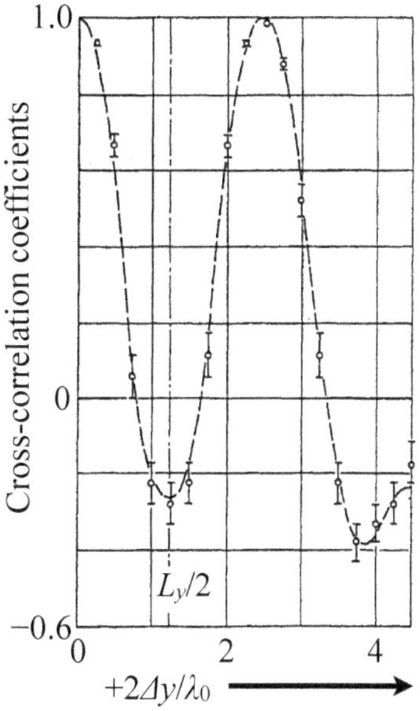

Fig. 12.16 Correlation coefficients along line $(L_x/2, y, L_z/2)$. y_A=2.72 (m). All data are shown with observational error range. Dotted line is calculated by Eq. 12.52 from [88] (Fig.6)

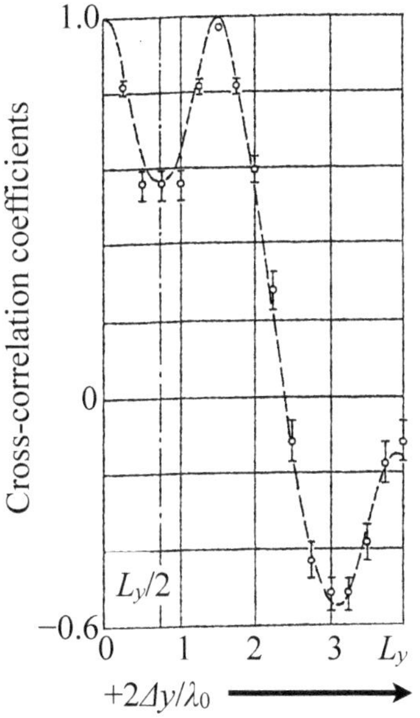

Fig. 12.17 Correlation coefficients along line $(L_x/2, y, L_z/2)$. y_A=3.4 (m). All other data and signatures as in Fig. 12.16 from [88] (Fig.5)

Figures 12.16 and 12.17 are examples of measurements of the cross-correlation coefficients, where the correlation of unity can be seen at the pairs of symmetrical points. These examples indicate that such a pair of reverberation sounds might be perceived as if a single source of the reverberation sounds is located in a free-field.

By taking the space average with respect to y_A on the y-axis,

$$C_F(y)|_{x_0,z_0} = \sum_{\omega_a < \omega_{lmn} < \omega_b} D_{pm} \cos \frac{m\pi}{L_y} y \tag{12.53}$$

can be derived, which is interpreted as the relationship between the power spectrum and auto-correlation function on the time domain. Namely, the two-point cross-correlation coefficients represent the **spatial auto-correlation function**, and D_{pm} shows the power spectrum in terms of the **angular distribution**[86].

12.3.3 Cross-Correlation Function in Two-Dimensional Random Sound Field

Similar to the cross correlation in a three-dimensional field, that for the two-dimensional one can be formulated such as

$$\overline{\mathrm{E}\left[p_1(n)p_2(n-m)\right]} = \frac{AB}{2}\frac{2}{\pi}\int_0^{\pi/2} \cos(kr\cos\theta)d\theta = \frac{AB}{2}\mathrm{J}_0(kr), \tag{12.54}$$

where $\mathrm{J}_0(x)$ denotes the **zero-th order Bessel function**, which is defined as[89]

$$\mathrm{J}_0(x) = \frac{1}{2\pi}\int_0^{2\pi} \mathrm{e}^{\mathrm{i}x\cos\theta} d\theta. \tag{12.55}$$

In particular,

$$C_{F_2}(kr) = \mathrm{J}_0(kr) \tag{12.56}$$

is called the **cross-correlation coefficients in a two-dimensional reverberation field**[84]. In addition,

$$C_{F_1}(kr) = \cos(kr) \tag{12.57}$$

is called the **cross-correlation coefficients in a one-dimensional reverberation field**.

Figure 12.18(b) shows an example of the cross-correlation coefficients in a two-dimensional reverberation field that was constructed as shown in Fig. 11.6. By comparing the results with those for a three-dimensional case all the walls are covered by reflective materials in the same room as shown in Fig. 12.18(a), it can be seen that the correlation coefficients given by the open plots for the two-dimensional case mostly follow the Bessel function indicated by the dotted line. On the other hand, the plots of solid circles in Fig. 12.18(b) are the results measured using a pair of microphones that were set perpendicular to the floor in the two-dimensional field. It can be confirmed that the sound field might be mostly composed of the one-dimensional waves along the line perpendicular to the floor .

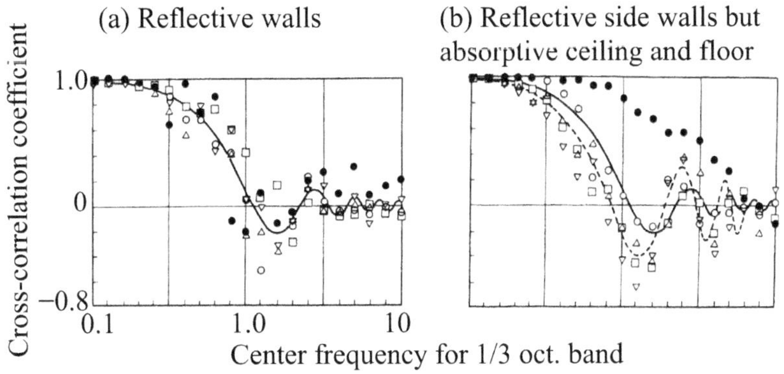

Fig. 12.18 Examples of two-point sound pressure correlation coefficients under different acoustical conditions in variable reverberation room. (a) reflective walls, (b) reflective side walls but absorptive ceiling and floor. Solid circle in figure (b): microphones located along line perpendicular to floor, open plots: microphones in plane parallel with floor. Solid line: $\sin kr/kr$, dotted line: $J_0(kr)$ from [90] (Fig.4)

12.3.4 Effect of Direction Sound on Spatial Correlation

The cross correlation in the sound field depends also on the **directivity of the source**. Rewrite the cross-correlation coefficients using the pair of impulse response records between the sound source and paired observation positions in the sound field. Suppose that $p_1(n)$ and $p_2(n)$ are the sound pressure signals at the positions, and similarly $h_1(n)$ and $h_2(n)$ are the impulse response records. The paired sound signals can be written as

$$p_1(n) = x(n) * h_1(n) \tag{12.58}$$

$$p_2(n) = x(n) * h_2(n), \tag{12.59}$$

where $x(n)$ denotes the source signal. Assuming that the source radiates white noise and starts at $n = 0$, the cross-correlation function becomes

$$\begin{aligned} C_F(m) &= \mathrm{E}[p_1(n)p_2(n-m)] \\ &= \mathrm{E}\left[\left(\sum_{q=0}^{n} x(n-q)h_1(q)\right)\left(\sum_{l=0}^{n-m} x(n-m-l)h_2(l)\right)\right] \\ &= \sum_{q=m}^{n} h_1(q)h_2(q-m) \end{aligned} \tag{12.60}$$

after taking the ensemble average due to the randomness of the white noise as

$$\begin{aligned} \mathrm{E}\left[x(l)x(m)\right] &= 1 \qquad l = m \\ &= 0 \qquad l \neq m. \end{aligned} \tag{12.61}$$

It can be seen that the cross-correlation function is also a function of the time n. By taking a long time n that exceeds the length of the impulse response record, the correlation function is independent of time, namely, the sound field reaches the steady state. The cross correlation might also be sensitive to the effect of direct sound on the sound field. If the sound pressure signal includes the early portion of the impulse response, the correlation becomes high. In contrast, if such a direct sound with its early echoes were removed from the receiving signal, the correlation might dramatically decreases; in particular, the receiving positions are close to the sound source[91][69].

Figure 12.19 demonstrates another example of the cross-correlation coefficients in a reverberation room[69]. The results actually indicate the effect of the **directivity**, i.e., the direct sound on the correlation even in a highly reverberant room. When

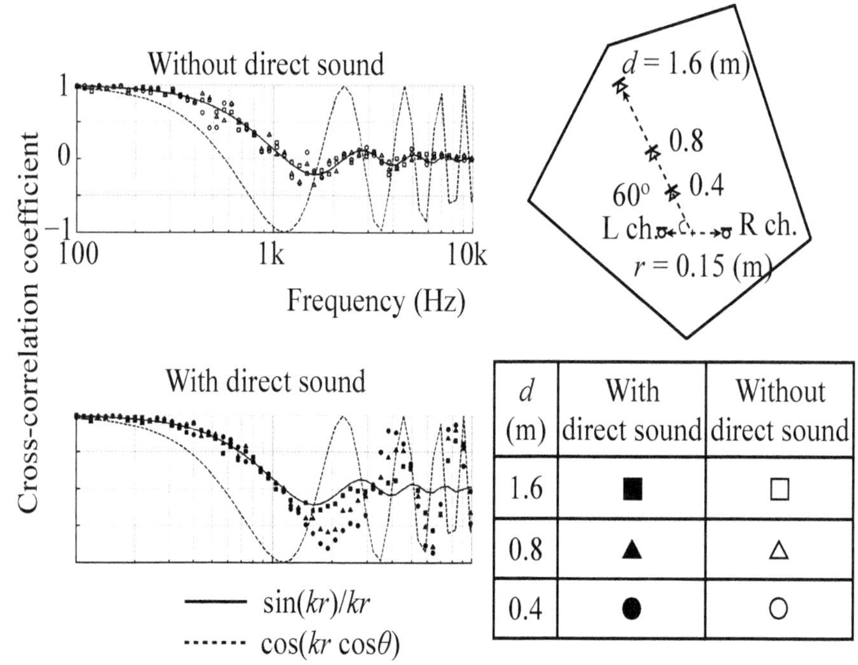

d (m)	With direct sound	Without direct sound
1.6	■	□
0.8	▲	△
0.4	●	○

Fig. 12.19 Examples of two-point sound pressure correlation coefficients with/without direct sound in reverberation room[69]

the loudspeaker faces the paired receiving positions, the correlation coefficient follows that for a free-field or one-dimensional field as shown by the closed plots as the frequency increases. However, if the direct sound is removed, the results follow those for a three-dimensional field as expected in a reverberation room. Perception of reverberation or a sound field in general is greatly sensitive to the direct sound followed by its early echoes[83][92]. Spatial correlation is also an important factor for sound field reproduction. Reproducing spatial correlation properties of original field has been also investigated[85][93].

Chapter 13
Room Transfer Function

This chapter describes room transfer functions in terms of poles and zeros. In general, a sound field in a room can be characterized by power response of a source located in the room. The resonant peaks of the responses due to the eigenfrequencies are observed even if the source locations were averaged throughout the room. This explains why eigenfrequencies are so significant in room acoustics. The power response could be controlled by using secondary sources closely located to the primary source, subject to the low modal overlap condition. In addition to the eigenfrequencies or poles, details of the phase characteristics of the transfer functions will be described, as the magnitude responses expressed by the modal density were in the previous chapters. Consequently, it will be shown how the propagation and reverberation phases are estimated according to the number of minimum and non-minimum-phase zeros distributed on the complex frequency plane instead of the eigenfrequencies.

13.1 Power Response of Point Source in Room

13.1.1 Sound Field in Room Expressed by Modal Functions

Following Eq. 10.49, suppose the **Helmholtz equation** for sound pressure P of sound field in a room is

$$\nabla^2 P(\mathbf{r}',\mathbf{r}) + \frac{\omega^2}{c^2}P(\mathbf{r}',\mathbf{r}) = -\mathrm{i}\omega\rho_0 Q_{0d}\delta(\mathbf{r}'-\mathbf{r}), \qquad (\mathrm{Pa/m^2}) \tag{13.1}$$

where $P(\mathbf{r}',\mathbf{r})$ denotes the sound pressure for a sinusoidal wave at the receiving position $\mathbf{r}$ for the point source located at $\mathbf{r}'$,

$$\begin{aligned} P(\mathbf{r}',\mathbf{r}) &= \frac{-\mathrm{i}\omega\rho_0 c^2 Q_0}{V}\sum_N \frac{\phi_N(\mathbf{r}')\phi_N(\mathbf{r})}{(\omega-\omega_{pN_1})(\omega-\omega_{pN_2})} \\ &\cong \frac{-\mathrm{i}\omega\rho_0 c^2 Q_0}{V}\sum_N \frac{\phi_N(\mathbf{r}')\phi_N(\mathbf{r})}{\omega^2-\omega_N^2-2\mathrm{i}\omega\delta_N} \qquad (\mathrm{Pa}) \end{aligned} \tag{13.2}$$

M. Tohyama: *Sound and Signals*, Signals and Communication Technology, pp. 243–304.
springerlink.com

$\phi_N(\mathbf{r})$ is the **orthogonal modal function** for the sound field with the angular eigenfrequency ω_N. Here the poles are located in the complex frequency plane such that

$$\omega_{p_{N_{1,2}}} = \mp\omega_{N_0} + \mathrm{i}\delta_N \quad (\omega_{N_0} \gg \delta_N, \delta_N \cong \delta) \tag{13.3}$$

$$\omega_{N_0}^2 = \omega_N^2 - \delta_N^2, \tag{13.4}$$

which represents the decaying characteristic δ_N of the sound field due to sound absorption through the walls. ω_{N_0} denotes the angular frequency of the free oscillation of the space.

13.1.2 Sound Power Response of Source and Energy Balance Equation

Recall that the radiation acoustic impedance of the source can be expressed as

$$\begin{aligned} Z_{A_{\mathrm{rad}}} &= \left.\frac{P(\mathbf{r}',\mathbf{r})}{Q_0}\right|_{\mathbf{r}\to\mathbf{r}'} \\ &\cong \frac{-\mathrm{i}\omega\rho_0 c^2}{V}\sum_N \frac{\phi_N^2(\mathbf{r}')}{\omega^2 - \omega_N^2 - \mathrm{i}\omega 2\delta_N} \\ &= R_{A_{\mathrm{rad}}} + \mathrm{i}X_{A_{\mathrm{rad}}} \quad (\mathrm{Pa\cdot s/m^3}) \end{aligned} \tag{13.5}$$

in the room. Then the **sound power output** of the source can be written as

$$W_X(\mathbf{r}',\omega) = \frac{1}{2}R_{A_{\mathrm{rad}}}Q_0^2, \quad (\mathrm{W}) \tag{13.6}$$

assuming that the sound source is a **constant velocity source** whose volume velocity is given by Eq. 10.51, namely

$$q(t) = \int_V Q_{0d}\delta(\mathbf{r}-\mathbf{r}')dv\mathrm{e}^{\mathrm{i}\omega t} = Q_0\mathrm{e}^{\mathrm{i}\omega t}. \quad (\mathrm{m^3/s}) \tag{13.7}$$

Therefore, the sound power output can be rewritten as

$$W_X(\mathbf{r}',\omega) \cong \frac{Q_0^2}{2}\frac{\omega^2\rho_0 c^2}{V}2\delta\sum_N \frac{\phi_N^2(\mathbf{r}')}{(\omega^2-\omega_N^2)^2 + 4\omega^2\delta^2} \quad (\mathrm{W}) \tag{13.8}$$

in a modal expansion form by assuming that

$$\delta_N \cong \delta \tag{13.9}$$

$$\omega_N >> \delta \quad (1/\mathrm{s}) \tag{13.10}$$

in the frequency band of interest. Moreover, the expression above can be rewritten as

$$W_X(\mathbf{r}',\omega) = 2\delta V \frac{<\overline{|p(\mathbf{r}',\omega)|^2}>/2}{\rho_0 c^2} \quad \text{(W)} \tag{13.11}$$

$$\frac{1}{2} <\overline{|p(\mathbf{r}',\omega)|^2}> \cong \frac{Q_0^2}{2} \frac{\omega^2 \rho_0^2 c^4}{V^2} \sum_N \frac{\phi_N^2(\mathbf{r}')}{(\omega^2-\omega_N^2)^2 + 4\omega^2\delta^2}, \quad (\text{Pa}^2) \tag{13.12}$$

which indicates the relationship between the sound power output and the **space-averaged mean square sound pressure** with respect to receiving positions. Recall the energy balance equation for a one-dimensional wave field, such as vibration of a string described by Eq. 5.78. Here the same type of formulation can be derived as the previous one. Namely, this is again the **energy-balance equation** for a three-dimensional sound field[94].

The decaying factor δ in the equation can be given by

$$2\delta = \frac{c\alpha S}{4V} \quad (1/\text{S}) \tag{13.13}$$

in terms of the geometrical acoustics as described by Eq. 11.84, where α is the averaged absorption coefficient for the walls and $S(\text{m}^2)$ is the area of the walls. By introducing the expression of δ into the energy balance equation, the sound-power output of the source can be rewritten as

$$W_X(\mathbf{r}',\omega) = \frac{1}{4} I \alpha S, \quad \text{(W)} \tag{13.14}$$

where

$$I = \frac{<\overline{|p(\mathbf{r}',\omega)|^2}>/2}{\rho_0 c} \quad (\text{W/m}^2) \tag{13.15}$$

denotes the **energy flow density** for a unit time interval or the **sound intensity**. The equation above says that the energy supplied to the sound field by the source in a unit time interval is equal to the sound energy absorbed by the walls in the interval of time. This is the principle of energy balance as previously stated in subsection 5.3.3.

In addition to the outcome, it can also be found that the energy flow density coming into the walls surrounding the room is equal to one quarter of that inside the room. Namely, the estimation of sound absorption in a room according to the geometrical acoustics can be represented based on the modal acoustics[94].

Following the energy-balance equation, the **space average of the mean square sound pressure** with respect to the receiving position can be written as

$$\frac{1}{2} <\overline{|p(\mathbf{r}',\omega)|^2}> = \frac{4\rho_0 c}{\alpha S} W_X(\mathbf{r}',\omega), \quad (\text{Pa}^2) \tag{13.16}$$

which indicates the spatial average increases in proportion to the sound power output or the **intrinsic acoustical impedance** of the medium (air) $\rho_0 c$, while it decreases as the sound absorption increases. Consequently, the sound power output of the source can be estimated by using the space average of the mean square sound pressure.

However, it is not easy to estimate the space average, when the frequency band width of sound radiated from the source becomes narrow as described in subsection 12.2.1.

The sound power output of a source depends on its position in a room. By taking the space average with respect to the sound source position in the room, the sound power output becomes

$$\frac{< W_X(\omega) >}{W_0} \cong \frac{4\pi c^3}{V} 2\delta \sum_N \frac{1}{(\omega^2 - \omega_N^2)^2 + 4\omega^2\delta^2}, \tag{13.17}$$

which is unique to the room where the sound source is located and indicates the eigenfrequencies are also significant factors for the sound power output. Namely, if estimation of the sound power output W_0 from a source in a free field is really desired, it might be extremely difficult in general, for a narrow-band source[95].

Figure 13.1 is an example for numerical calculation of sound power output from a point source of sinusoidal waves in a rectangular reverberation room[94] following the next equation:

$$\frac{< W_X(\omega) >}{W_0} \cong \frac{8\pi c^3}{V} \sum_N \frac{\delta}{(\omega^2 - \omega_N^2)^2 + 4\omega^2\delta^2} \tag{13.18}$$

where

$$W_0 = \frac{\rho_0 \omega^2 Q_0^2}{8\pi c}. \qquad \text{(W)} \tag{13.19}$$

following Eq. 9.45. It can be seen that the sound power output in a room might greatly differ from that in a free field. If the frequency of sound radiated from the source is the same as the eigenfrequency in the room, the sound power output becomes large, while it decreases for sounds of different frequencies. The eigenfrequencies are sparsely found at low frequencies, in particular in a small room. Consequently, the variance in the power response due to the frequency seems very large in general.

Look at the solid line of the lower panel in Fig. 13.1. The peak levels of the power response increase as the frequency decreases. This shows that the peak levels decrease in inverse proportion to the square of the frequency, even if the reverberation time were independent of the frequency. On the other hand, as the frequency increases (over about 250 Hz), such high levels of resonance peaks cannot be seen any longer. It can be interpreted that **overlap of the modal response** becomes significant as the modal density increases in proportion to the square of the frequency. In other words, the decrease of the peak levels inversely proportional to the frequency can be cancelled out by the increase of **modal overlap**. Consequently, the power response randomly varies around $W_X/W_0 \cong 1$ as the frequency increases.

However, when the reverberation time decreases, the sound power output seems to approach that for a free field, as shown in the solid line of the upper panel in Fig. 13.1. Namely, if the effect of resonance on the power response becomes weak by decreasing the reverberation time or increasing the sound absorption, it can be

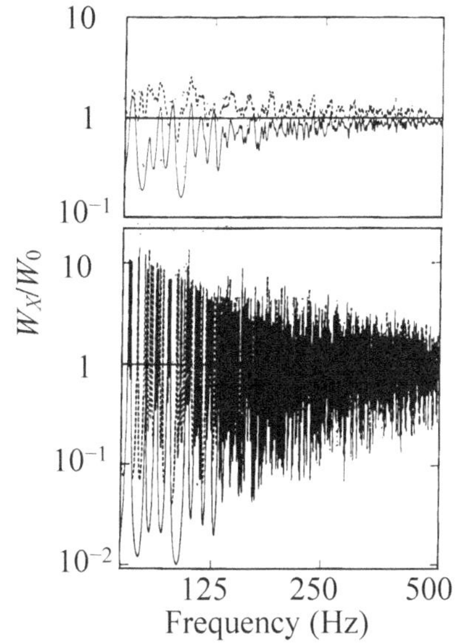

Fig. 13.1 Sound power output from source averaged with source position, calculated following Eq. 13.17. W_0 is sound power output of point source in free field (W), T_R is reverberation time in reverberation room in which sound source is located (s), room volume is 200 (m^3), and ratios of room dimensions are $1 : 2^{1/3} : 4^{1/3}$. (Upper panel) $T_R = 1$ (s); (lower panel)$T_R = 16$ (s), broken lines, axial, tangential and oblique modes, solid lines, oblique modes only from [94] (Fig. 1)

expected that a smoother frequency response of the sound power output (power response of a source) might be achieved even in a small room.

As shown in Fig. 13.1, the sound power output differs markedly from the output in free field, under the condition that the frequency of the source is low or the room reverberation time is long. The extent of the difference depends in a complicated way on the frequency of the source and on the room. Moreover, as shown by the broken lines in Fig. 13.1, the axial and tangential wave modes are contained in the sound fields of a rectangular room in the very low-frequency bands. Note that in such a mixed field the condition specified by Eq. 13.9 cannot be assumed, and thus the energy-balance principle as shown in Eq. 13.12 is not accurately applicable any longer.

13.2 Estimation of Space Average for Mean Square Sound Pressure

The sound power output of a source in a room can be estimated by using the space average of the mean square sound pressure in the room. It is difficult to estimate the space average, but if the sound source radiates random noise, the estimation error of the space average could be reduced. In particular, by locating the receiving positions on the floor or on the edge or at the corner, the estimation variance can be reduced.

13.2.1 Increase of Mean Square Sound Pressure on Wall

Consider the mean square sound pressure averaged on the plane parallel with the x-wall in a rectangular room. Suppose that a random noise source is located in the room. The mean square sound pressure along the center-line y perpendicular to the wall can be expressed as

$$\frac{1}{2}\overline{|p(y)|^2} = \sum_{m=0}^{m_{Max}} D_{pm} \cos^2 \frac{m\pi}{L_y} y \tag{13.20}$$

following Eq. 12.51. Therefore, by assuming that a uniform distribution of the angular power spectrum D_{pm}, i.e., equally probable incidence of the waves into the line y, the sound pressure distribution can be written as

$$\frac{1}{2}\overline{|p(y)|^2} \cong \frac{1}{k_0}\int_0^{k_0} \cos^2 kydk = \frac{1}{2}\left(1 + \frac{\sin 2ky}{2ky}\right), \tag{13.21}$$

which is called the **interference pattern** of a random sound field due to a plane wall[96]. It can be seen that the sound pressure increases by 3 dB on the wall as illustrated by Fig. 13.2; however, such an interference effect becomes weak as the distance from the wall increases.

The same type of effect can be seen for the edge or the corner of the room. Figure 13.3 illustrates measured results of the **interference patterns near the**

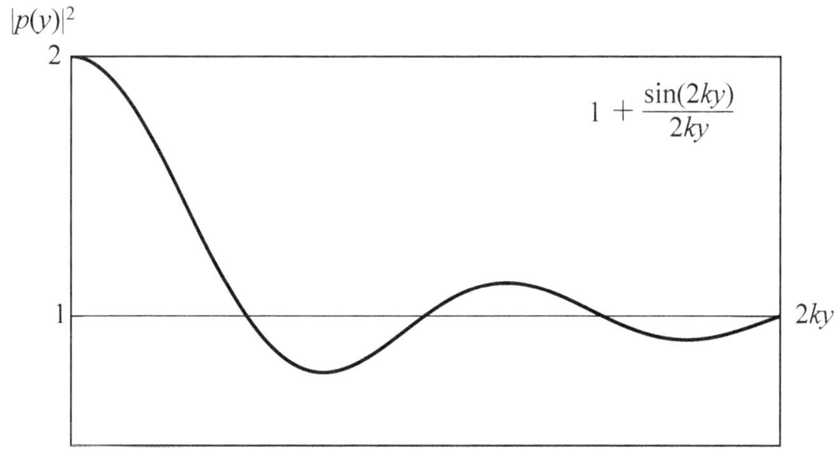

Fig. 13.2 Sound pressure increase on wall in random sound field

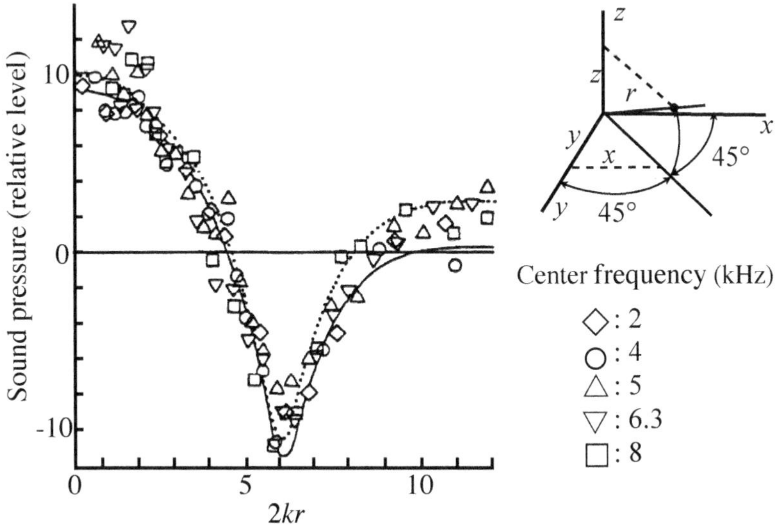

Fig. 13.3 Interference patterns of mean square sound pressure in dB near corner (frequency band: $\nu_c \pm 5$ (Hz)) solid line: calculated under rigid wall condition[96], dotted line: estimated under non-rigid wall and lossy air condition[98] from [97] (Fig.2)

corner in the rectangular reverberation room where a narrow-band noise source is located[97]. It can be seen that sound pressure increases by 9 dB at the corner. Similarly to the pattern for the plane wall, the increasing effect of the sound pressure can be seen at the positions close to the corner.

The interference patterns stated above can be formulated by using the modal functions in the rectangular reverberation room[98]. Set the sound pressure position $\mathbf{r}'(x', y', z')$ and receiving position $\mathbf{r}(x, y, z)$, and assume that the decaying factor δ_N is independent of the modal number and frequencies. The mean square sound pressure can be represented by

$$\frac{1}{2}\overline{|p(\mathbf{r}',\mathbf{r})|^2} \cong C \sum_{\omega_N in \Delta} T_N(\mathbf{r}')R_N(\mathbf{r}) \tag{13.22}$$

$$T_N(\mathbf{r}) = R_N(\mathbf{r}) = \cos^2\frac{l\pi}{L_x}x\cos^2\frac{m\pi}{L_y}y\cos^2\frac{n\pi}{L_z}z \tag{13.23}$$

$$\omega_N = c\sqrt{\left(\frac{l\pi}{L_x}\right)^2 + \left(\frac{m\pi}{L_y}\right)^2 + \left(\frac{n\pi}{L_z}\right)^2}, \tag{13.24}$$

where L_x, L_y and L_z (m) denote the lengths of the sides, Δ is the frequency band of noise, l, m, and n denote positive integers, $\sum_{\omega_N in \Delta}$ expresses the summation for the oblique waves in the frequency band, c is the speed of sound, and C is a constant.

Thus the space average with respect to the receiving positions throughout the room can be rewritten as

$$\frac{1}{2} < \overline{|p(\mathbf{r}')|^2} >_V = \frac{1}{2V}\int_V |p^2(\mathbf{r}',\mathbf{r})| d\mathbf{r} \cong \frac{C}{8} \sum_{\omega_N in \Delta} T_N(\mathbf{r}'), \tag{13.25}$$

where $V(\mathrm{m}^3)$ shows the room volume. Similarly the average only on the xy-wall $z_m = 0$ or $z_m = L_z$ becomes

$$\frac{1}{2} < \overline{|p(\mathbf{r}')|^2} >_{z_m=0} = \frac{1}{2S_{bxy}}\int_{\mathbf{r}onxy} \overline{|p(\mathbf{r}',\mathbf{r})|^2} d\mathbf{r} = \frac{C}{4} \sum_{\omega_N in \Delta} T_N(\mathbf{r}'), \tag{13.26}$$

where $S_{bxy}(\mathrm{m}^2)$ is the surface area for the wall. This is the same as that for the yz-wall ($x_m = 0$ or $x_m = L_x$) or zx-wall ($y_m = 0$ or $y_m = L_y$). Moreover, if the average is taken only on a single edge in the room, the averaged sound pressure is given by

$$\frac{1}{2} < \overline{|p(\mathbf{r}')|^2} >_{y_m=z_m=0} = \frac{C}{2} \sum_{\omega_N in \Delta} T_N(\mathbf{r}'). \tag{13.27}$$

Similarly, by taking mean square sound pressure on the corner instead of averaging, the mean square sound pressure is obtained by

$$\frac{1}{2}\overline{|p(\mathbf{r}',\mathbf{r})|^2}_{x_m=y_m=z_m=0} = C \sum_{\omega_N in \Delta} T_N(\mathbf{r}'). \tag{13.28}$$

Consequently, the next relationship holds well for the averaged ones:

$$\begin{aligned} < \overline{|p(\mathbf{r}')|^2} >_V &= \frac{1}{2} < \overline{|p(\mathbf{r}')|^2} >_{z_m=0} \\ &= \frac{1}{4} < \overline{|p(\mathbf{r}')|^2} >_{y_m=z_m=0} \\ &= \frac{1}{8}\overline{|p(\mathbf{r}',\mathbf{r})|^2}_{x_m=y_m=z_m=0}. \end{aligned} \tag{13.29}$$

This outcome indicates that the space-averaged mean square sound pressure in the room can be estimated by taking the averaged one only for a single wall in the room, or along a single edge, or by simply taking the sound pressure on a corner instead of averaging through the entire room[99][100].

13.2.2 Variances in Mean Square Sound Pressure on Wall

As stated in the previous subsection, the space average for mean square sound pressure can be estimated by using the average on a single wall in the room. In particular, if the variances in the distribution on the wall could be smaller than those for the distribution in the room, it can be expected that the number of samples required for estimating the average within a reasonable variance might be reduced[99].

Figure 13.4 shows arrangements of the sound source and receiving positions. The variance observed under condition A indicates the variance when source and receiving positions are randomly chosen inside the room. Similarly the variances under condition B and C show those for randomly taking only the receiving positions on the floor and on the edge, respectively. Namely, the variances under conditions A-C include the effects of both the source and receiving positions. In contrast, the variances observed under conditions D-G represent the variances due to the source positions only.

Conditions	Source positions	Microphone positions	Notation of normalized space variance
A	Different positions throughout room	Different positions throughout room	$\sigma^2{}_A$
B	Different positions throughout room	Different positions on floor	$\sigma^2{}_B$
C	Different positions throughout room	Different positions on edge	$\sigma^2{}_C$
D	Different positions throughout room	Space average throughout room	$\sigma^2{}_D$
E	Different positions throughout room	Space average on floor	$\sigma^2{}_E$
F	Different positions throughout room	Space average on edge	$\sigma^2{}_F$
G	Different positions throughout room	Fixed at corner	$\sigma^2{}_G$

Fig. 13.4 Conditions for both point source position and microphone position (seven conditions, A-G in rectangular reverberation room ($11\text{m}^L, 8.8\text{m}^W, 6.6\text{m}^H$)) from [99] (Table 1)

Calculated results indicating the space variances, under the conditions specified in Fig. 13.4, are shown in Fig. 13.5. Room dimensions used for the calculation were $L_x = 11$ (m), $l_y = 8.8$ (m), and $L_z = 6.6$ (m), and the center frequencies for a narrow-band noise source were the center frequencies for the 1/3-octave bands from 31.5-250 Hz. The frequency bandwidth for the noise source was 10 Hz. Space variances calculated by using the mean square sound pressure data obtained at each cross point in the divided meshes in the room or at the boundaries of the room. Each side of the meshes was the shortest wavelength in each frequency band.

The results indicate that under conditions A, B, C, and G, the next relation holds well:

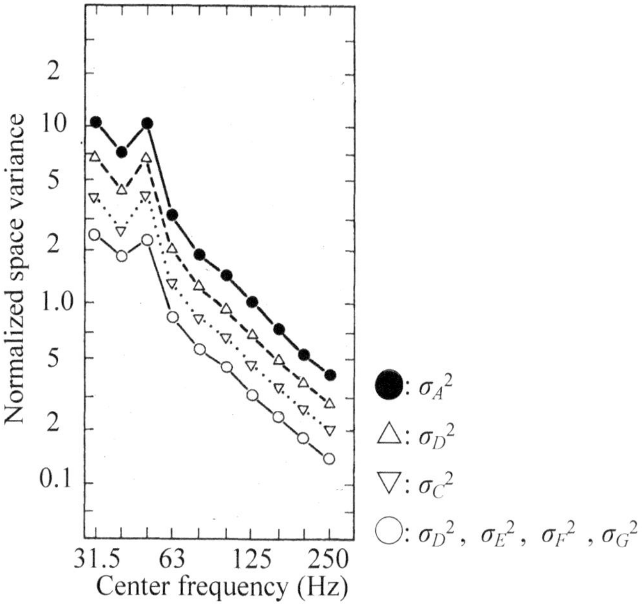

Fig. 13.5 Normalized space variance in mean square sound pressure under conditions A-G specified in Fig. 13.4, source frequency band: $\nu_c \pm 5$ Hz from [99] (Fig.2)

$$\sigma_G^2 < \sigma_C^2 < \sigma_B^2 < \sigma_A^2. \qquad (13.30)$$

However, under the conditions D, E, F, and G,

$$\sigma_D^2 = \sigma_E^2 = \sigma_F^2 = \sigma_G^2 \qquad (13.31)$$

holds well. The amount of variances that depends on source positions $\sigma_D^2, \sigma_E^2, \sigma_F^2$ and σ_G^2 is approximately equal to 1/3 of the amount of σ_A^2. This outcome corresponds to the result described in reference[101][102]. Sampling the mean square sound pressure data at the corner is equivalent to averaging receiver positions through the entire space in a rectangular reverberation room.

Consequently, it is possible to reduce the number of independent samples needed for estimating the space average inside the room, or sound power measurements of a source, that is, by sampling mean square sound pressure data on the floor (condition B), on the edge (condition C), or at the corner (condition G). Under condition G, the receiver position fixed at the corner, the number of samples (number of source positions) becomes 1/2-1/3 the number of samples needed under condition A. Figure 13.6 shows examples of power measurements made using the **corner method**[99]. These experiments were performed in the rectangular reverberation room. The true power level (reference: 1 pW) for sound source L_w is given by

$$L_w = L_{w_c} - 9, \qquad \text{(dB)} \qquad (13.32)$$

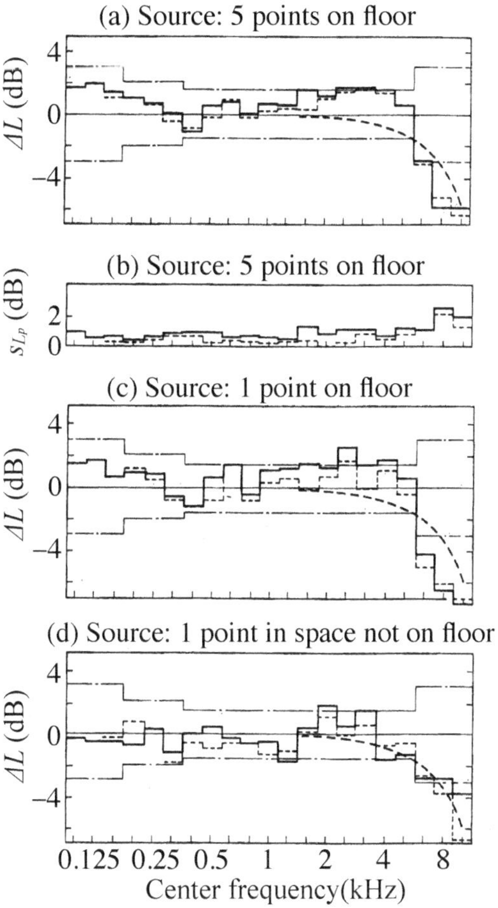

Fig. 13.6 Level differences ΔL of power levels measured by corner method (in rectangular reverberation room) from power level in free field (a), (c), (d), and standard deviation in sound pressure levels (b). Frequency band for sound sources: solid line, 10 Hz (ν_c: 100-10,000 Hz), dotted line 32 Hz (160-10,000 Hz); broken line shows level differences corrected from interference pattern near corner; distance between microphone and corner: 10 mm, dash-dotted line is the uncertainty in determining sound power levels for sound sources in reverberation rooms by ISO-3741,2, and s_{L_p} is the standard deviation in sound-pressure levels, from [99] (Fig.3)

where L_{W_c} is the power level measured by the corner method. Level differences $L_{W_c} - L_{W_f}$ where L_{W_f} shows the power level in a free field reflect in Fig. 13.6 (a) and (c) the condition that all the source positions are sampled on the floor. Thus the results obtained by the corner method are favourable below 1 kHz.

At high frequencies, however, the results by the corner method are not acceptable. It seems mainly due to the distance between the microphone and the corner point because the microphone is not imbedded in the corner but is placed as close as possible to the corner. A 1/2-inch microphone was used in this experiment and the distance between the microphone and the corner point is about 10 mm. The broken line curves in Fig. 13.6 show the corrected level difference calculated from the interference pattern at the point (10 mm away from the corner point) following Fig. 13.3. The correction, however, is not so effective at high frequencies. Therefore, at high frequencies, it seems necessary to imbed the microphone in the corner.

13.3 Active Power Minimization of Source

13.3.1 Active Power Minimization of Source in Closed Space

The sound power output of paired sources was already described in subsection 9.3.3, where the power output was changed according to the phase relationship between the pair of sources. This indicates that the output of a source can be controlled in a closed space by using secondary sources. These secondary sources have a certain coherency with the original (primary) source. This power-reduction method is called **active power minimization**[103].

Suppose that a point source of a sinusoidal wave is located in a rectangular reverberant room at position $\mathbf{r}'_1$. The sound-power output of the source can be formulated as

$$\frac{W_X(\mathbf{r}'_1,\omega)}{W_0/Q_1^2} \cong \frac{4\pi c^3}{V} 2\delta \sum_N \frac{Q_1^2 \phi_N^2(\mathbf{r}'_1)}{(\omega^2-\omega_N^2)^2+4\omega^2\delta^2} = \frac{4\pi c^3}{V} 2\delta Q_1^2 E_1(\mathbf{r}'_1,\omega) \qquad (\mathrm{m}^6/\mathrm{s}^2) \tag{13.33}$$

where W_0 is the sound-power output of the source in a free field such as

$$W_0 = \frac{\rho_0 \omega^2 Q_1^2}{8\pi c}, \qquad (\mathrm{W}) \tag{13.34}$$

and its volume velocity is given by $q_1 = Q_1 e^{i\omega t}$, ϕ_N denotes the eigenfunction of the space assumed as a real function here, and $\delta_N \cong \delta$ is also assumed. To simplify the representation above, introduce the next expression:

$$E_1(\mathbf{r}'_1,\omega) = \sum_N \frac{\phi_N^2(\mathbf{r}'_1)}{(\omega^2-\omega_N^2)^2+4\omega^2\delta^2} \qquad (\mathrm{s}^4)$$

$$= D_1^2(\mathbf{r}'_1,\omega) + R_1^2(\mathbf{r}'_1,\omega) \tag{13.35}$$

$$D_1^2(\mathbf{r}'_1,\omega) = \frac{\phi_{N_p}^2(\mathbf{r}'_1)}{(\omega^2-\omega_{N_p}^2)^2+4\omega^2\delta^2}, \tag{13.36}$$

where ω_{N_p} denotes the nearest eigenfrequency to the source frequency, i.e., D_1^2 expresses the resonance response contributed from the nearest resonance, and the second term R_1^2 expresses the remainder contributed by the other modes[23][24][66].

Now by assuming that a secondary point source $q_2 = Q_2 e^{i\omega t}$ is located at $\mathbf{r}'_2$ and by extending the magnitude of the secondary source to be a negative one (that is equivalent to the phase of π), the total power response TPR from the primary and secondary sources can be written as

$$\begin{aligned} TPR &= \sum_N \frac{(Q_1\phi_N(\mathbf{r}'_1) + Q_2\phi_N(\mathbf{r}'_2))^2}{(\omega^2-\omega_N^2)^2 + 4\omega^2\delta^2} \\ &\cong \frac{(Q_1\phi_{N_p}(\mathbf{r}'_1) + Q_2\phi_{N_p}(\mathbf{r}'_2))^2}{(\omega^2-\omega_{N_p}^2)^2 + 4\omega^2\delta^2} + Q_1^2R_1^2 + Q_2^2R_2^2 \end{aligned} \tag{13.37}$$

where

$$\begin{aligned} E_2(\mathbf{r}'_2,\omega) &= \sum_N \frac{\phi_N^2(\mathbf{r}'_2)}{(\omega^2-\omega_N^2)^2 + 4\omega^2\delta^2} \\ &= D_2^2(\mathbf{r}'_2,\omega) + R_2^2(\mathbf{r}'_2,\omega) \end{aligned} \tag{13.38}$$

$$D_2^2(\mathbf{r}'_2,\omega) = \frac{\phi_{N_p}^2(\mathbf{r}'_2)}{(\omega^2-\omega_{N_p}^2)^2 + 4\omega^2\delta^2}. \tag{13.39}$$

Consequently, the magnitude $Q_{2_{\min}}$ that minimizes TPR and the minimum power response MPR achieved by that are given by

$$Q_{2_{\min}} = \frac{-Q_1\phi_{N_p}(\mathbf{r}'_1)\phi_{N_p}(\mathbf{r}'_2)}{\phi_{N_p}^2(\mathbf{r}'_2) + R_2^2[(\omega^2-\omega_{N_p}^2)^2 + 4\omega^2\delta^2]} \quad (\mathrm{m}^3/\mathrm{s}) \tag{13.40}$$

$$MPR \cong Q_1^2\left(E_1 - D_1^2\left(1 - \frac{R_2^2}{D_2^2}\right)\right) \cong Q_1^2\left(E_1 - D_1^2\frac{D_2^2}{E_2}\right). \quad (\mathrm{m}^6\cdot\mathrm{s}^2) \tag{13.41}$$

The MPR is possibly smaller than that for the primary source only. However, MPR is greatly sensitive to the acoustical conditions of surroundings, including the position of the secondary source. If

$$\frac{D_2^2(\mathbf{r}'_2,\omega)}{E_2(\mathbf{r}'_2,\omega)} \cong 1 \tag{13.42}$$

could be assumed at the position of the Secondary source, then the greatest reduction would be achieved. Namely, the MPR might be reduced to the remainder of the power response of the primary source. In contrast, when

$$\frac{D_2^2(\mathbf{r}'_2,\omega)}{E_2(\mathbf{r}'_2,\omega)} \cong 0 \tag{13.43}$$

can be assumed at the position of the secondary source, the power minimization effect may not occur any longer. In other words, active power minimization is possible in the sound field where the peaks and troughs are separately observable, while it cannot be expected in a sound field where most of those peaks and troughs overlap. However, it is possible to obtain the minimization effect if the secondary source is located closely enough to the primary one so that $E_1(\mathbf{r}'_1,\omega) \cong E_2(\mathbf{r}'_2,\omega)$ holds well.

Figure 13.7(a) is an example of calculation assuming a rectangular reverberation room[66][104]. Only oblique wave modes were taken into account for simplification. In the results, some of the resonance response peaks were reduced. The formulation stated above for *MPR* and *TPR* can be extended into cases that include secondary sources. Figure 13.7(b) illustrates *MPR* that was achieved by using two secondary sources. A larger reduction can be expected than that obtained by using a single secondary source only. In particular, the total power output is reduced at the frequency where the two dominant modes overlap. Generally, *N* secondary sources are necessary to control the resonance peak in which *N* modes overlap.

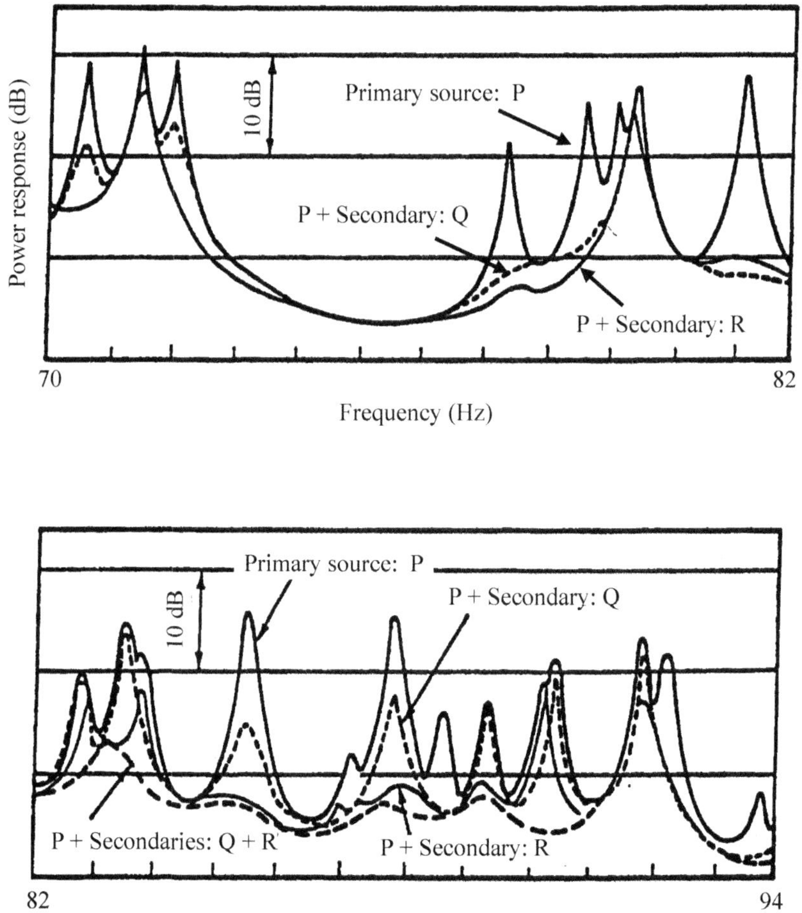

Fig. 13.7 Numerical results for minimum power response of primary source and additional sources in rectangular reverberation room. (Upper panel) One additional source. (Lower panel) Two additional sources are included from [104] (Fig.1)

Figure 13.8(a) shows power responses of pure tone sources in a reverberation room[66][104]. The resonance peaks in the power response are greatly reduced by using a secondary source that is located at a point far from the primary source, as shown in Fig. 13.9. Power reduction was observed, although the correspondence to theoretical calculations was not confirmed numerically. The amplitude and phase (in-phase or anti-phase) of the secondary source are controlled in order to minimize the space-averaged mean square sound pressure. The power response of the sources can be obtained from the averaged mean square sound pressure following the energy balance equation as stated subsection 13.1.2. The space-averaged sound pressure was estimated from the observed data by using 6 microphones that were located randomly as shown in Fig. 13.9.

The sound pressure responses are illustrated by Fig. 13.8 (b), where the microphone is located at a corner. A corner is a suitable microphone location when the modal overlap is very small or when uncorrelated modes are excited in the sound field, because the resonance peaks observed in the sound pressure response at the

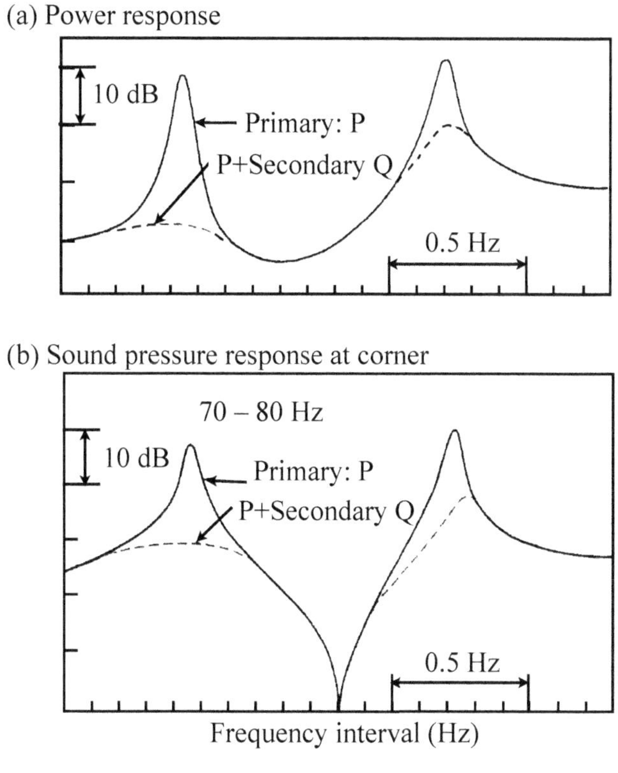

Fig. 13.8 Experimental results of active power minimization for pure tone source, using additional source in reverberation room. (a) Power response. (b) Sound pressure response at corner from [104] (Fig.2)

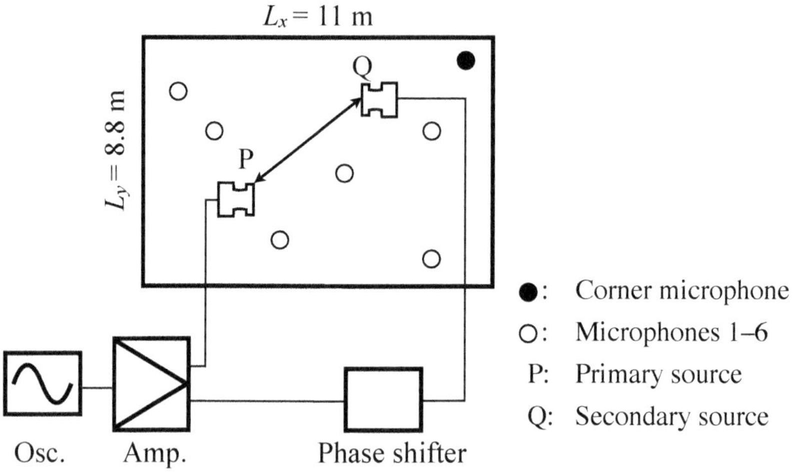

Fig. 13.9 Experimental arrangement in reverberation room from [104] (Fig.3)

corner correspond to those of the power response that is estimated by using the space-averaged sound pressure as described in subsection 13.1.2. Thus, the power response peaks can be minimized, if the amplitude and phase of the secondary source are controlled in order to minimize the resonance peaks of the sound pressure response data observed at the corner.

13.3.2 Estimation of MPR in Terms of Modal Overlap

As stated above, the *MPR* has been evaluated by using the dominant term and the remainder in the power response; however, it is difficult to estimate precisely the remainder under practical conditions. The space-and-frequency-averaged remainder, however, can be evaluated by using the **modal overlap** of the sound field[66]. Thus, it is possible to obtain a theoretical estimate of the expected (averaged) *MPR*. Recall that the power response of the primary source can be represented by

$$E_1(\mathbf{r}'_1,\omega) = D_1^2(\mathbf{r}'_1,\omega) + R_1^2(\mathbf{r}'_1,\omega). \tag{13.44}$$

Then, take the space average with respect to the source position such that

$$< D_1^2(\omega) > = \frac{1}{(\omega^2-\omega_{N_p}^2)^2+4\omega^2\delta^2} \cong \frac{1}{4\omega_{N_p}^2\delta^2} \tag{13.45}$$

$$< R_1^2(\omega) > \cong \sum_N \frac{1}{(\omega^2-\omega_N^2)^2+4\omega^2\delta^2}$$

$$\cong \int_0^\infty \frac{n(x)}{(\omega^2-x^2)^2+4x^2\delta^2}dx$$

$$\cong \frac{V}{2\pi^2 c^3} \frac{1}{4\delta^2} \int_0^{+\infty} \frac{dx}{1+(\frac{\omega-x}{\delta})^2}$$
$$= \frac{V}{2\pi^2 c^3} \frac{1}{4\delta} \int_{-\infty}^{+\infty} \frac{d\xi}{1+\xi^2}$$
$$= \frac{V}{8\pi c^3 \delta}, \tag{13.46}$$

where

$$n(x) = \frac{Vx^2}{2\pi^2 c^3} \tag{13.47}$$

and $n(x)$ denotes the **modal density** where x is interpreted as the angular frequency. Consequently,

$$\frac{< R_1^2(\omega) >}{< D_1^2(\omega) >} \cong \frac{V\omega^2}{2\pi^2 c^3}\pi\delta = n(\omega)\pi\delta = n(\omega)B_M = M(\omega) \tag{13.48}$$

is derived, where $M(\omega)$ is the **modal overlap** and $B_M = \pi\delta$ is called the **modal bandwidth** or **equivalent bandwidth**. The reason B_M is called the modal bandwidth is according to

$$\int_0^{\infty} < D_1^2(\omega) > d\omega = \int_0^{\infty} \frac{1}{(\omega^2-\omega_N^2)^2+4\omega^2\delta^2} d\omega$$
$$\cong \frac{\pi\delta}{4\omega_N^2\delta^2} \cong < D_1^2(\omega_N) > \pi\delta = < D_1^2(\omega_N) > B_M. \tag{13.49}$$

Therefore, the modal overlap denotes the number of modes within the modal bandwidth.

According to the expressions above, the relationship

$$\frac{< D_1^2(\omega) >}{< E_1(\omega) >} = \frac{< D_1^2(\omega) >}{< D_1^2(\omega) > + < R_1^2(\omega) >} \cong \frac{1}{1+M(\omega)} \cong \frac{< D_2^2(\omega) >}{< E_2(\omega) >} \tag{13.50}$$

can be obtained. Thus, by taking the space average of

$$\frac{< MPR >}{Q_1^2 < E_1 >} = 1 - \frac{< D_1^2 >< D_2^2 >}{< E_1 >< E_2 >} \cong 1 - \left(\frac{1}{1+M(\omega)}\right)^2 \tag{13.51}$$

can be derived. The modal overlap is about $M(\omega) \cong 0.25$ under the condition where the numerical calculation was made as shown in Fig. 13.7. Therefore, the reduction effect by using a single secondary source amounting to about 4.4 dB can be expected[66], detailed study can also be seen in reference[103][105].

13.4 Representation of Transfer Functions

Transfer functions are complex functions defined on a **complex frequency plane**. They are represented by **poles and zeros**. Consider the distribution of the poles and zeros on the complex frequency plane in the following sections in this chapter.

13.4.1 Frequency Characteristics for Single-Degree-of-Freedom System

Recall Eq. 2.19. A vibration system whose **impulse response** is represented by a decaying sinusoidal function such as

$$h(t) = Ae^{-\delta_0 t}\sin(\omega_d t + \phi) \tag{13.52}$$

is called a **single-degree-of-freedom system**. Here ω_d is the **angular frequency for the damped free oscillation** that is equal to the eigenfrequency if the **decaying factor** (or **damping constant**) δ_0 were equal to zero. Note that $2\delta_0$ is also called the damping constant sometimes. Expressing the damping constant $2\delta_0$ by the reverberation time T_R can be written as

$$2\delta_0 = \frac{\ln 10^6}{T_R} \cong \frac{13.8}{T_R}. \qquad (1/\mathrm{s}) \tag{13.53}$$

The **free oscillation of a single-degree-of-freedom system** can be obtained as a solution for the differential equation

$$M\frac{d^2x(t)}{dt^2} + R\frac{dx(t)}{dt} + Kx(t) = 0, \qquad (\mathrm{N}) \tag{13.54}$$

which can be interpreted as the free oscillation of a mass and a spring as already described in Chapter 2. The frequency of free oscillation is derived as the solution of the quadratic equation

$$\omega_s^2 M - \mathrm{i}\omega_s R - K = 0. \tag{13.55}$$

Namely, the complex frequency of the free oscillation is expressed as

$$\omega_s = \pm\sqrt{\omega_0^2 - \delta_0^2} + \mathrm{i}\delta_0 = \pm\omega_d + \mathrm{i}\delta_0 \qquad (1/\mathrm{s}) \tag{13.56}$$

where

$$\omega_0 = \sqrt{K/M}, \qquad \delta_0 = R/2M. \quad (1/\mathrm{s}) \tag{13.57}$$

As mentioned previously in Chapter 2, the frequency of the oscillation, which means the real part of the **complex frequency**, becomes low as the damping factor increases. Therefore, the condition $\omega_0 > \delta_0$ is normally required for determining the frequency of free oscillation of a vibrating system.

By taking Fourier transform of the impulse response,

$$H(\omega) = \frac{A}{4\pi}\left(\frac{e^{-i\phi}}{\omega - \omega_{s_1}} - \frac{e^{i\phi}}{\omega - \omega_{s_2}}\right) \tag{13.58}$$

is derived; this is called the**frequency characteristic** of the vibrating system, where

$$\omega_{s_1} = -\omega_d + i\delta_0 \tag{13.59}$$

$$\omega_{s_2} = +\omega_d + i\delta_0. \tag{13.60}$$

In addition, by extending the frequency into the complex variable $\omega_s = \omega + i\delta$, the function of ω_s is called the **transfer function** defined on the **complex frequency plane**. However, $\omega_{s_1} = \omega_{p_1}$ and $\omega_{s_2} = \omega_{p_2}$ are denoted by the **poles** where the transfer functions are not defined. The poles are also called **singularities** of the transfer function. The poles or singularities are located above the **real frequency line (axis)** for sinusoidal vibrations represented by using $e^{i\omega t}$ as shown in Fig. 13.10. The

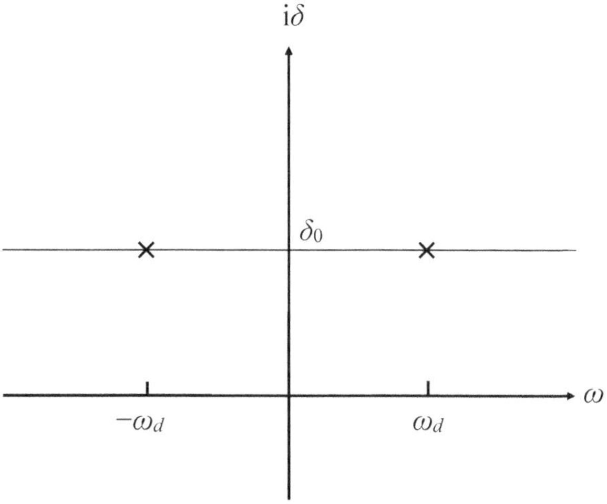

Fig. 13.10 Example of pair of poles on complex frequency plane

distance between the poles and real frequency line corresponds to the damping constant, and thus the distance increases with the damping. In contrast, if the damping is small, the distance is also short, and thus the poles are located very close to the real frequency axis.

The magnitude of frequency characteristics is called the **magnitude frequency response**, while the angle of frequency characteristics is called the **phase frequency response**. The magnitude response is a function of the frequency, so it takes its maximum at the frequency, which is called the **resonance frequency** and is close to the eigenfrequency. The frequency characteristics can be approximated as

$$H(\omega) \cong \frac{A}{4\pi}\left(\frac{(e^{-i\phi}-e^{i\phi})\omega_d-(\omega_{p_2}e^{-i\phi}-\omega_{p_1}e^{i\phi})}{(\omega-\omega_{p_1})(\omega-\omega_{p_2})}\right)=\frac{A}{4\pi}\frac{N(\omega)}{D(\omega)} \tag{13.61}$$

subject to $\omega \cong \omega_d$. Therefore, the **resonance frequency** at which the magnitude becomes its maximum is given by

$$\omega_M = \sqrt{\omega_d^2-\delta_0^2} = \sqrt{\omega_0^2-2\delta_0^2} \cong \omega_0, \qquad \text{(rad/s)} \tag{13.62}$$

indicating that the denominator becomes the minimum at the frequency. Consequently, there are three kinds of frequencies that represent the single-freedom-of-vibration system: **eigenfrequency** ω_0, **frequency of free oscillation** ω_d, and **resonance frequency** ω_M, where $\omega_0 > \omega_d > \omega_M$ generally holds well[5](See Appendix).

The frequency characteristics around the resonance frequency are sometimes simply called the **resonance response**. The resonance response can be written as

$$H(\omega) = \frac{A}{4\pi}H_N(\omega)H_D(\omega) \tag{13.63}$$

$$H_D(\omega) = \frac{1}{(\omega-\omega_{p_1})(\omega-\omega_{p_2})}. \tag{13.64}$$

The **half-power bandwidth**, which is similar to the equivalent bandwidth defined by Eq. 13.49 can be defined as follows. By setting

$$|H_D(\omega_M)|^2 \cong \frac{1}{4\delta_0^2\omega_M^2}, \tag{13.65}$$

the frequency ω_B at which the squared magnitude becomes

$$|H_D(\omega_B)|^2 = \frac{1}{2}|H_D(\omega_M)|^2 \tag{13.66}$$

is given by

$$\omega_B \cong \omega_M \pm \delta_0, \tag{13.67}$$

where

$$|H_D(\omega)|^2 = \frac{1}{(\omega^2-\omega_d^2-\delta_0^2)^2+4\delta_0^2\omega^2} = \frac{1}{(\omega^2-\omega_0^2)^2+4\delta_0^2\omega^2} \tag{13.68}$$

and δ_0 is called the **half-power bandwidth**. The **equivalent bandwidth** can be written as $B_M = \pi\delta_0$ by using the **half-power bandwidth**. Note here that $\omega^2 - \omega_d - \delta_0^2 = \omega^2 - \omega_0^2$. Thus, the modal expansion form of the transfer function can be interpreted as nothing but superposition of the responses of single-degree-of-freedom systems subject to $\delta_N \cong \delta_0$ where δ_N denotes the modal damping constant.

13.4.2 Residues and Zeros of Transfer Function

As stated in the previous subsection, the transfer function of a single degree-of-freedom system can be characterized by the poles. However, the transfer function of a **multi-degree-of-freedom system**, such as the **room transfer function**, can contain the **zeros** as well as the **poles**. The **occurrence of zeros** depends on the sign of **residues** of the poles[23][24].

Consider a sound field in a room that is surrounded by hard walls, and set the N-th pole such that

$$\omega_{p_N} = \omega_{d_N} + \mathrm{i}\delta_N \cong \omega_N + \mathrm{i}\delta_N, \tag{13.69}$$

where $\omega_N > \delta_N > 0$ and $\omega_{d_N} \cong \omega_N$ are assumed, ω_N denotes the angular eigenfrequency for the sound field surrounded by the rigid walls without sound absorption, and ω_{d_N} denotes the angular frequency of the N-th free oscillation. Figure 13.11 is an image of the distribution for the poles and zeros in the complex frequency

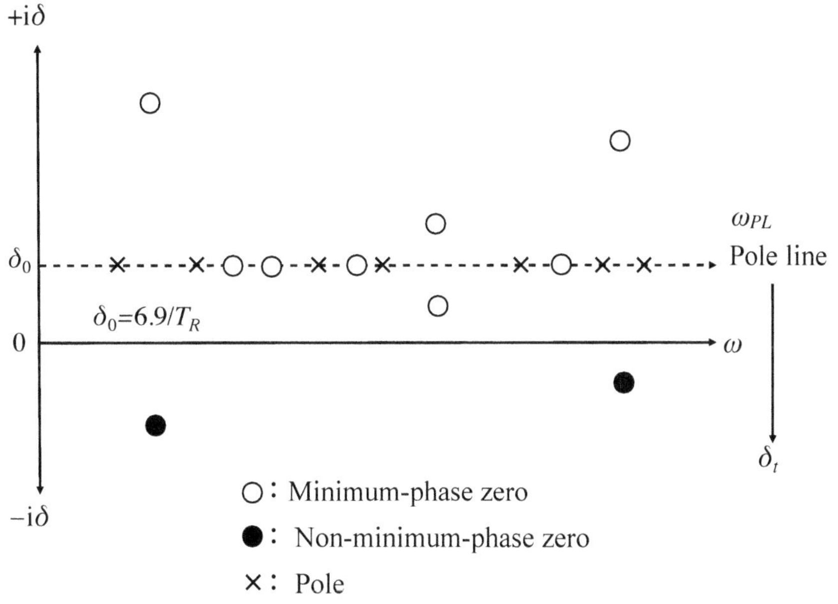

Fig. 13.11 Pole-zero pattern in complex-frequency plane

plane. The horizontal axis shows the real frequency, while the vertical one corresponds to the imaginary part of the complex frequency, i.e., the **damping constant**. The **poles** are located above the real frequency axis, while the **zeros** are distributed above and below the frequency axis. The line connecting the poles is called the **pole line**, which is a parallel with the real frequency axis subject to the damping constant being independent of the frequency. It will be shown that the transfer function is symmetric with respect to the pole line.

Now consider the occurrence of zeros between two adjacent poles on the pole line. Define the transfer function that has two poles for $\omega_A < \omega < \omega_B$ as

$$H(\omega_{PL}) = \frac{A}{\omega_{PL} - \omega_{PA}} + \frac{B}{\omega_{PL} - \omega_{PB}}, \tag{13.70}$$

where

$$\omega_{PA} = \omega_A + i\delta_0 \tag{13.71}$$
$$\omega_{PB} = \omega_B + i\delta_0 \tag{13.72}$$
$$\omega_{PL} = \omega + i\delta_0 \tag{13.73}$$

and A, B are called the **residues** for the respective poles and are assumed to be real numbers. Figure 13.12 is a schematic of zeros occurring, which indicates that a zero

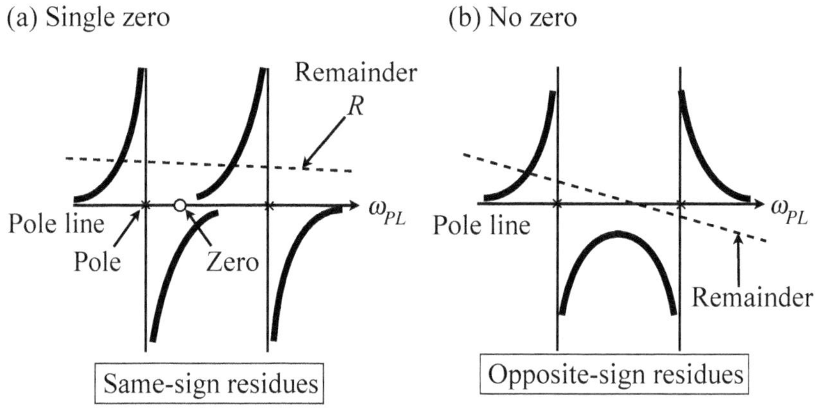

Fig. 13.12 Possibility of formation of zero in interval of two adjacent poles, depending on relative signs of residues from [106] (Fig.4)

occurs on the pole line between the two poles with the same sign residues, while no zeros occur between poles with the residues of opposite signs[23][24][106].

The **transfer function for a multi-degree-of-freedom system** can be expressed as superposition of resonance and out-of-resonance responses as already stated in subsection 13.3. Now reconsider the occurrence of zeros between two adjacent poles for a multi-degree-of-freedom system. For that purpose, define the transfer function as

$$H(\omega_{PL_s}) = \frac{A}{\omega_{PL_s} - \omega_A} + \frac{B}{\omega_{PL_s} - \omega_B} + R(\omega_{PL_s}), \tag{13.74}$$

where the complex frequency is extended into the complex frequency plane from the pole line as

$$\omega_{PL_s} = \omega_{PL} \pm i\delta_t \tag{13.75}$$

and δ_t shows the distance from the pole line. $R(\omega_{PL_s})$ is called the **remainder function** assuming that

$$R(\omega_{PL}) \cong \text{const} \qquad (\omega_A < \omega < \omega_B). \tag{13.76}$$

Look again at Fig. 13.12. As stated previously, there is a zero on the pole line between the two adjacent poles with the same sign residues even if the remainder function exists. There are three cases of zero occurring, however, for the poles of opposite-sign residues including the remainder function. Namely, no zero, **double zero**, and a **symmetric pair of zeros** are possible as illustrated Fig. 13.13. These zero locations can be formulated as follows[107].

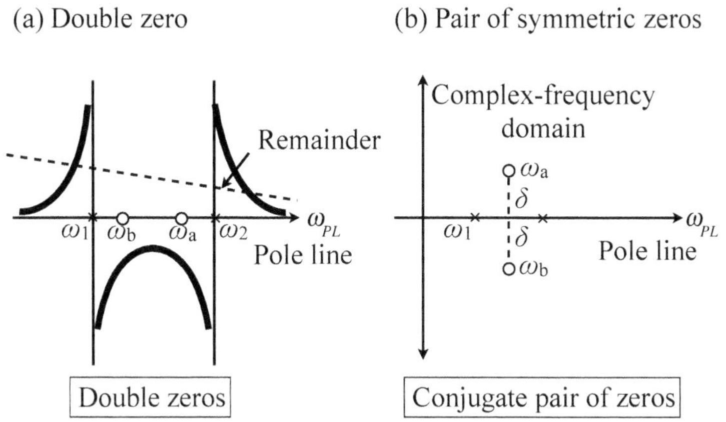

Fig. 13.13 Zeros from opposite-sign residues and remainder from [106] (Fig.5)

By approximating the remainder function to be almost constant, the transfer function can be approximated by[23]

$$H(\omega) \cong \frac{A}{\omega - \omega_A} + \frac{B}{\omega - \omega_B} + R, \tag{13.77}$$

where ω_{PL} is simply written as ω. First suppose that $R \cong 0$. A zero,

$$\omega_z = \omega_0 + \Delta\omega \frac{A - B}{A + B} \tag{13.78}$$

is obtained as a solution of the equation

$$H(\omega_z) = \frac{A}{\omega_z - \omega_A} + \frac{B}{\omega_z - \omega_B} = 0, \tag{13.79}$$

where

$$\omega_0 = \frac{\omega_A + \omega_B}{2}, \qquad \Delta\omega = \frac{\omega_B - \omega_A}{2} \quad (\omega_B > \omega_A)$$
$$\omega_B = \omega_0 + \Delta\omega, \quad \omega_A = \omega_0 - \Delta\omega.$$

If A and B have the same sign residue,

$$\left|\frac{A-B}{A+B}\right| < 1 \tag{13.80}$$

holds well and consequently the zero is located on the pole line between the poles.

In contrast, suppose that A and B are with the opposite sign and consider the zero on the pole line for the equation such that

$$\frac{A}{\omega_z - \omega_A} + \frac{B}{\omega_z - \omega_B} + R = 0. \tag{13.81}$$

The equation above can be rewritten as

$$\frac{A}{\hat{\omega} + \Delta\omega} + \frac{B}{\hat{\omega} - \Delta\omega} + R = 0 \tag{13.82}$$

by introducing the variables

$$\hat{\omega} = \omega - \omega_0, \quad \omega - \omega_A = \hat{\omega} + \Delta\omega, \quad \omega - \omega_B = \hat{\omega} - \Delta\omega. \tag{13.83}$$

The solutions of the quadratic equation above are given by

$$\omega_z = \omega_0 + \frac{-(A+B) \pm \sqrt{(A+B)^2 + 4R\cdot(R\cdot(\Delta\omega)^2 + (A-B)\Delta\omega)}}{2R}. \tag{13.84}$$

For simplicity, suppose that $|A| = |B| = A > 0$. Assuming that the residues are with the same sign, the zeros are given by

$$\omega_{z_1} = \omega_0 + \sqrt{\frac{A^2}{R^2} + (\Delta\omega)^2} + \frac{-A}{R} \tag{13.85}$$

$$\omega_{z_2} = \omega_0 - \sqrt{\frac{A^2}{R^2} + (\Delta\omega)^2} + \frac{-A}{R}, \tag{13.86}$$

one of which is the zero on the pole line between the poles[23][24][107]. Namely, if $A/R > 0$, then ω_{z_1}, is the corresponding zero. On the other hand, ω_{z_2} is the zero when $A/R < 0$. This is because

$$-\Delta\omega < \sqrt{\frac{A^2}{R^2} + (\Delta\omega)^2} + \frac{-A}{R} < \Delta\omega \tag{13.87}$$

holds well when $A/R > 0$.

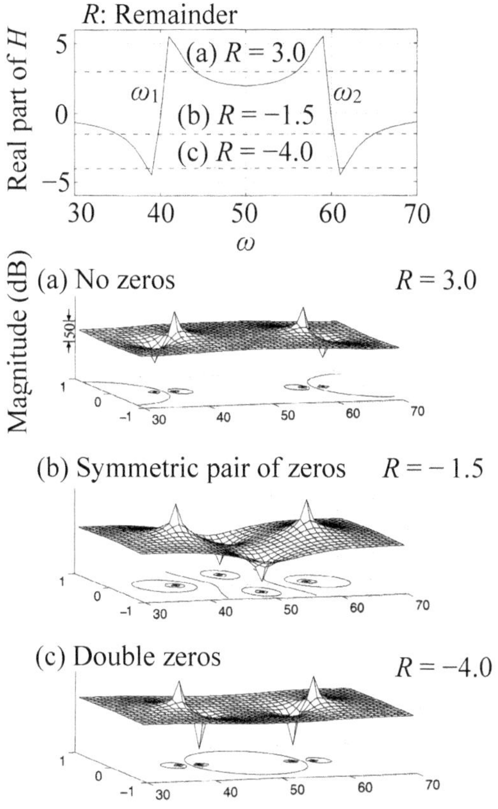

Fig. 13.14 Occurrence of zeros under opposite sign residues from [46] (Fig.7.4.4)

Figure 13.14 indicates the three cases of zero occurring for the opposite sign residues. Recall that $|A| = |B| > 0$ for the opposite sign residues, and the zeros are rewritten as

$$\omega_z = \omega_0 \pm \Delta\omega\sqrt{1 + \frac{2A}{\Delta\omega R}}. \tag{13.88}$$

This indicates the three cases: (a) no zero for $R > 0$
(b) a **symmetric pair of zeros** for $R < 0$ and $1 + \frac{2A}{\Delta\omega R} < 0$ that is given by a pair of complex numbers such that

$$\omega_z = \omega_0 \pm i\Delta\omega\sqrt{-\left(1+\frac{2A}{\Delta\omega R}\right)}; \tag{13.89}$$

and
(c) **double zero** on the pole line for $R < 0$ and

$$1 > 1 + \frac{2A}{\Delta\omega R} > 0, \tag{13.90}$$

which coincide when $1 + \frac{2A}{\Delta\omega R} = 0$. Similar analysis can be done when $A < 0$ and the condition of the sign of R is converted[107].

Both minimum and non-minimum-phase zeros can be contained in the transfer function. A **non-minimum-phase zeros**, however, is produced as one of a pair of zeros located at equal distances to the pole line from each other. This is because the transfer function has a symmetrical form in a complex domain with respect pole line, assuming real residues. One member of this pair of zeros should be non-minimum phase in a slightly damped system, because the pole line runs just above (below) the real-frequency axis assuming $e^{i\omega t}$ ($e^{-i\omega t}$) time dependency. Note, however, that no clear phase jump (because of counter balancing of phase behavior due to the symmetry of the pair of zeros) is observed on the real frequency axis near the non-minimum-phase zero when the damping of the transfer function is very small, which, therefore, does produce phase characteristics consistent with minimum-phase behavior. The symmetric location of the pair of "conjugate-like" zeros at equal distance above and below the pole line (approximately equal to the real frequency axis) cancels their phase effects[107]. Phase responses for minimum-phase or non-minimum-phase transfer functions will be described in following sections in detail.

13.5 Sign of Residues of Transfer Function

The phase characteristics of the transfer function can be determined according to the poles and zeros as well as the magnitude response[23][24]. The phase changes by $-\pi$ at the **pole**, while it recovers π at the **zero** (**minimum phase zero**). Suppose that there are $N_p(\omega)$ poles and $N_z(\omega)$ zeros below the angular frequency ω. The **accumulated phase progression** from 0 to ω can be represented by

$$\Phi(\omega) = -\pi N_p(\omega) + \pi N_z(\omega) = -\pi(N_p(\omega) - N_z(\omega)), \tag{13.91}$$

which is called **accumulated phase** characteristics. The number of zeros depends on the sign change of the residues. By neglecting the case of double zeros, the accumulated phase can be approximated as[23][24]

$$\Phi(\omega) \cong -\pi P_{sc} N_p(\omega), \tag{13.92}$$

where the number of zeros is expressed as

$$N_z(\omega) = (1 - P_{sc})N_p(\omega) \tag{13.93}$$

and P_{sc} denotes the **probability of the residue sign change**. Note that there is no phase effect on the pole line by a symmetric pair of zeros with respect to the pole line.

13.5.1 Transfer Function Phase for One-Dimensional Waves

Now consider the probability of a sign change for the residues. Recall the transfer function can be expressed as the modal expansion using the orthogonal functions as described in subsection 10.1.4, and assume that the transfer function can be written as

$$H(x', x, \omega) = C \sum_n \frac{\sin(\frac{n\pi}{L}x')\sin(\frac{n\pi}{L}x)}{(\omega + \omega_n)(\omega - \omega_n)}, \tag{13.94}$$

where ω denotes the angular frequency on the pole line, $\omega_n = c(n\pi/L)$, c is the speed of sound, L gives the length of the one-dimensional system, x', x are the source and observation points, respectively, and C is a constant.

The probability of sign changes of the numerator depends on the number of **nodes** that are located between the source and receiving positions, namely,

$$\sin k_n x_z = 0 \tag{13.95}$$

is satisfied where $k_n = \omega_n/c$. Consider the modal patterns corresponding to the poles as shown in Figure 13.15. If the number of nodes increases as the pole changes from the n-th to the $(n+1)$-th pole, the sign of the numerator is converted. The sign can normally be expected to remain the same despite the poles changing, if the distance

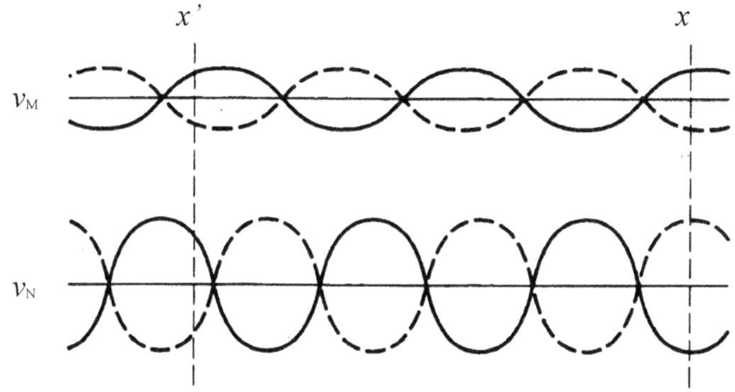

Fig. 13.15 Standing wave pattern in one-dimensional system showing phase advances as nodal pattern changes from [24] (Fig.1)

between the source and receiving positions is sufficiently short. Consequently, a zero could be located at every interval between adjacent poles[23][24]. That is, such node occurrences correspond to the process of losing the zeros that can be represented by the probability P_{sc}.

The **number of nodes** can be estimated by N_n such that

$$\mathrm{Int}(k\frac{|x-x'|}{\pi}) = N_n(k), \tag{13.96}$$

where $\mathrm{Int}(x)$ denotes the maximal integer that does not exceed x. By introducing this number of nodes, the accumulated phase between the source x' and receiving x positions can be estimated using the continuous function

$$\Phi(k) = -\pi N_n(k) \cong -k|x-x'| = -kr. \tag{13.97}$$

This phase difference is called the **propagation phase**, which is equivalent to that between a pair of two locations where a progressive plane wave travels[23][24]. Note that the propagation phase can be rewritten as

$$\frac{\partial\Phi(k)}{\partial k} = -r, \tag{13.98}$$

which indicates the slope of the phase progression between a pair of two frequencies of sound waves is also constant and is equal to the distance r from the source position.

The propagation phase can also be derived according to the **acoustic transfer impedance** such that

$$H(x',x,\omega) = \frac{i\rho_0 c}{S}\frac{\sin kx' \sin k(L-x)}{\sin kL}, \qquad (\mathrm{Pa\cdot s/m^3}) \tag{13.99}$$

which was defined by Eq. 7.27. The poles k_p and zeros k_z' and k_z are located at

$$k_p = \frac{l\pi}{L}, \quad k_z' = \frac{m\pi}{x'}, \quad k_z = \frac{n\pi}{L-x}, \tag{13.100}$$

where l, m, and n are positive integers. Therefore, the accumulated phase can be expressed as

$$\Phi(k) \cong -\pi\left(\frac{k}{\pi/L} - \frac{k}{\pi/x'} - \frac{k}{\pi/(L-x)}\right) = -k(x-x') = -kr, \tag{13.101}$$

which indicates the propagation phase when $x > x'$.

The propagation phase that might be observed between the source and receiving positions is reminiscent of the **two-point correlation** coefficients in a sound field that was dealt with in section 12.3. Now consider the phase average change in an interval of two adjacent poles. Suppose the phase progression at the wavenumber k on average is[79]

$$\Phi_1(r,k) = -\pi P_{sc_1}(r,k). \tag{13.102}$$

Recalling Eq. 12.57, the phase can be estimated by

$$\Phi_1(r,k) = -\cos^{-1} C_{F_1}(kr); \tag{13.103}$$

namely,

$$P_{sc_1} = \frac{1}{\pi}\cos^{-1} C_{F_1}(kr), \tag{13.104}$$

where $C_{F_1}(kr)$ denotes the two-point correlation coefficient of the sound field.

By introducing the correlation coefficient of a one-dimensional wave travelling system given by Eq. 12.57, the accumulated phase can be approximated as

$$\Phi_1(r,k) = -\Delta kr = -\frac{\pi}{L}r \tag{13.105}$$

where

$$P_{sc_1} = \frac{r}{L} \qquad (r \le L) \tag{13.106}$$

and L denotes the size of the one-dimensional space of interest.

The phase characteristics described above for one-dimensional systems can also be confirmed by numerical calculations[108]. Recall the transfer function by the modal expansion form written by Eq. 10.50 and approximate it such that

$$H(x',x,\omega) \cong K\sum_n \frac{\sin(\frac{n\pi}{L}x')\sin(\frac{n\pi}{L}x)}{\omega-\omega_n}. \tag{13.107}$$

for one-dimensional vibrating systems. The numbers of poles and zeros follow the integral formulation on the complex frequency domain[109],

$$\frac{1}{2\pi \mathrm{i}}\int_C \frac{H'(\omega_s)}{H(\omega_s)}d\omega_s = N_{z_{in}} - N_p, \tag{13.108}$$

where ω_s is the complex angular frequency, C denotes the contour on which the integration is taken in the complex frequency domain, and N_p and $N_{z_{in}}$ are the numbers of the poles and zeros inside the region surrounded by the contour, respectively. According to Fig. 13.16, by taking the contour C_2, the integration formula becomes

$$\frac{1}{2\pi \mathrm{i}}\int_{C_2} \frac{H'(\omega_s)}{H(\omega_s)}d\omega_s = N_z^- - N_p \tag{13.109}$$

where N_z^- is the numbers of the zeros inside the region surrounded by the contour above the real-frequency axis. In contrast, it is written as

$$\frac{1}{2\pi \mathrm{i}}\int_{C_1} \frac{H'(\omega_s)}{H(\omega_s)}d\omega_s = N_z^+ \tag{13.110}$$

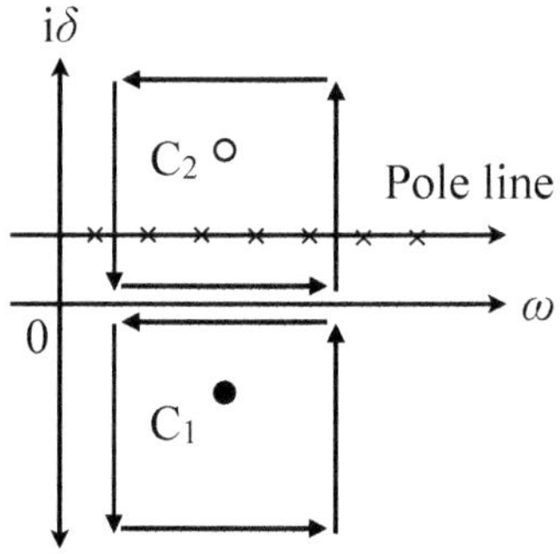

Fig. 13.16 Contours of integration for detecting zeros on complex frequency plane from [108] (Fig.1)

when following the contour C_1, where N_z^+ is the numbers of the zeros inside the region surrounded by the contour below the real-frequency axis. Therefore, the number of zeros can be numerically estimated, if the number of poles is known.

Consider an acoustical tube as an example of a one-dimensional wave travelling system. Figure 13.17 illustrates examples of numerical calculation of the accumulated phase between the source and receiving positions as shown in Fig. 13.18. Figures 13.17(a) and (b) show the results corresponding to sound waves of 500 Hz for two different source locations. Similarly, Fig. 13.17(c) shows the average results (of 2652 samples) when the pair of source and receiving positions are randomly taken in the pipe, and the frequency is also randomly sampled in the range up to 500 Hz. In all cases, the accumulated phase that is estimated by

$$\Phi(\omega) = -\pi(N_p(\omega) - N_z(\omega)) \tag{13.111}$$

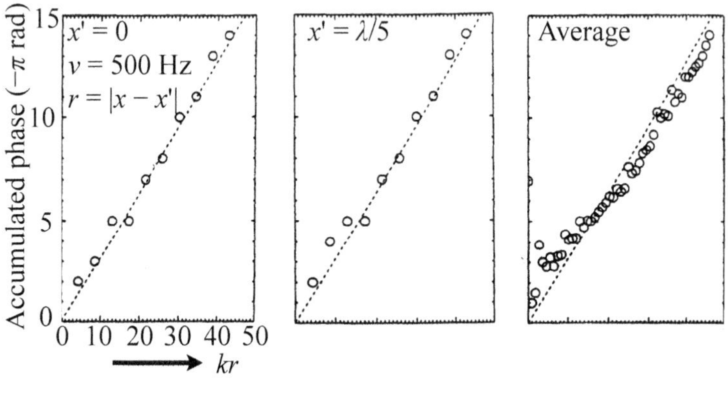

Fig. 13.17 Examples of accumulated phase for one-dimensional system as shown in Fig. 13.18 from [108] (Fig.3)

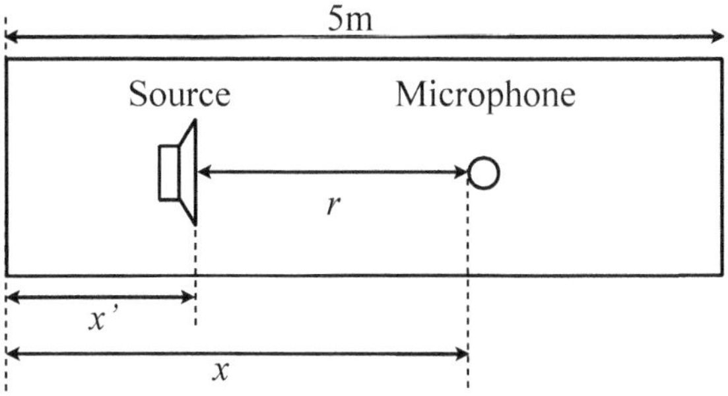

Fig. 13.18 Model for one-dimensional wave-travelling system

follows the propagation phase $-kr$, where r shows the distance from the source to the receiving position.

As stated above, the phase might be simply estimated by using the **propagation phase** for one-dimensional systems, and the probability of a residue sign change is proportional to the distance between the source and receiving positions. However, the transfer functions are not as simple as that any longer for two-dimensional cases.

13.5.2 Transfer Function Phase for Two-Dimensional Waves

Recall the phase accumulation stated in the previous subsection. By introducing the correlation coefficient in a two-dimensional reverberation field,

$$C_{F_2}(kr) = \mathrm{J}_0(kr), \tag{13.112}$$

which was described in subsection 12.3.1. The probability of a sign change can be estimated by

$$P_{sc_2} = \frac{1}{\pi}\cos^{-1}(\mathrm{J}_0(kr)). \tag{13.113}$$

The accumulated phase that is expected when $P_{sc_2} = 1/2$ is called the **reverberation phase**, in contrast to the **propagation phase**[23][24]. That condition occurs when $C_{F_2}(kr) = \mathrm{J}_0(kr) = 0$, and thus the zero of the Bessel function $\mathrm{J}_0(kr)$ shows the condition under which the reverberation phase might be observed between the source and receiving positions. The distance $kr \cong 2.405$ corresponds to the estimate $kr \cong 2$ in references[23][24].

Figure 13.19 shows the two curves of $\mathrm{J}_0(kr)$ and $(1/\pi)\cos^{-1}(\mathrm{J}_0(kr))$. The probability remains around 1/2 without approaching unity, different from that in one-dimensional cases. By introducing $P_{sc_2} = 1/2$ into the probability, the accumulated phase up to k can be estimated by

$$\Phi(k) \cong -\frac{\pi}{2}N_p(k) \tag{13.114}$$

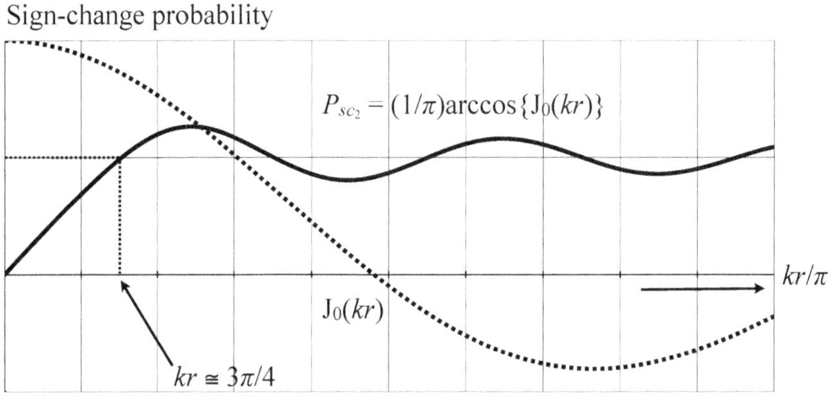

Fig. 13.19 Two-point spatial correlation and probability of residue sign changes for two-dimensional reverberation sound field

where $N_p(k)$ denotes the number of poles up to the wavenumber k. The reverberation phase does not follow the propagation phase any longer; it even exceeds the propagation phase. This is because the number of poles is not linearly proportional to the wavenumber in the two-dimensional field, and in addition, the probability of residue sign changes remains around 1/2 even if the distance from the source to the receiving position becomes longer than $kr \cong 2.405$.

The outcome that has been derived can be numerically confirmed[23][24][110]. Recall that the transfer function can be written in the modal expansion form as

$$H(\mathbf{x}', \mathbf{x}, \omega) = \sum_{l,m} \frac{1}{\Lambda_{lm}} \frac{\cos(\frac{l\pi}{L_x}x')\cos(\frac{m\pi}{L_y}y')\cos(\frac{l\pi}{L_x}x)\cos(\frac{m\pi}{L_y}y)}{(\omega - \omega_{lm_1})(\omega - \omega_{lm_2})} \quad (13.115)$$

$$\omega_{lm_{1,2}} = \mp c\sqrt{\left(\frac{l\pi}{L_x}\right)^2 + \left(\frac{m\pi}{L_y}\right)^2} + i\delta_0 = \mp\omega_{lm} + i\delta_0 \quad (13.116)$$

where

$$\begin{aligned} \frac{\Lambda_{lm}}{L_x L_y} &= \frac{1}{L_x L_y}\int_S \cos^2(\frac{l\pi}{L_x}x)\cos^2(\frac{m\pi}{L_y}y)dxdy \\ &= \frac{1}{4} \qquad l \neq 0, \quad m \neq 0 \\ &\ \frac{1}{2} \qquad l = 0, \quad \text{or} \quad m = 0, \end{aligned} \quad (13.117)$$

(x', y') and (x, y) are the source and receiving positions, respectively, L_x and L_y are the lengths of the sides of the rectangular boundary, and c is the sound speed in the medium. Figure 13.20 shows the probability that an adjacent pair of poles has opposite sign residues, when the source and receiving positions are sampled in the

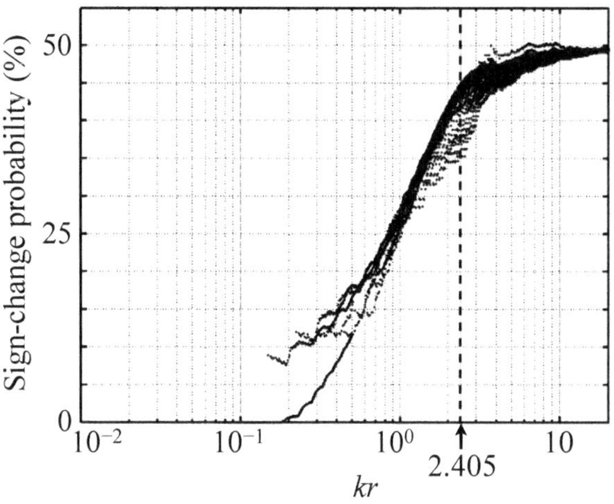

Fig. 13.20 Probability of residue sign changes in 2-D space where $L_x = 4.5$ and $L_y = 4.5 \times 2^{1/3}$[69]

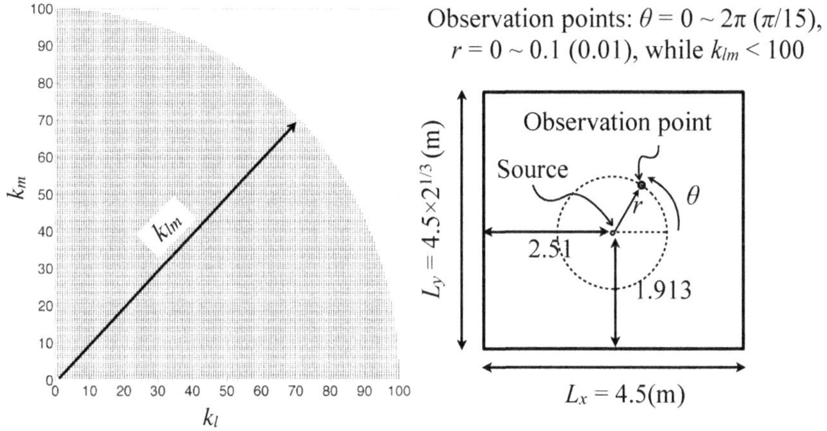

Fig. 13.21 Conditions of calculation for Fig. 13.20

space following Fig. 13.21 in the frequency range where the wavenumber is smaller than 100. It can be seen that the probability approaches 1/2 after kr exceeds about 2. Here, r is the distance between the source and receiver positions.

The accumulated phase can be numerically estimated for two-dimensional space as well as the one-dimensional field[110]. For comparison with a theoretical estimation, introduce the simplified formula such that

$$H(\mathbf{x}',\mathbf{x},\omega) = \sum_{l,m} \frac{A_{lm}}{\omega - \omega_{lm} - \mathrm{i}\delta_0} \tag{13.118}$$

$$A_{lm} = 1 \quad f_{lm}(\mathbf{x}')f_{lm}(\mathbf{x}) > 0$$
$$-1 \quad f_{lm}(\mathbf{x}')f_{lm}(\mathbf{x}) < 0 \tag{13.119}$$

$$f_{lm}(\mathbf{x}) = \cos(\frac{l\pi}{L_x}x)\cos(\frac{m\pi}{L_y}y). \tag{13.120}$$

This is called the **residue sign model**, while the modal expansion including the orthogonal functions is called the **wave theoretic model** such that

$$H(\mathbf{x}',\mathbf{x},\omega) \cong \sum_{l,m} \frac{1}{\Lambda_{lm}} \frac{f_{lm}(\mathbf{x}')f_{lm}(\mathbf{x})}{\omega - \omega_{lm} - \mathrm{i}\delta_0}. \tag{13.121}$$

Suppose the two-dimensional space where $L_x = 4.5$ (m), $L_y = L_x \times 2^{1/3}$ (m), and the frequency interval of interest is 0-300 Hz, and the distance between the pole line and real-frequency axis is set to $\delta_0 = 0.25$ on the complex-frequency plane. Figure 13.22 illustrates the averaged numbers of zeros Fig. 13.22 (a) and corresponding phase accumulation (b), where 50 samples of the source and receiver positions are randomly taken, keeping the constant distance r in the frequency range.

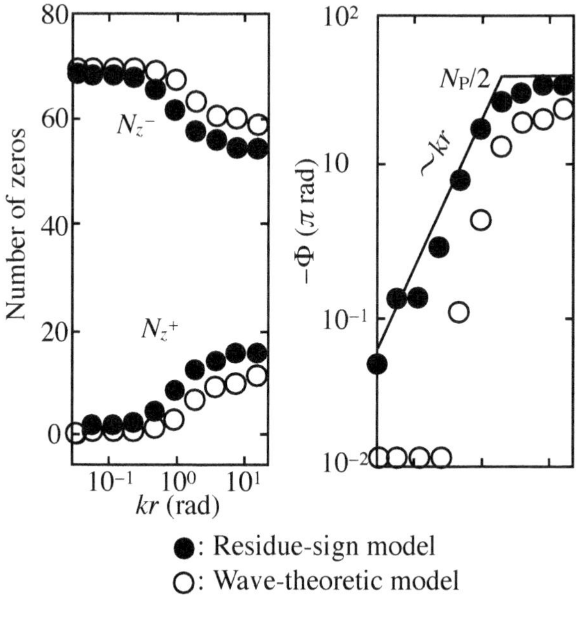

Fig. 13.22 Number of zeros detected by two models (a) and accumulated phases estimated by zeros (b), N_z^+: number of non-minimum-phase zeros, N_z^-: number of minimum-phase zeros, N_p: number of poles from [110] (Fig.4)

The number of zeros is estimated according to integration formula described by Eq. 13.108 where contour C_1 or C_2 is taken as shown in Fig. 13.16.

The zeros distributed above (below) the real frequency axis are called **minimum-phase zeros** assuming that $e^{i\omega t}$ ($e^{-i\omega t}$) time dependency, in contrast to the **non-minimum-phase zeros** below (above) the real-frequency axis. The reason for the zeros are named **minimum phase** is that zeros above (below) the real-frequency axis compensate for the phase lag due to the poles. In contrast, the **non-minimum phase** zeros below (above) the frequency line also add to the phase lag. Figure 13.22 (a) shows that the number of non-minimum-phase zeros N_z^+ is equal to zero when kr is short, while it increases with kr. Consequently, the number of minimum-phase zeros N_z^- decreases, keeping the total number of zeros equal to that for the poles. The dependence of the distribution of zeros on kr is similar for the two models, i.e., the residue sign and wave theoretic models, but the number of zeros is a little different.

Figure 13.22 (b) presents the calculated results of the accumulated phase following

$$\Phi(\omega) = -\pi(N_p(\omega) - N_z^-(\omega) + N_z^+(\omega)). \qquad (13.122)$$

The estimated phase by the residue model indicates the reverberation phase is given by $-N_p\pi/2$ after kr exceeds about 2.4, but the phase is already greater than the propagation phase even before $kr \cong 2$, despite the phase seeming to be proportional to kr[23][24]. However, the results according to the wave theoretic model are smaller than those by the residue model. This is probably because the double zeros, neglected in the estimation of $-N_p\pi/2$, cannot discarded for the wave theoretic model.

Figure 13.23 is the estimation for the on-line zeros located on the pole line subject to no off-line zeros being close to the pole line (in the pole line $\pm\delta_0$), and the contours being taken every 0.2 Hz between adjacent poles as shown in Fig. 13.24[110]. Figure 13.23 (a) is the zeros estimated by the residue model, while Fig. 13.23 (b) presents those by the wave theoretic model. It can be seen that the zeros are mostly single on-line zeros for the residue model. In contrast, for the wave model, single or double on-line zeros are distributed as well as the off-line zeros. This difference in the number of double (on-line) zeros explains why a smaller phase is accumulated for the wave model than for the residue model.

However, both of the estimation by two models are a little smaller than the theoretical estimate by $-N_p\pi/2$. This might be because of the double-zero effect and may also be due to the whole distribution of the zeros on the complex frequency plane. Figures 13.25 (a) and (b) are the examples of distributions of zeros that were detected following the contours as shown in Fig. 13.26. The distribution is symmetric with respect to the pole line, as expected from the models, and the zeros are concentrated around the pole line. In particular, the number of on-line zeros decreases and approaches $-N_p/2$ as kr increases, for the residue model; however, it does not noticeably decreases as kr increases for the wave model. On the whole, it can be confirmed that the number of off-line zeros increases with kr.

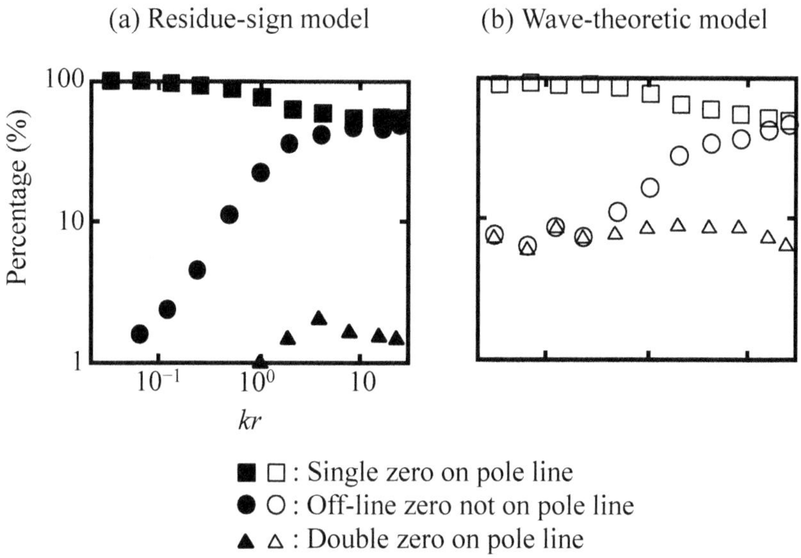

Fig. 13.23 Classification of zeros with respect to pole line from [110] (Fig.11)

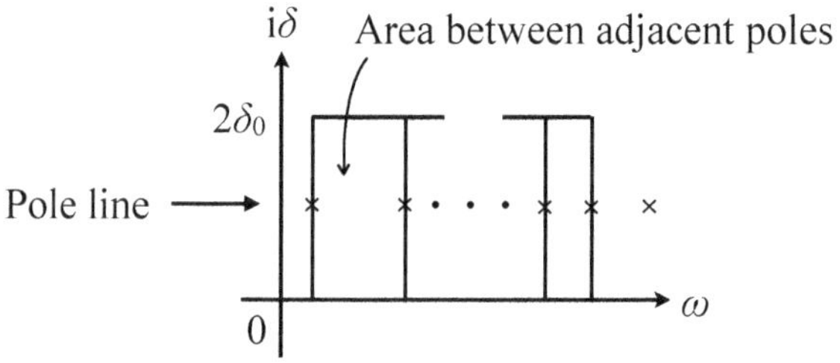

Fig. 13.24 Contours on complex frequency plane from [110] (Fig.10)

As stated above, the reverberation phase can be mostly confirmed by the numerical calculation, but the results displayed in Fig. 13.22 are still not intuitively understood. It seems natural to expect that the directional sound from the source might be dominant even in highly reverberant space if the receiver is located close to the source. In other words, it is quite likely to see the propagation phase in the sound field close to the sound source even in a reverberation field. However, there is no region where the propagation phase might be observable even when kr is very small as in Fig. 13.22. This fact suggests that neither of the theoretical models might be adequate for a sound field close to the sound source.

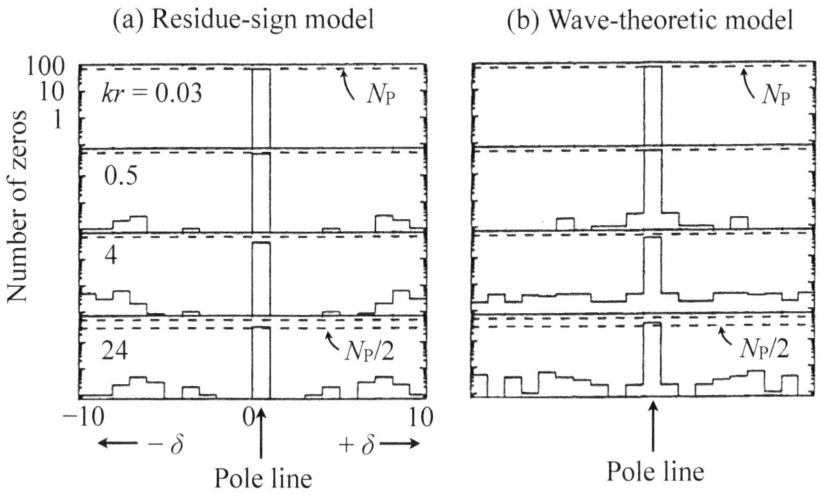

Fig. 13.25 Distribution of zeros on complex frequency plane[110] (Fig.6)

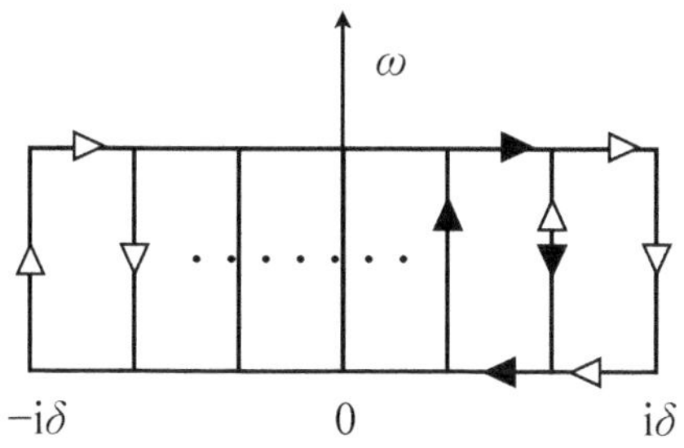

Fig. 13.26 Integration contours used for Fig. 13.25 from [110] (Fig.5)

13.5.3 Phase of Transfer Functions for Three-Dimensional Systems

Recall the correlation coefficient between two points in a three dimensional space defined by Eq. 12.50:

$$C_{F_3}(kr) = \frac{\sin kr}{kr}. \tag{13.123}$$

The probability of a sign change can be estimated by

$$P_{sc_3} = \frac{1}{\pi}\cos^{-1}\left(\frac{\sin kr}{kr}\right). \tag{13.124}$$

Therefore, the reverberation phase occurs when $\sin kr/kr = 0$ namely $kr = \pi$. Figure 13.27 shows the two curves of the spatial correlation coefficient and the residue-sign-change probability. Similarly, Fig. 13.28 shows the calculated results of the residue-sign-change probability in the three-dimensional reverberant space following Fig. 13.29 in the frequency range where the wavenumber is smaller than 20[69]. The probability remains around 1/2, as does that for the two-dimensional space as shown in Fig. 13.20, when kr exceeds π. That is, the probability reaches 1/2 when $kr = \pi/2$ for one-dimensional systems, $kr \cong 3\pi/4 \cong 2.4$ for two-dimensional fields, or $kr = \pi$ for three-dimensional spaces.

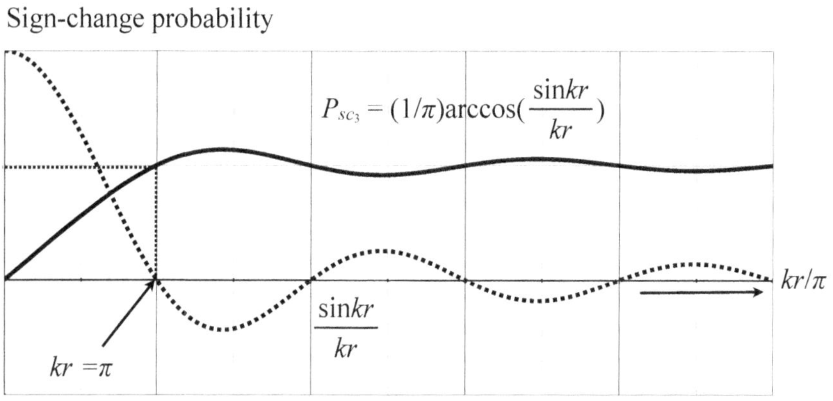

Fig. 13.27 Spatial correlation coefficient and residue-sign-change probability for three-dimensional reverberation space

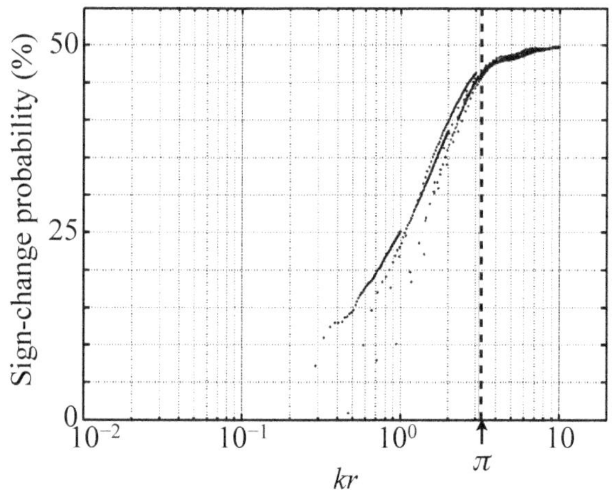

Fig. 13.28 Calculated residue-sign-change probability according to the wave model in three dimensional reverberation field[69]

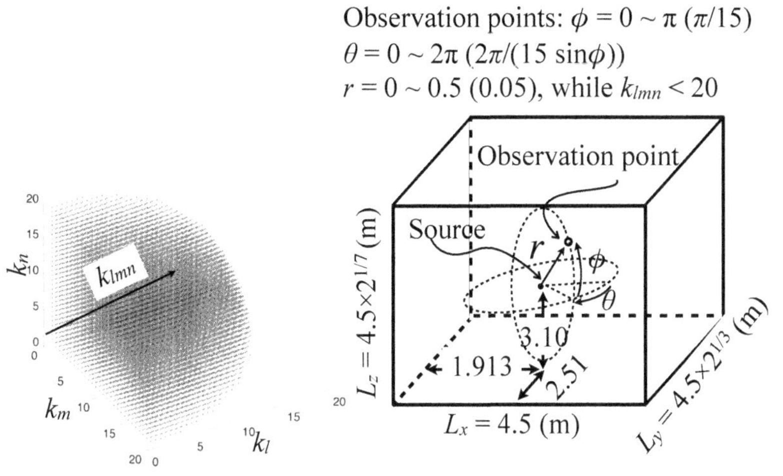

Fig. 13.29 Conditions of calculation for Fig. 13.28

13.6 Coherent Field, Propagation, and Reverberation Phase

In subsection 13.5.2, the propagation phase could not be confirmed by numerical calculations for two-dimensional wave-travelling systems according to the residue-sign or wave-theoretic model. However, the condition of kr, i.e., the propagation phase might break down can be interpreted in terms of the **coherent field** according to the modal wave theory[78]. In this section, the condition of the distance from the source for the coherent field in a reverberation field will be described.

13.6.1 Coherent Field in Three-Dimensional Reverberation Space

As mentioned earlier, it is natural to expect that the propagation phase might be observable if the receiving position is put close to the sound source. Recall the Green function defined by Eq. 10.70 where $\Lambda_N = 8$, assuming oblique waves only. It can be intuitively understood that the summation can be approximated by integration with respect to the continuous variable of $k_N = k'$.

For that approximation, it is necessary to impose the condition that the phase shift of between $\mathrm{e}^{-\mathrm{i}\mathbf{k}_N\cdot\mathbf{R}_P}$ and $\mathrm{e}^{-\mathrm{i}\mathbf{k}_{N+1}\cdot\mathbf{R}_P}$ might remain within $\pi/2$, i.e., the probability of a sign change of residues is smaller than 1/2. This is equivalent to assuming the modal response might be positively superposed, and thus it corresponds to the **break down of the propagation phase** (or **occurrence of the reverberation phase**). Recall that the modal response can be decomposed into resonant and non-resonant portions as mentioned in subsection 13.3.1. Now suppose that the summation can be well approximated by taking only the resonant response, and set $k' - \Delta k' < k' < k' + \Delta k'$ so that $k' \cong k$ might hold well.

Consider the volume of the spherical shell cut from the wavenumber space in which $\mathbf{k}_N$ corresponding to the eigenfrequencies are arranged on the latices as shown in Fig. 10.1. Assuming that $k' >> \Delta k'$, the volume of the shell Δ can be estimated by

$$\Delta_3 = 4\pi k'^2 \cdot dk' \qquad (1/\text{m}^3), \tag{13.125}$$

and thus the density of the eigenfrequencies contained in the volume becomes

$$n_{\Delta_3}(k') \cong \frac{8n_3(k')dk'}{\Delta_3} \cong \frac{V}{\pi^3}, \tag{13.126}$$

where $n_3(k')$ denotes the modal density of the oblique wave modes in the wavenumber space such that

$$n_3(k') \cong \frac{Vk'^2}{2\pi^2} \tag{13.127}$$

and V denote the room volume.

By again cutting out a small portion from the spherical shell so that the volume of the small portion might be

$$d\Delta_3 = k'^2 \sin\theta d\theta d\phi dk', \tag{13.128}$$

the number of eigenfrequencies contained in the small portion is given by

$$dn_{\Delta_3}(k') = \frac{V}{\pi^3} d\Delta_3 = \frac{V}{\pi^3} k'^2 \sin\theta d\theta d\phi dk'. \tag{13.129}$$

Now by converting the summation into integration using the modal density in the small portion of the spherical shell stated above, the Green function can be rewritten as[78]

$$\begin{aligned}
G(\mathbf{x}', \mathbf{x}, k) &= \frac{Q_{0_3}}{8V} \sum_{N=-\infty}^{+\infty} \sum_{P=1}^{8} \frac{\mathrm{e}^{-\mathrm{i}\mathbf{k}_N \cdot \mathbf{R}_P}}{k_N^2 - k^2} \\
&\cong \frac{Q_{0_3}}{8\pi^3} \int_0^{2\pi} d\phi \int_0^{\pi} \mathrm{e}^{-\mathrm{i}k' r\cos\theta} \sin\theta d\theta \int_0^{+\infty} \frac{k'^2}{(k'+k)(k'-k)} dk' \\
&= \frac{Q_{0_3}}{4\pi^2} \int_0^{\infty} \frac{\mathrm{e}^{-\mathrm{i}k'r} - \mathrm{e}^{\mathrm{i}k'r}}{-\mathrm{i}k'r} \frac{k'^2}{(k'+k)(k'-k)} dk' \\
&\cong \frac{Q_{0_3}}{8\pi^2} \int_{-\infty}^{+\infty} \frac{\mathrm{e}^{-\mathrm{i}k'r} - \mathrm{e}^{\mathrm{i}k'r}}{-\mathrm{i}r} \frac{dk'}{k'-k} \\
&= \lim_{\beta \to 0} \frac{Q_{0_3}}{8\pi^2} \frac{-1}{\mathrm{i}r} \int_{-\infty}^{+\infty} \frac{\mathrm{e}^{-\mathrm{i}k'r}}{k' - (k - \mathrm{i}\beta)} dk' \\
&\cong \frac{Q_{0_3}}{4\pi r} \mathrm{e}^{-\mathrm{i}kr} \qquad r > 0. \qquad (\text{m}^2/\text{s})
\end{aligned} \tag{13.130}$$

This result indicates the spherical wave from the source to the receiving position.

Here recall the condition that was imposed on $\mathbf{R}_P$ for approximating the summation by integration, i.e., for deriving the spherical wave with a propagation phase in a three-dimensional room. The number of eigenfrequencies in the width of $2\Delta k'$ is given by

$$n_3(k') \cdot 2\Delta k' \cong \frac{Vk'^2}{\pi^2}\Delta k'. \tag{13.131}$$

Assuming $k' >> \Delta k'$ so that all the eigenfrequencies can be regarded as distributed on the surface of the the average distance between the adjacent eigenfrequencies $\Delta k_{N_{3_{AV}}}$ can be expressed as

$$\Delta k_{N_{3_{AV}}} \cong \sqrt{2}\sqrt{\frac{\frac{\pi}{2}k'^2}{\frac{V}{\pi^2}k'^2\Delta k'}} = \sqrt{\frac{\pi^3}{V\Delta k'}}. \tag{13.132}$$

Consequently, the condition that was assumed, $\Delta k_{N_{AV}} \cdot r < \pi/2$ corresponding to the **range of the propagation phase**, can be rewritten as

$$\sqrt{\frac{\pi^3}{V\Delta k'}}r < \frac{\pi}{2}. \tag{13.133}$$

Recall the modal bandwidth given by Eq. 13.49 that represents the spread of the resonant response on the frequency axis. The condition above gives the limit of **coherent field**:

$$\sqrt{\frac{\pi^3}{V\Delta k'}}R_{c_3} = \frac{\pi}{2} \tag{13.134}$$

or

$$R_{c_3} = \sqrt{\frac{V\Delta k'}{4\pi}} = \sqrt{\frac{V\delta_0}{8c}} \cong \sqrt{\frac{A_3}{64}}, \qquad \text{(m)} \tag{13.135}$$

where $c\Delta k' \cong \pi\delta_0/2$, $\delta_0 \cong 6.9/T_{R_3}$, and $T_{R_3} \cong 0.163V/A_3$ that is given by Eq. 11.43. The range of the coherent field is proportional to the room volume or equivalent absorption area.

13.6.2 Coherent Field in Two-Dimensional Reverberation Space

Similarly to that for three-dimensional space stated in the previous subsection, the coherent field can be derived for two-dimensional space. Introducing the ring

$$\Delta_2 = 2\pi k' dk' \tag{13.136}$$

instead of the spherical shell, the modal density in the ring is given by

$$n_{\Delta_2}(k') \cong \frac{4n_2(k')dk'}{\Delta_2} = \frac{S_2}{\pi^2}, \tag{13.137}$$

where $n_2(k')$ denotes the modal density for the two-dimensional space, $n_2(k') = S_2k'/2\pi$, and S_2 is the area of the field. Consequently, the number of eigenfrequencies contained in the small portion of the ring can be expressed as

$$dn_{\Delta_2}(k') = \frac{S_2}{\pi^2}d\Delta_2 = \frac{S_2}{\pi^2}k'dk'd\phi, \qquad (13.138)$$

where

$$d\Delta_2 = k'dk'd\phi. \qquad (13.139)$$

By introducing the number of eigenfrequencies in the small portion stated above, the Green function defined for the two-dimensional space can be rewritten as[89]

$$\begin{aligned} G(\mathbf{x}',\mathbf{x},k) &= \frac{Q_{0_2}}{4S_2}\sum_{N=-\infty}^{+\infty}\sum_{P=1}^{4}\frac{\mathrm{e}^{-\mathrm{i}\mathbf{k}_N\cdot\mathbf{R}_P}}{k_N^2-k^2} \\ &\cong \frac{Q_{0_2}}{2\pi}\frac{1}{\pi}\int_{-\frac{\pi}{2}}^{+\frac{\pi}{2}}\mathrm{e}^{-\mathrm{i}k'r\cos\phi}d\phi\int_0^{+\infty}\frac{k'dk'}{k'^2-k^2} \\ &\cong \frac{Q_{0_2}}{2\pi}\frac{1}{\pi}\int_{-\frac{\pi}{2}+\mathrm{i}\infty}^{+\frac{\pi}{2}-\mathrm{i}\infty}\mathrm{e}^{-\mathrm{i}k'r\cos\phi}d\phi\int_0^{+\infty}\frac{k'dk'}{k'^2-k^2} \\ &= \frac{Q_{0_2}}{2\pi}\int_0^{+\infty}\frac{\mathrm{H}_0^{(2)}(k'r)k'}{k'^2-k^2}dk' \\ &\cong \frac{Q_{0_2}}{4\pi}\int_{-\infty}^{+\infty}\frac{\mathrm{H}_0^{(2)}(k'r)}{k'-k}dk' \\ &= -\mathrm{i}\frac{Q_{0_2}}{2}\mathrm{H}_0^{(2)}(kr). \quad (\mathrm{m}^2/\mathrm{s}) \end{aligned} \qquad (13.140)$$

This represents the directional wave from the source in the two-dimensional space, where $H_0^{(2)}(kr)$ is called the second kind of **Hankel function**[89],

$$\mathrm{H}_0^{(2)}(kr) = \mathrm{J}_0(kr) - \mathrm{i}\mathrm{N}_0(kr), \qquad (13.141)$$

the real part is the **Bessel function**, and the imaginary part denotes the **Neumann function**[89]. The Neumann function is expressed as

$$\pi N_0(r) = 2\ln\frac{\gamma r}{2}J_0(r) + 4\left(J_2(r) - \frac{1}{2}J_4(r) + \frac{1}{3}J_6(r) - \cdots\right) \qquad (13.142)$$

and thus shows the singularity due to the logarithmic function when $r \to 0$ and γ is **Euler's constant**, i.e.[89],

$$\gamma \cong 0.57721\cdots. \qquad (13.143)$$

As described above, the propagation phase due to the direction sound from the source can also be observed in two-dimensional space. However, the **singularity** that appears in the limit when $r \to 0$ is different from that for three-dimensional space. Suppose that the Green function for the sound field close to the source in the two-dimensional space is written as

$$\Phi(r) = \Psi(r)\mathrm{e}^{-ikr}. \qquad (\mathrm{m}^2/\mathrm{s}) \tag{13.144}$$

Assuming that the strength of the source is unity $(\mathrm{m}^2/\mathrm{s})$,

$$\int_{-\pi}^{\pi}\left(\frac{-\partial\Phi(r)}{\partial r}\right) r d\theta = 1 \tag{13.145}$$

holds well for the Green function above in the limit when $r \to 0$. Namely, according to

$$\int_{-\pi}^{\pi}\left(\frac{-\partial\Psi(r)}{\partial r}\mathrm{e}^{-ikr}\right) r d\theta + \int_{-\pi}^{\pi}\left(\mathrm{i}k\Psi(r)\mathrm{e}^{-ikr}\right) r d\theta = 1, \tag{13.146}$$

the relation

$$\int_{-\pi}^{\pi}\left(-\frac{\partial\Phi(r)}{\partial r} r d\theta\right) = 1 \to -\frac{\partial\Psi(r)}{\partial r}2\pi r = 1 \qquad (r \to 0) \tag{13.147}$$

holds well where $r\Psi(r) \to 0 (r \to 0)$ is assumed. Consequently, the singularity is represented by the function $\Psi(r)$, where

$$\Psi(r) = -\frac{1}{2\pi}\ln r, \tag{13.148}$$

in two-dimensional space. This might be interpreted as the difference between the sources for **spherical** and **cylindrical waves**.

Similar to that in a three-dimensional field, the **range of the coherent** can also be derived for two-dimensional space. According to the following relations

$$n_2(k')2\Delta k' \cong \frac{S_2 k'}{\pi}\Delta k' \tag{13.149}$$

$$\Delta k_{N_{Av}} \cong \frac{\pi k'/2}{n_2(k')2\Delta k'} = \frac{\pi^2}{2S_2\Delta k'} \tag{13.150}$$

$$\frac{\pi^2}{2S_2\Delta k'}R_{c_2} = \frac{\pi}{2}, \tag{13.151}$$

the range of the distance from the source in the coherent field is given by

$$R_{c_2} = \frac{S_2\Delta k'}{\pi} = \frac{S_2\delta_{0_2}}{2c} \cong \frac{A_2}{12.6}, \qquad (\mathrm{m}) \tag{13.152}$$

where $\delta_{0_2} \cong 6.9/T_{R_2}$, T_{R_2} is given by Eq. 11.68, $A_2 = -\ln(1-\alpha_2)L_2$, α_2 denotes the averaged absorption coefficient of the space, L_2 is the length of the circumference of the space, and S_2 is the area of the space.

13.6.3 Coherent Field in One-Dimensional Reverberation Space

As described in subsection 13.5.1, the phase is represented by the propagation phase in one-dimensional space. However, following the definition of the **coherent field**, i.e., the directional wave is dominant and the phase characterisitic is that of a propagation phase like the plane wave, the range for the one-dimensional space can be also derived. Similarly to the previous discussions, the Green function for the one-dimensional space can be written as

$$\begin{aligned}
G(\mathbf{x}',\mathbf{x},k) &= \frac{Q_{0_1}}{2L_x}\sum_{N=-\infty}^{+\infty}\sum_{P=1}^{2}\frac{\mathrm{e}^{-\mathrm{i}\mathbf{k}_N\cdot\mathbf{R}_P}}{k_N^2-k^2} \\
&\cong \frac{Q_{0_1}}{2\pi}\left[\mathrm{e}^{-\mathrm{i}k'r}+\mathrm{e}^{\mathrm{i}k'r}\right]\int_0^{+\infty}\frac{dk'}{k'^2-k^2} \\
&\cong \frac{Q_{0_1}}{4\pi k}\int_{-\infty}^{+\infty}\left[\mathrm{e}^{-\mathrm{i}k'r}+\mathrm{e}^{\mathrm{i}k'r}\right]\frac{dk'}{k'-k} \\
&= -\mathrm{i}\frac{Q_{0_1}}{2k}\mathrm{e}^{-\mathrm{i}kr} \qquad r>0, \qquad (\mathrm{m}^2/\mathrm{s})
\end{aligned} \tag{13.153}$$

where

$$\Delta_1 = 2dk' \tag{13.154}$$

$$n_{\Delta_1}(k') \cong \frac{2\frac{L_x}{\pi}dk'}{2dk'} = \frac{L_x}{\pi} \tag{13.155}$$

and Q_{0_1} $(\mathrm{m/s})$ denotes the strength of the source.

Consequently, the range of the coherent field is given by

$$R_{c_1} = \frac{L_x}{2}, \qquad (\mathrm{m}) \tag{13.156}$$

according to

$$n_1(k')2\Delta k' \cong \frac{L_x}{\pi}2\Delta k' \tag{13.157}$$

$$\Delta k_{N_{Av}} \cong \frac{2\Delta k'}{n_1(k')2\Delta k'} = \frac{\pi}{L_x} \tag{13.158}$$

$$\frac{\pi}{L_x}R_{c_1} = \frac{\pi}{2}. \tag{13.159}$$

The range is independent of the reverberation time in the one-dimensional space. This can be interpreted as the outcome according to the phase itself being a propagation phase in the one-dimensional space.

As described above, the propagation phase might be observed even in two- or three-dimensional space according to the wave theoretic analysis, although such characteristics could not be confirmed by numerical calculations. Results obtained using experimental records in a reverberation room or an small echoic room will be presented in the following section.

13.7 Phase Responses and Distribution of Zeros for Transfer Functions in Three-Dimensional Reverberant Space

Recall the transfer function in a modal expansion form such that

$$H(\mathbf{x}',\mathbf{x},\omega) = \sum_N \frac{\phi_N(\mathbf{x}')\phi_N(\mathbf{x})}{(\omega-\omega_{N_1})(\omega-\omega_{N_2})} \tag{13.160}$$

$$\omega_{N_{1,2}} \cong \mp\omega_{N_0} + \mathrm{i}\delta_0 \tag{13.161}$$

$$\omega_{N_0}^2 \cong \omega_N^2 - \delta_0^2 \tag{13.162}$$

according to Eq. 13.116, where ϕ_N is the normalized orthogonal function of the space with its eigenfrequency ω_N. When the source and the observer are in the same location, all the residues are positive. The poles thus interlace with zeros and the numbers of poles and zeros are equal. As the source and receiver move apart, the zeros migrate (poles do not move)[23][24]. Some move above the pole line, an equal number moves symmetrically below the line, and the remainder stays on the line. This migration of zeros can produce the **propagation phase**, which will be analyzed using experimental data in this section.

If the observer is located in the coherent field where the probability of residue sign changes is around 1/2 away from the source, the possible number of zeros below the real-frequency axis (non-minimum-phase zeros) could thus be estimated using $1/4N_p$, where N_p denotes the number of poles and the case for double zeros is neglected[23][24][110]. These zeros produce the **reverberation phase**; however, this possible number of zeros is reduced as $\delta_0 \cong 6.9/T_R$, which indicates the distance between the pole line and the real-frequency axis, increases.

Occurrences of the zeros are explained in subsection 13.4.2 under the **low-modal overlap condition** where the individual modal responses are observed separately. The distribution of zeros far from the pole line on the complex frequency plane, namely for the **high-modal overlap condition**, will be described in this section. Consequently, the effects of the reverberation time on the phase will be developed using experimental data.

13.7.1 Phase Response and Impulse Response Records in Reverberation Room and Exponential Time Windowing

Normally, the frequency characteristics of a system are defined on the real-frequency axis. However, the transfer function is defined on the complex-frequency plane except for the singularities. **Exponential time windowing** is a possible method for transfer function analysis in a wide area of the complex-frequency plane[107]. Namely, when the exponential time function is applied to the impulse response record of a linear system of interest, the observation-frequency line moves into the complex-frequency plane apart from the real-frequency axis. In other words, if an exponential function such as $w(t) = e^{-\delta' t}$, where δ' is a positive number, the observation line moves to $\delta = -\delta'$. This is equivalent to the pole line moving farther away by δ' from the original pole line. Therefore, it can be interpreted that the exponential windowing represents the effects of the change of the reverberation time on the frequency-response function.

Figure 13.30 shows the measured impulse responses taken from the original data that were recorded in a reverberation room whose room volume is 189 (m^3). The reverberation time is reduced to around 1 second by the exponential windowing. It can be seen that the arrival time is delayed and the energy of direct sound is decreased when the sound source distance (SSD) increases. Figure 13.31 plots samples of phase responses obtained using narrow-band analysis as the sound source distance increases. It can be seen that the phase reaches the reverberation phase when kr becomes longer than about π, as expected in subsection 13.5.3, theoretically. Both results by numerical calculation for two-dimensional systems and by experimental analysis in three-dimensional space confirmed that the propagation phase breaks down when kr exceeds around 2.405 and π respectively.

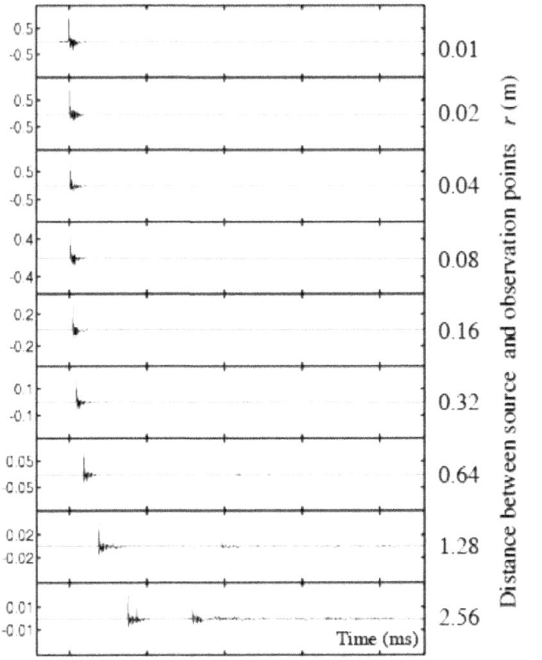

Fig. 13.30 Display of impulse responses measured in reverberant space from [70] (Fig. 10)

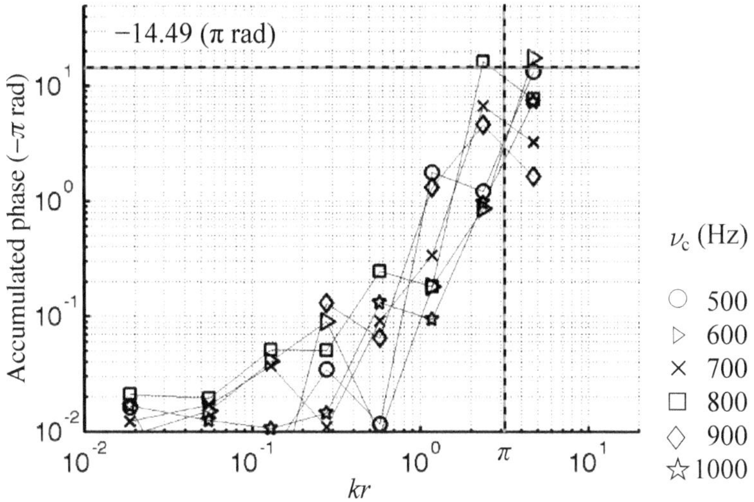

Fig. 13.31 Samples of accumulated phase responses in reverberant space from [60] (Fig. 10)

The reverberation phase after the propagation phase breaks down must decrease from the maximum $N_p\pi/2$ according to the reverberation time. Such a reverberation effect can be seen in the distribution of non-minimum-phase zeros on the complex-frequency plane[109].

13.7.2 Non-minimum-Phase Zeros and Reverberation Phase

Recall the image of distribution of the poles and zeros on the complex-frequency plane. Figure 13.32 shows a schematic of the **accumulated phase** due to the poles and zeros, and with the magnitude response. Here, Fig. 13.32 (c) gives the geometric image of the phase due to the zeros when the observation frequency passes on the real-frequency axis. It can be intuitively understood as shown in Fig. 13.32 (b) that the accumulated phase can be estimated by Eq. 13.122.

As described in subsection 13.4.2, the zeros are distributed symmetrically with respect to the pole line. The effects of the reverberation time (or the damping) on the distribution of the zeros, and therefore the effects on the phase, in particular the reverberation phase, are determined by the distribution of the non-minimum-phase zeros on the complex-frequency plane. This is because the distance between the pole line and the real-frequency axis depends on the reverberation time.

Suppose that the transfer function of a linear system is expressed by the complex function

$$H(\omega_s) = H_r(\omega_s) + \mathrm{i}H_i(\omega_s) \tag{13.163}$$

where $\omega_s = \omega + \mathrm{i}\delta$ denotes the complex frequency. By taking the inverse Fourier transform of $H(\omega)$ observed on the real-frequency axis, the impulse response $h(t)$

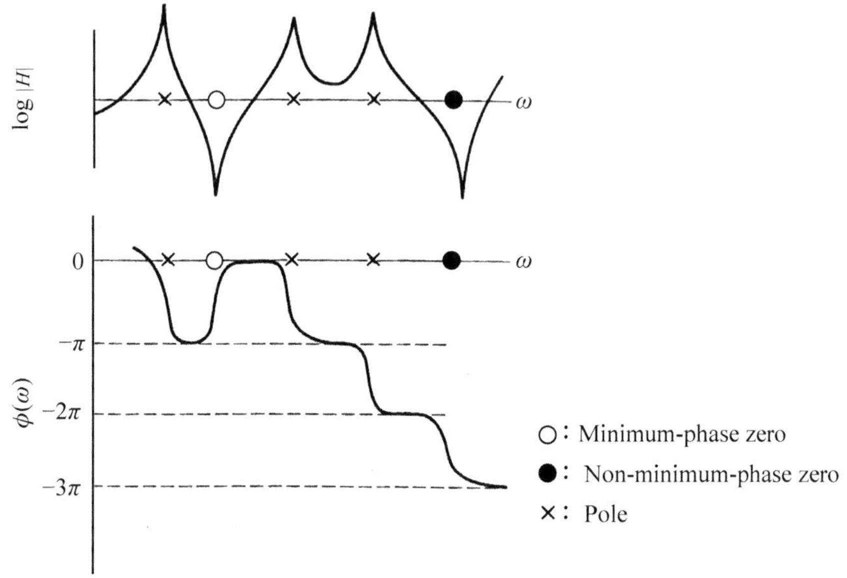

Fig. 13.32 Poles, zeros, and accumulated phase from [109] (Fig.2)

from the source to receiving positions can be derived. Assuming that $H(\omega)$ can be regarded as a random process for a three-dimensional field [73][74][111], the causal part ($t \geq 0$) of the inverse Fourier transform of the real part $H_r(\omega)$ is $h_r(t) = h(t)/2$ and thus ensemble average for $h^2(t)$ with respect to the source and observation points can be expressed by

$$W_{Av}(t) = \mathrm{E}[h_r^2(t)] = Ce^{-t/\tau} \tag{13.164}$$

$$\tau = \frac{1}{2\delta_0} = \frac{T_R}{6\ln 10} \cong \frac{T_R}{13.8} \tag{13.165}$$

where T_R is the reverberation time in the field and C is a constant. Here the reverberation sound field is modeled by superposing random plane waves as described in subsection 12.2.1, and therefore the frequency responses are regarded as a random process.

According to the random process theory [79], the expected number of **zero crossings** per unit increases in the frequency is given by

$$n_z = 2\left(\frac{\int_0^\infty t^2 W_{Av}(t)dt}{\int_0^\infty W_{Av}(t)dt}\right)^{1/2} = \frac{\sqrt{2}}{\delta_t} \tag{13.166}$$

where δ_t shows the distance from the pole line. Therefore, if the real and imaginary parts of $H(\omega_s)$ are statistically uncorrelated[73], the density of the zeros should be proportional to the squared inverse of the distance from the pole line in the complex-frequency domain. The distribution of non-minimum-phase zeros must, therefore, decrease inversely as the damping increases[107]:

$$N_z^+ \to \int_{\delta_t}^{\infty} \frac{dx}{x^2} = \frac{1}{\delta_t}. \tag{13.167}$$

This result is expected to hold for vibrating systems or a sound field of high modal overlap, which is the normal situation in room acoustics.

A possible candidate that gives the probabilistic density function for the distribution of the zeros might be a **Cauchy distribution** such that

$$w_{Cau}(\eta_t) = \frac{4/\pi}{1+4\eta_t^2} \tag{13.168}$$

$$\eta_t = \frac{\delta_t}{\Delta\omega_{N_{Av}}} \tag{13.169}$$

$$\Delta\omega_{N_{Av}} \cong \frac{1}{n_{V3D}(\omega)}, \tag{13.170}$$

where $\Delta\omega_{N_{Av}}$ corresponds the average pole spacing. For a high modal overlap, the Cauchy distribution reduces to

$$w_{Cau}(\eta_t) = \frac{1}{\pi\eta_t^2}. \tag{13.171}$$

Thus, the number of non-minimum-phase zeros in frequency interval $\Delta\omega$ is

$$N_z^+(\eta_0, \Delta\omega) \cong \frac{N_p(\Delta\omega)}{4} \int_{\eta_0}^{\infty} \frac{1}{\pi x^2} dx = \frac{n_{V3D}(\omega)\Delta\omega/4}{\pi\eta_0} = \frac{\Delta\omega/4}{\pi\delta_0}, \tag{13.172}$$

where $\eta_0 = \delta_0/\Delta\omega_{N_{Av}} \cong n_{V3D}(\omega)\delta_0$, $N_p(\Delta\omega)$ denotes the number of poles in frequency interval $\Delta\omega$ or $N_p(\omega) \cong n_{V3D}(\omega)\Delta\omega$. The density of the non-minimum-phase zeros is independent of the frequency under the high modal-overlap conditions, while the number of poles increases with the frequency. Here the probability of occurrences of the double-zeros is neglected in the estimation of the number of non-minimum-phase zeros.

The Cauchy distribution can be mostly confirmed for a two-dimensional field as described in subsection 13.5.2. Figure 13.33 illustrates the results of counting the non-minimum-phase zeros by numerical calculations according to the two models: residue-sign and wave-theoretic models. Figure 13.33 (a) is the contour used for integration on the frequency plane. Both of results in Figs. 13.33 (b) and (c) show that the numbers of zeros approach the theoretical estimates by the Cauchy distribution when kr exceeds about 2.

As described above, the density of the non-minimum-phase zeros is estimated by

$$n_z^+(\eta_0, \omega) \cong \frac{n_{V3D}(\omega)}{4} \int_{\eta_0}^{\infty} \frac{1}{\pi x^2} dx = \frac{1}{2}\frac{1}{2\pi\delta_0} = \frac{n_{max}}{2} \tag{13.173}$$

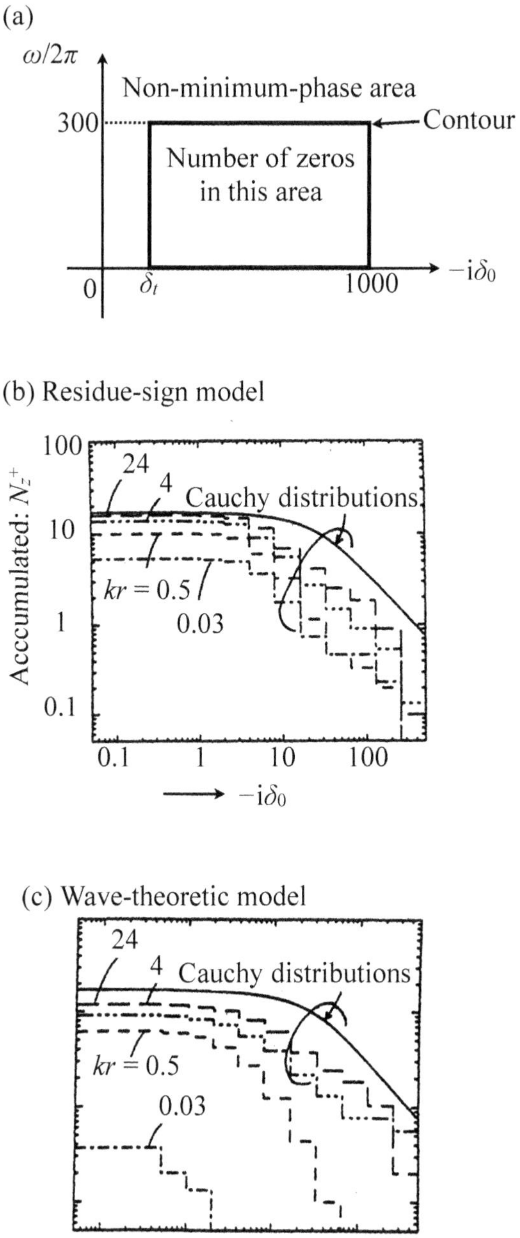

Fig. 13.33 Number of non-minimum-phase zeros following residue-sign model (b) or wave-theoretic model (c) in area illustrated by Fig. (a) from [110] (Fig.8)

for the high-modal overlap condition, where

$$n_{max} = \frac{1}{2\pi\delta_0} \cong \frac{1}{\pi}\frac{T_R}{13.8} \quad \text{(s/rad)} \tag{13.174}$$

indicates the density of the **maximal amplitude** of the frequency characteristics in a reverberation field under a high modal-overlap condition[73]. It is interesting to see that the density of non-minimum-phase zeros is given by the density of the maxima independent of the frequency, but it depends on the reverberation time, as do most room acoustic parameters.

The reverberation phase can be estimated by using the number of non-minimum-phase zeros. The **phase accumulation** every $\Delta\omega$ is expressed as

$$\begin{aligned}\Delta\Phi(\eta_0,\omega,\Delta\omega) &\cong -\pi(n_{V_{3D}}(\omega) - n_z^-(\eta_0,\omega) + n_z^+(\eta_0,\omega))\Delta\omega \\ &= -2\pi n_z^+(\eta_0,\omega)\Delta\omega = -\pi n_{max}(\eta_0,\omega)\Delta\omega \end{aligned} \tag{13.175}$$

independent of frequency where $n_{V_{3D}}(\omega) = n_z^+(\omega) + n_z^-(\omega)$. This outcome indicates that the reverberation phase characteristics become those of a **linear phase** under the high modal overlap condition in reverberation space.

Figure 13.34 shows the experimental arrangements for the impulse response measurements in an echoic room. The averaged number of zeros below the observation-frequency line in the lower half plane of the complex frequency domain is illustrated by Fig. 13.35 and Fig. 13.36. The results clearly show that the distribution of zeros follows the solid lines, following the Cauchy distribution, that increases in inverse

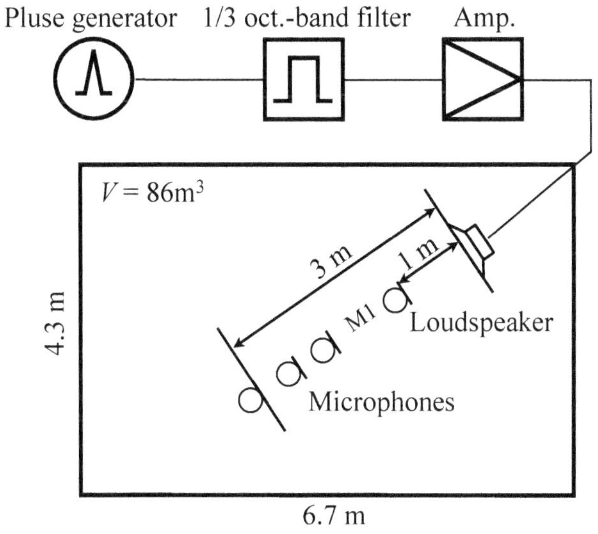

Fig. 13.34 Experimental setup for impulse response measurements in echoic room

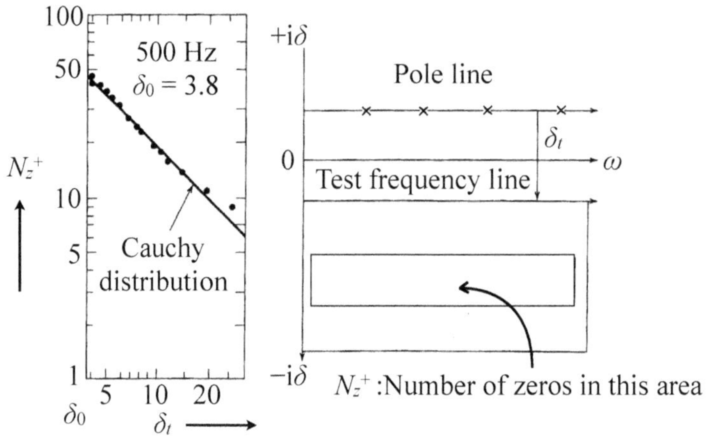

Fig. 13.35 Distribution of zeros of transfer function (Number of non-minimum-phase zeros in 500-Hz 1/1 octave band) from [109] (Fig.6)

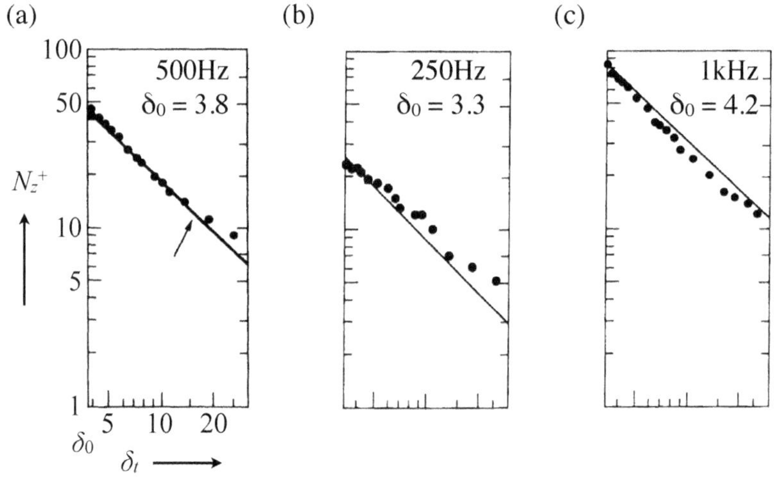

Fig. 13.36 Distribution samples of non-minimum-phase zeros from [109] (Fig.6)

proportion to the distance from the pole line. The total number of non-minimum-phase zeros is estimated to be 48 in the octave band centered at 500 Hz[109].

Figure 13.37 presents the results of the magnitude and phase responses at the observation-frequency lines whose distances from the pole line are indicated by δ_t in the figure[112]. The volume of the room where the impulse responses were recorded is 86 (m^3), and reverberation time is about 1.8 s, thus, the distance between the pole line and the real-frequency axis is about 3.8 (1/s). The responses shown in Fig. 13.37 were obtained for the microphone position M1 in Fig. 13.34. The distance

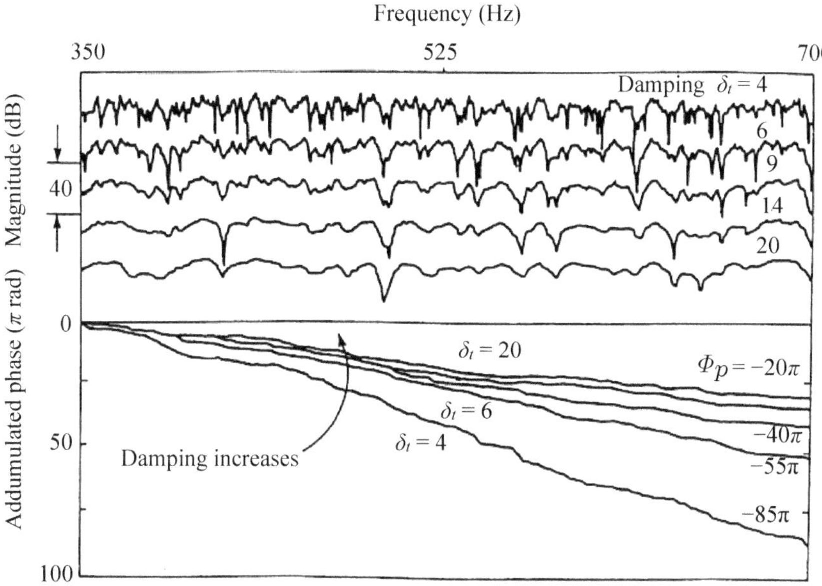

Fig. 13.37 Magnitude and phase of transfer functions from [112] (Fig.6)

between the pole line and the observation-frequency line was changed by applying an exponential time-window to the impulse response record[107].

The amplitude curves in Fig. 13.37 for different damping factors are displayed vertically. The steep peaks cannot be clearly seen as the observation-frequency line moves far from the pole line. This is because the effects of the poles on the magnitude response become weak as the distance from the pole line becomes far. On the other hands, deep troughs can be seen even if the distance is far from the pole line, although the number of dips decreases as the distance increases. This can be interpreted as the deep troughs being due to the zeros that are closely located to the observation-frequency line, but those zeros are located far from the real-frequency axis. However, the density of the zeros decreases in squared inverse proportion to the distance from the pole line, and therefore the total number of dips decreases overall.

The phase curves are plotted in the lower part of Fig. 13.37. The end points of the phase Φ_P in the figure denote the theoretical estimates for the accumulated phase from the number of non-minimum-phase zeros according to the Cauchy distribution. The reverberation-phase trend is is predictable, although the fluctuations from the trend are included, since the transfer function is not averaged in the space but taken at the position M1. However, it can be seen that the reverberation phase follows mostly the trend of linear phases predicted by the number of non-minimum phase zeros, and decreases to zero as the limit as the damping increases.

By applying the Cauchy distribution to the distribution of zeros, the number of non-minimum-phase zeros can be estimated as

$$N_z^+(\eta_0, \Delta\omega) \cong \int_0^\omega \frac{n_{V3D}(\omega)}{4} d\omega \int_{\eta_0}^\infty w_{Cau}(x)dx \tag{13.176}$$

and thus the reverberation phase is expressed as

$$\begin{aligned}\Phi(\eta_0, \omega) &\cong -\pi(N_{V3D}(\omega) - N_z^-(\eta_0, \omega) + N_z^+(\eta_0, \omega)) \\ &= -2\pi N_z^+(\eta_0, \omega) &(13.177)\\ N_z^-(\eta_0, \omega) &= N_{V3D}(\omega) - N_z^+(\eta_0, \omega). &(13.178)\end{aligned}$$

Namely, the decrease of the reverberation phase from the maximum $-N_p(\omega)\pi/2$ due to the damping effect could be formulated by using the number of non-minimum-phase zeros. In addition, the reverberation phase in the interval $d\omega$ could also be formulated such that[107][112]

$$d\Phi(\eta_t, \omega) \cong n_{V3D}(\omega)[\frac{\pi}{2} - \tan^{-1}(2\eta_t)]d\omega. \tag{13.179}$$

Consequently, the local fluctuating behavior from the linear-phase trend can be described in terms of the **group delay** as

$$\frac{d\Phi(\eta_t, \omega)}{d\omega} \cong -\tau_\infty \left[1 - \frac{1}{3}\left(\frac{2M(\omega)}{\pi}\right)^{-2} + \frac{1}{5}\left(\frac{2M}{\pi}\right)^{-4} - \frac{1}{7}\left(\frac{2M}{\pi}\right)^{-6} + \ldots\right], \tag{13.180}$$

where

$$\begin{aligned}\tau_\infty &= \frac{1}{2\delta_t} \quad \text{(s)} &(13.181)\\ M(\omega) &= \pi n_p(\omega)\delta_t = \pi\eta_t. &(13.182)\end{aligned}$$

As shown above, the group delay is defined by

$$\tau(\eta_t, \omega) = -\frac{d\Phi(\eta_t, \omega)}{d\omega}, \tag{13.183}$$

which indicates the center of an energy time wave. The group delay for the reverberant space under the high modal overlap can be rewritten as

$$\begin{aligned}h^2(t) &\cong Ce^{-t/2\delta_0} &(13.184)\\ \tau &= \sqrt{X - Y^2} = \frac{1}{2\delta_0} &(13.185)\\ X &= 2\delta_0 \int_0^\infty t^2 h^2(t)dt &(13.186)\\ Y &= 2\delta_0 \int_0^\infty h^2(t)dt. &(13.187)\end{aligned}$$

where $\delta_0 = \delta_t$. Namely, the group delay shows the **standard deviation of the delay time** as indicating the center of the energy time response, i.e., the time delay of the response. The process where the group delay approaches the limit when the modal overlap becomes high is shown in Fig. 13.38 following the power series of expansion using the modal overlap above. The group delay comes close to the limit when the modal overlap exceeds 2 or 3[73].

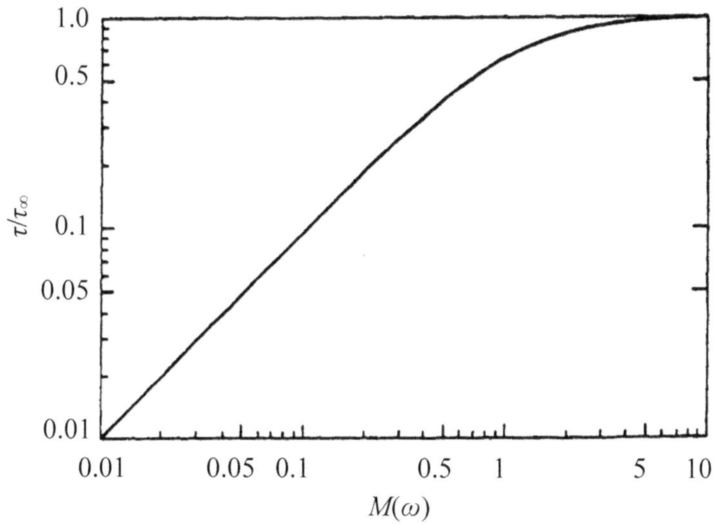

Fig. 13.38 Group delay and modal overlap in reverberation phase from [109] (Fig.4)

The fluctuation from the linear phase can be visualized by using the group delay. Figure 13.39 shows examples of the group delay samples that were calculated from the phase shown in Fig. 13.37[112]. The cause of the phase fluctuation is the non-minimum-phase zeros for the high modal-overlap condition. Assuming the group-delay sequences as random process, Fig. 13.40 illustrates the variances of the group-delay sequences[112]. The variances decreases as the damping increases independent of the frequencies. This outcome indicates that variances are mainly due to the zeros closely located to the observation-frequency line, i.e., the variances depend on the density of the non-minimum-phase zeros.

The properties due to the random sequences can also be displayed by using the auto-correlation sequence, i.e., by the normalized covariances. Figure 13.41 presents examples of the auto-correlation sequences of the group-delay pulsive trains[112]. Here, 4 transfer functions as shown in Fig. 13.34 and 5 conditions of the damping (observation-frequency-lines) as shown in Fig. 13.37 were taken, and therefore 20 curves are plotted in each frequency band. The horizontal axis is normalized according to the density of the non-minimum-phase zeros. The auto-correlation curves of the group delay mainly depend on the density of the non-minimum-phase zeros and are mostly independent of the frequency band. Detailed analysis of the group delay sequences for the room transfer functions can be seen in reference [112].

13.7.3 Minimum-Phase Zeros and Propagation Phase

Recall Fig. 13.31. As described in the previous section, the reverberation phase could be predicted according to the number of non-minimum-phase zeros. In the figure, the number of those zeros indicates correspondence to the reverberation phase in each frequency interval. However, the propagation phase, which might be observed in the coherent field, still is not confirmed from the figure. This is probably because the propagation phase might be produced by the **minimum-phase zeros** instead of the non-minimum-phase zeros. Therefore, the phase analysis for the minimum-phase component of the room transfer functions will be described in this subsection.

The **impulse response** of a linear system can be decomposed into the **minimum-phase** and **all-pass** components. The minimum-phase component has the minimum-phase zeros only, except the poles, in contrast to the all-pass part having the poles and non-minimum-phase zeros. Figure 13.42 is a schematic of such decomposition of the impulse response. The **minimum-phase** part of the response keeps the original magnitude frequency response, but it has a phase different from the original, so the newly created phase by the minimum-phase component has no phase accumulation even in the whole frequency interval of interest. Namely, the phase returns to its initial position after the frequency reaches to the end point of the interval. In contrast, the **all-pass** part has a constant magnitude response (normally to unity) but a different phase from the original. Therefore, the sum of the two phases must be equal to the original.

The **propagation phase** might be produced by zero migration on the pole line, when the receiver moves far from the source[23][24]. The image is inspired by the propagation phase governs the phase characteristic of one-dimensional wave travelling systems. Actually, all the zeros could be minimum phase for the transfer functions of one-dimensional systems. However, note that main cause of the propagation phase must be imbalance between the numbers of poles and zeros, and thus such an imbalance could be cancelled out in the whole-frequency interval. Namely, the propagation phase might be observed only in narrow-band frequencies where the imbalance remains. These local properties of the phase for minimum-phase systems do not contradict that fact there is no phase accumulation of the minimum-phase systems in the whole-frequency range. Figure 13.43 illustrates the phase characteristics for the minimum-phase components of the impulse responses shown in Fig. 13.30 recorded in the reverberation room. The room volume is 189 (m^3), and the reverberation time is reduced to about 1 s by applying exponential windowing to the original impulse-response records. All of the accumulated phase characteristics return to their initial phases of 0 at the end-frequency point. Therefore, it seems almost impossible to detect the propagation phase properties from those phase response; however, it could be possible to perform **linear regression analysis**[2][46] of narrow-band-phase frequency characteristics[60].

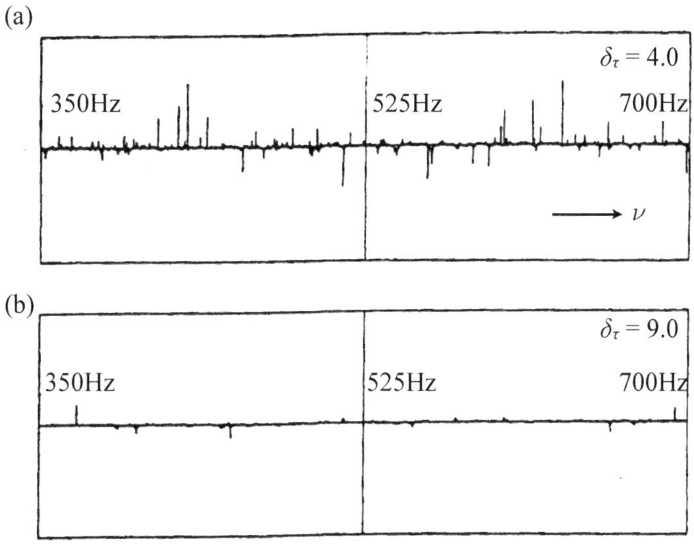

Fig. 13.39 Group delay samples under different damping conditions where δ_t is distance from pole line from [112] (Fig.7)

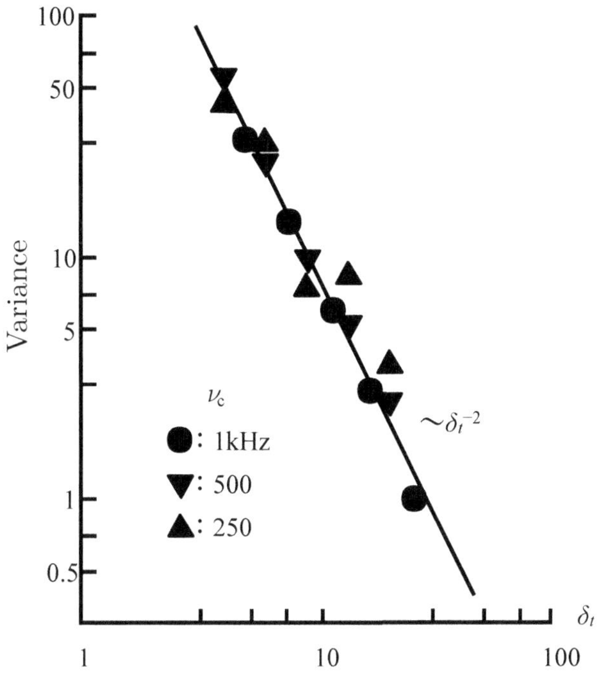

Fig. 13.40 Variances of group delay under different damping conditions from [112] (Fig. 9)

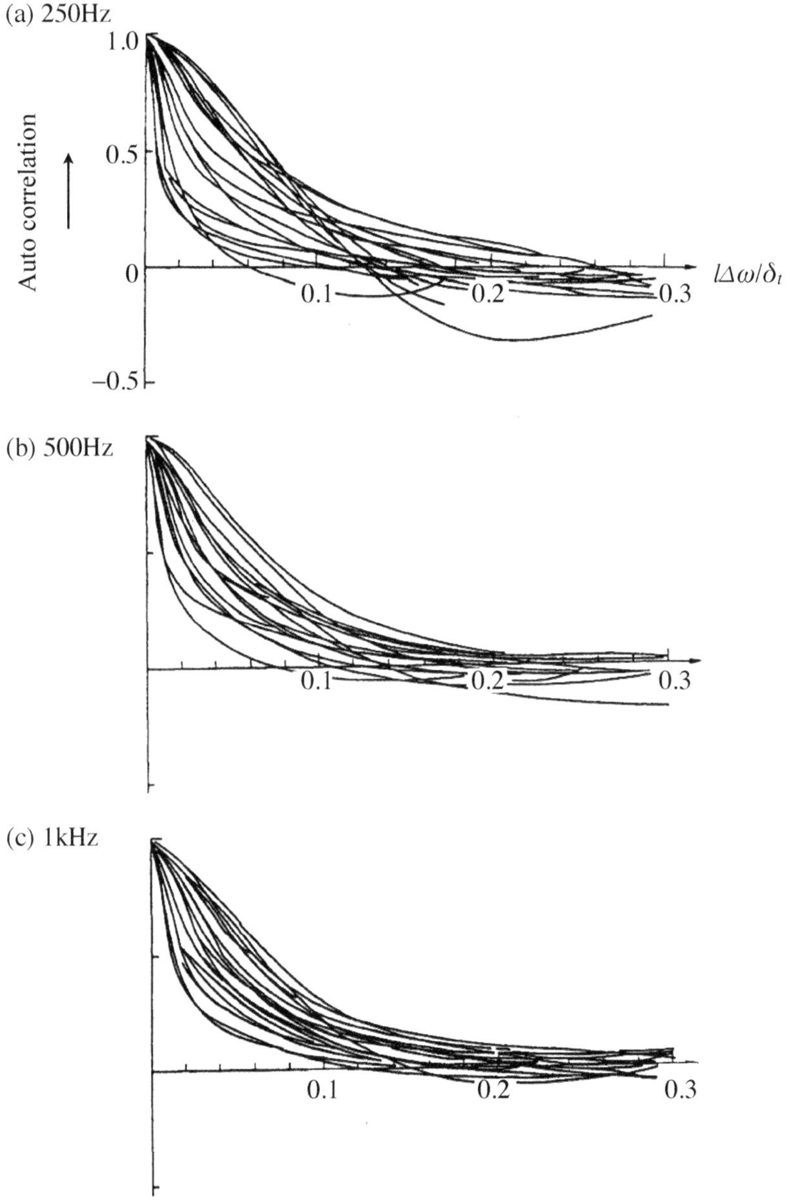

Fig. 13.41 Correlation functions of group delay. Here $l\Delta\omega$ denotes lag for correlation functions on frequency axis, where $\Delta\omega$ is angular frequency sampling interval and l is number of sampling data points in lag: (a) at 250 Hz (1/1 oct. band); (b) at 500 Hz (1/1 oct. band); (c) at 1000 Hz (1/1 oct. band) from [112] (Fig. 8).

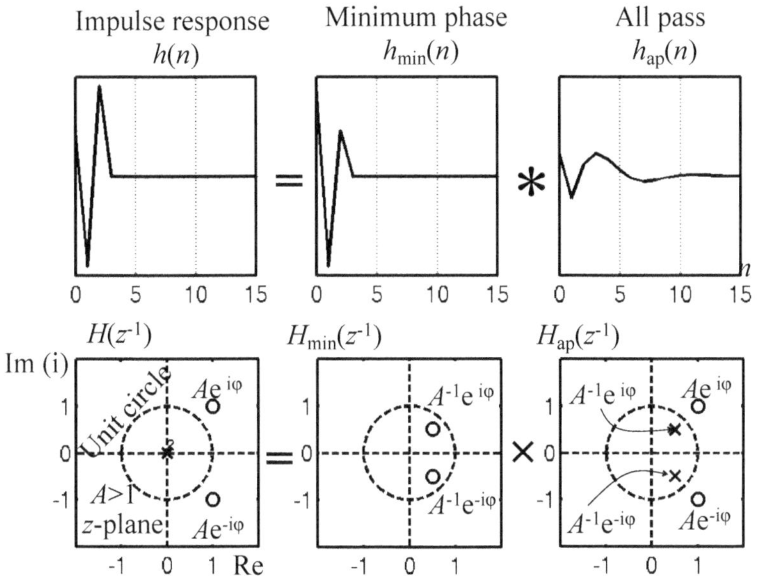

Fig. 13.42 Illustration of impulse-response decomposition into minimum-phase and all-pass components, (a) impulse response, (b) minimum-phase component, (c)all-pass component, top: time response; bottom: poles and zeros on z plane; open circle: zero; cross: pole from [113] (Fig. 3)

Figure 13.44 shows the linear regression analysis of (normalized) minimum-phase characteristics against wavenumber k, where $k_{200} = 2\pi \cdot 200/c$ and c is the sound speed[60]. The increasing gradient of the regression line (**phase trend**) with increasing r can be observed. Figure 13.45 plots the results from evaluating the gradients of the regression lines with r. The dotted line in this figure represents the **propagation phase**, i.e., r. The limit, which can be considered to be the distance from the source, in the gradient of the phase regression line in the minimum-phase component is around $r - r_0 \cong 0.7$ (m).

As described in subsection 13.6.1, the **range of the coherent field** can be estimated by Eq. 13.135. By introducing the conditions of the room, such as $V = 189(\text{m}^3)$ and $T_R = 1$ (s), the range becomes $R_{c_3} \cong 0.68$ (m). The **critical distance**, which was defined in subsection 11.4.3, also shows the distance from the source for the region where the direct sound is dominant according to the energy criterion. The critical distance given by Eq. 11.52 yields $r_c \cong 0.77$ (m). The coherent field presumed from the propagation phase of minimum-phase characteristics ranges mostly within the critical distance.

The propagation phase itself is a local property of the phase characteristics that are the outcome of the local imbalance of the numbers of poles ans zeros. However,

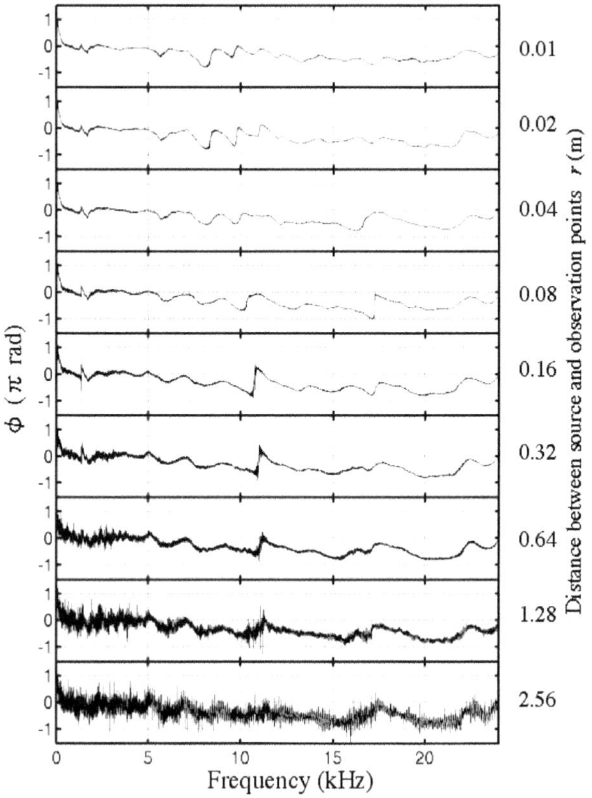

Fig. 13.43 Phase characteristics of minimum-phase components for impulse responses shown in Fig. 13.30 from [60] (Fig.2)

again the fluctuations, such as local-micro characteristics, can be visualized by using the group delay sequences, as were the reverberation phase characteristics. In Fig. 13.45, not only the increasing gradient for the regression line but also the variance in deviation from the propagation phase with increasing r can be seen. Figure 13.46 plots the variance in deviation from the propagation phase of minimum-phase characteristics by using the group-delay sequences[60]. It is interesting to see that the variances again are independent of the frequency bands but depend on distance from the source within the coherent field, and thus approach the limit when the distance exceeds the coherent length.

The minimum-phase component of the transfer function has two types of zeros: the original minimum-phase zeros and the converted ones from the original non-minimum-phase ones. This can be understood by recalling the decomposition

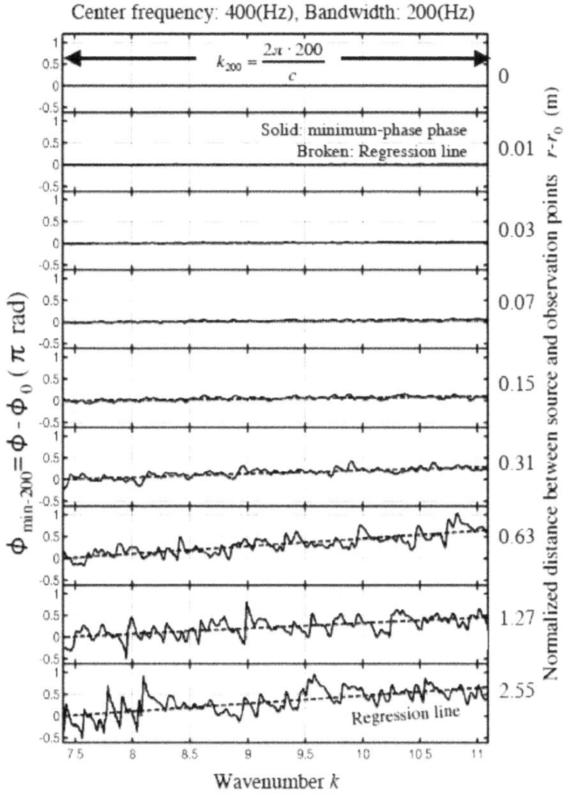

Fig. 13.44 Linear regression analysis for narrow-band minimum-phase characteristics normalized by accumulated phase at $r_0 = 0.01$ (m) from [60] (Fig.5)

schematic of the transfer function into minimum-phase and all-pass components illustrated in Fig. 13.42. As stated previously, migration of the minimum-phase zeros produces the propagation phase; however, an increase in the number of converted minimum-phase zeros does not contribute to constructing the propagation phase, but makes only the fluctuations from the propagation phase. This might explain why the variances in the fluctuations from the propagation phase are independent of the frequencies because the numbers of non-minimum-phase zeros are almost independent of the frequencies as described in the previous subsection 13.7.2.

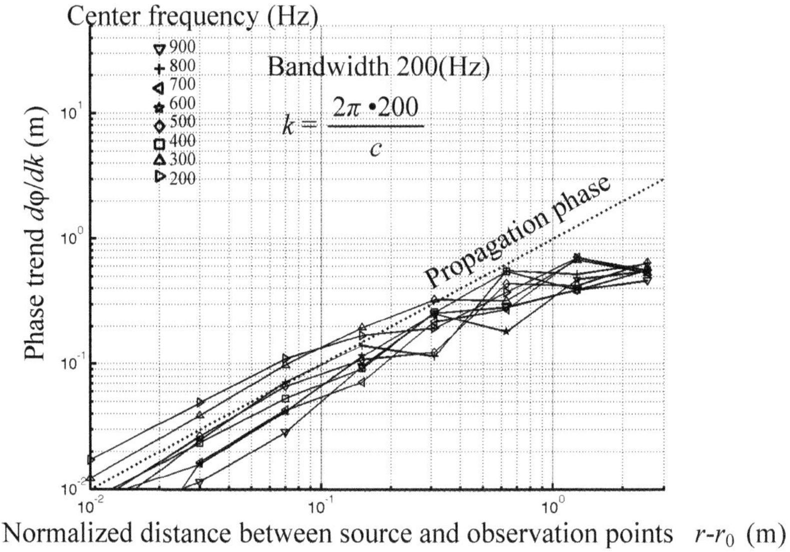

Fig. 13.45 Slopes of the regression lines (phase trends) for narrow-band phase as shown in Fig. 13.44 from [60] (Fig.6)

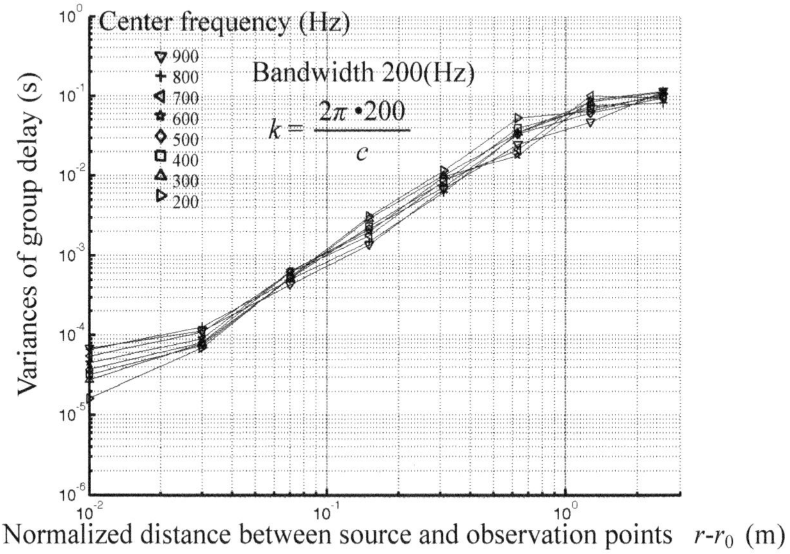

Fig. 13.46 Group-delay variances from propagation phase from [60] (Fig.7)

Chapter 14
Representation of Sound in Time and Frequency Plane

This chapter describes signal analysis and synthesis according to the correspondence between the time and frequency region. Speech intelligibility is sensitive to temporal narrow-band envelopes. To represent the envelopes, the magnitude and phase information are important, respectively. When taking a medium-sized window-length for analysis and synthesis, such as 20 - 60 ms, the magnitude spectral information is crucial. In contrast, for shorter or longer frame lengths, the phase is significant instead. Under moderate frame lengths, a speech waveform can be represented by spectral peak selection, and the temporally changing fundamental frequencies of sound are tracked by the auto-correlation analysis of the selected spectral peaks along the frequency axis. Sound that is made of the harmonic structure of spectral components can be expressed as clustered line-spectral components around the spectral peaks including the time envelope. In contrast, a transient signal in a short time period can be characterized by the clustered time series according to the correspondence between the time and frequency region. Following the correspondence, a modulated waveform can be interpreted in terms of the magnitude and phase for the complex time region. Consequently, the minimum- and non-minimum-phase concepts can also be applied to the modulated signals between the envelopes and carrier components.

14.1 Magnitude- or Phase-Spectral Information for Intelligible Speech

The discrete Fourier transformation (DFT) is a general principle for signal analysis and/or synthesis using sinusoidal signals that are determined according to the frame length of observation. However, the significance of magnitude or phase spectral information for constructing intelligible speech, with regard to the observation frame lengths to be taken, has been of research interest[13]. The magnitude spectrum has been considered important in almost all types of applications of speech processing, while the phase spectrum has received less attention.

An experimental approach similar to that in reference [114] is applied to a spoken sentence and random noise in this section[13]. From these signals, two new signals

M. Tohyama: *Sound and Signals*, Signals and Communication Technology, pp. 305–349.
springerlink.com

are created by a cross-wise combination of the magnitude and phase spectra of the speech and noise signals. These two hybrid signals are made for a wide range of window lengths.

14.1.1 Test Materials and Signal Processing

Synthesized hybrid (magnitude- or phase only) speech signals were obtained using female-spoken speech and random-noise samples, as shown in Fig. 14.1. Sentence intelligibility for the two hybrid signals, as a function of the window length used in the DFT analysis and reconstruction, was estimated using listening tests.

The original speech signals were everyday sentences spoken by two female speakers. All of the speech materials were in Japanese and digitized at a sampling rate of 16 kHz. The speech and random-noise pairs were analyzed using DFT (Fig. 14.1), where a rectangular window function was applied to cut signals into frames, and a 50% overlapping window was applied. Two hybrid signals were synthesized by inverse DFT on the frame-wise basis using the magnitude spectrum of the speech (or the noise) and the phase spectrum of the noise (or the speech). The first type will be referred to as magnitude-spectral speech (MSS) and the second type as phase-spectral speech (PSS). A triangular window, with a frame length equal to the rectangular window used for the analysis, was used for reconstructing the hybrid signals.

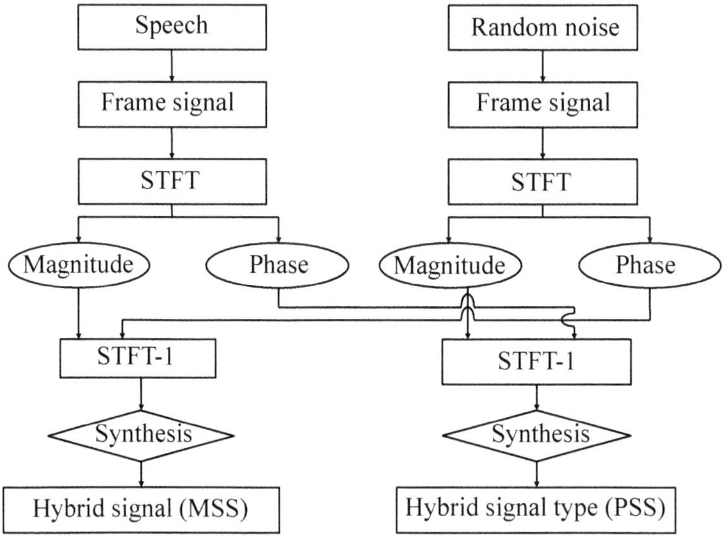

Fig. 14.1 Method for deriving two types of hybrid signals from speech and random noise from [13] (Fig.1)

The listeners were seven native speakers of Japanese. The total set of materials, which consisted of 192 processed sentences (6 sentences for each of 16 frame lengths and 2 types of hybrid signals), were presented in random order through headphones under diotic listening conditions at an individually preferred level. The **diotic listening condition** means that a subject listens to the same signals through both ears simultaneously. Each subject was asked to write down the sentences as they listened. A sentence was considered intelligible only if the complete sentence was written down correctly.

14.1.2 Speech Intelligibility Scores for Synthesized Speech Materials

Figure 14.2 shows the sentence intelligibility scores (with the standard deviation) for each signal type and frame length[13]. Each data point is based on an average for six presentations (sentences) to seven listeners.

Note that the frequency resolution of DFT is given by $1/N$ (Hz), where N denotes the frame length in seconds without spectral interpolation. For the shorter time frames, the results in the figure suggest that a frequency resolution finer than 250 Hz (frame length longer than 4 ms) is needed to obtain intelligible speech from the spectral magnitude. For the longer time frames, the temporal resolution required to obtain intelligible speech from the magnitude spectrum should be better than about 128 ms. It can be intuitively understood that appropriate temporal and spectral resolutions might be necessary to construct intelligible speech materials. Interestingly, where the magnitude spectrum fails in reproducing intelligible speech, the phase spectrum (partly) takes over this role[115][116]. This outcome can be interpreted as that the temporal properties or signal dynamics represented by the envelopes can be expressed as the very local characteristics of the phase spectrum, such as the group delay. In other words, the phase spectra with a fine spectral resolution will allow a partial reconstruction of the **narrow-band temporal envelope**.

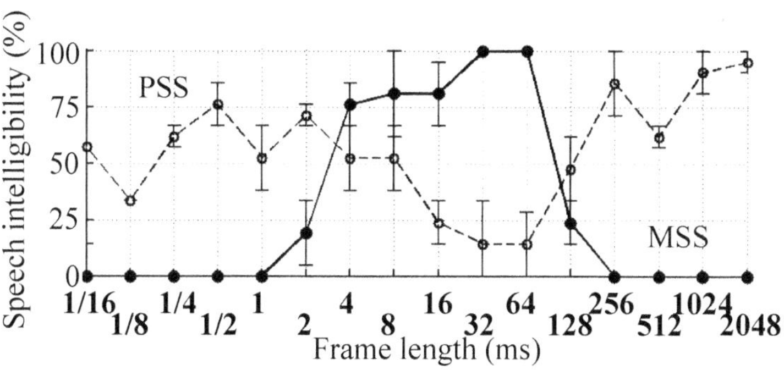

Fig. 14.2 Sentence intelligibility for PSS and MSS, as function of frame length used in DFT procedure from [13] (Fig.2)

Signal envelopes can be represented by **Hilbert envelopes**. Suppose a real signal $x(n)$ with a length N and its Fourier transform $X(k)$. By introducing the modified **causal spectrum** such that

$$\begin{aligned}\hat{X}(k) &= 2X(k) && 0<k<N/2 \\ &= 0 && N/2<k\leq N-1 \\ &= X(k) && k=0, \quad \text{or} \quad k=N/2\end{aligned} \tag{14.1}$$

and taking the inverse Fourier transform of the causal spectrum, the complex signal

$$z(n) = x(n) + \mathrm{i}y(n) \tag{14.2}$$

can be obtained instead of the original real signal $x(n)$. The complex signal $z(n)$ above is called **analytic representation** of a real signal $x(n)$, and the spectrum $\hat{X}(k)$ is called the **causal** or **single-sided spectrum**.

The relationship between the real original signal and the imaginary part newly created corresponds to that between the real and imaginary parts of the spectrum of a **causal** or **single-sided signal**. Namely, if the real part of the spectrum is known for a causal real signal $x(n)$, the imaginary part can be derived from the real part, and vice versa, subject to $x(0)$ being known. Figure 14.3 is an example of causal signals, which can be decomposed into even and odd sequences. The even sequence makes the real part of the spectrum of the causal signal, while the odd one produces the imaginary part of the spectrum. The odd (even) part can be constructed from the even (odd) part for a causal signal. This explains why the real (imaginary) part of a spectrum can be derived from the imaginary (real) part of the spectrum of the real and causal signal.

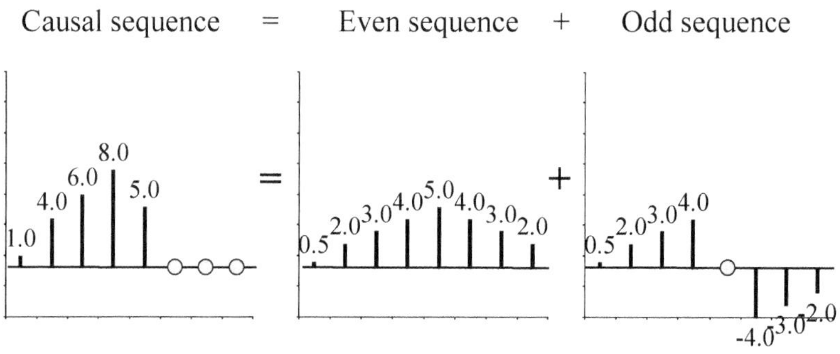

Fig. 14.3 Decomposition of causal sequence into even and odd sequences.

The magnitude of the analytic signal waveform is called **instantaneous magnitude** or a **Hilbert envelope**, in contrast to the magnitude of the complex spectrum for the causal signal. The angle of the complex signal (analytic signal) is called the **instantaneous phase**, similar to the phase spectrum for the causal signal. The envelopes of speech signal waveforms are closely related to intelligibility[11][12][13].

14.1.3 Narrow-Band Envelopes of Speech Materials

An interesting question is to what extent the narrow-band envelopes are preserved for the two types of hybrid signals. Here the narrow-band envelopes indicate the Hilbert envelopes that were derived for every sub band signal of speech, e.g., every 1/4 octave-band filtered signals. Figure 14.4 presents the cross-correlation coefficients between squared narrow-band envelopes of the hybrid signals and the original speech for each of four 1/4-octave bands. Figure 14.4(a) is just replication of the intelligibility results in Fig. 14.2. The correspondence between the intelligibility data and the narrow-band temporal envelopes confirms that the preservation of the narrow-band temporal envelopes is closely related to speech intelligibility. The cross-correlation coefficients are defined between two variables X and Y as

$$C_F(0) = \frac{\mathrm{E}[(X-\mu_X)(Y-\mu_Y)]}{\sqrt{\mathrm{E}[(X-\mu_X)^2]\mathrm{E}[(Y-\mu_Y)^2]}} \tag{14.3}$$

which means the normalized **covariance** between the two quantities. E[*] denotes taking the ensemble average of *, and μ_X denotes E[X].

It might be interesting to see the cross-over points in the figures. The correlation data for MSS and PSS show two cross-over points. The cross-over at a frame length of about 256 ms is almost independent of the frequency band considered, as can be

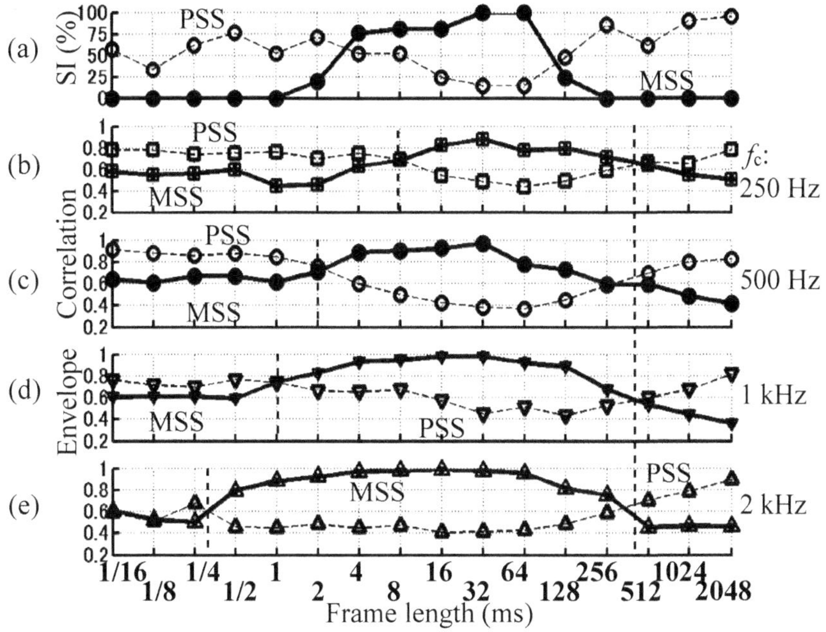

Fig. 14.4 Sentence intelligibility (a) and examples of envelope-correlation analysis (b)-(e) for MSS and PSS from [13] (Fig. 4)

seen by the vertical dotted line through figures. Since the observed decrease in the correlation for MSS toward long frame lengths reflects the loss of time resolution required for representing the temporal envelope, this cross-over point is supposed to be related to the dominant frequency of the envelope modulation. The corresponding cross-over point in the intelligibility data is considerably lower, suggesting that the speech envelope includes slow modulations, which are included in the correlation values, but contribute little to speech intelligibility. The cross-over point, 256 ms, corresponds to a modulation frequency of 4 Hz.

The cross-over point is frequency dependent as shown by the vertical dotted lines in each of the figures. This frequency dependency might be due to the limited frequency resolution associated with a short frame length for DFT analysis. A certain loss of frequency resolution, which can be represented by the inverse of the frame length in the DFT, will have less effect for higher center frequencies. Thus, to recover 1/4-octave band envelopes from the magnitude spectrum indicated by MSS in Fig. 14.4, the frame length used in the DFT should provide an adequate degree of frequency resolution, related to the width of the frequency-band considered. Therefore, shorter frames are allowed toward higher center frequencies.

The frame-length dependency of intelligibility for the synthesized hybrid signals represents the temporal and spectral properties preserved for intelligible speech. Such requirements can be translated into the conditions for frame-wise magnitude and phase spectral analysis from the viewpoint of narrow-band envelope recovery. For constructing such intelligible MSS, the frame length must be within 4 to 256 ms as shown by Fig. 14.4. For longer time frames ($>$ 256 ms), the temporal resolution is insufficient to follow the relevant envelope modulations, and for shorter frames ($<$ 4 ms), the frequency resolution becomes insufficient (this appears to depend on the center frequency of a band).

However, the PSS data surprisingly indicate that the envelope is (partly) reconstructed for longer time frames than 256 ms, and even for very short time frames. In the following subsections, recovery of narrow-band envelopes from the phase spectrum will be described.

14.1.4 Recovery of Narrow-Band Envelopes from Phase Information under Long Time Window

As already mentioned subsection 2.5.5, the importance of the phase spectrum is well illustrated by the difference between an amplitude- and a quasi-frequency-modulated (AM and QFM) sinusoid. The phase of the two side-band components determine the temporal envelope: essentially flat in the QFM case and modulated in the AM case. Figures 14.5(a) and 14.5(b) show a stationary random noise and a noise modulated by a co-sinusoidal function, respectively. The corresponding magnitude and phase spectra are shown in the middle and bottom rows in the figure. The (normalized) envelope-modulation frequency is given by $2(1/N)$, where N denotes the signal length, and DFT analysis was applied to the whole signal length. There are no clear indications of the envelope frequency in the magnitude and phase spectra, however.

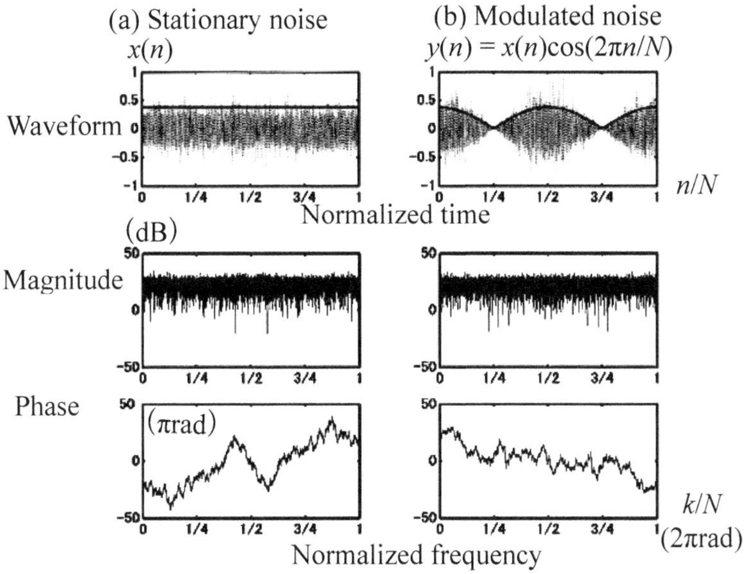

Fig. 14.5 Examples of stationary random noise (a) and modulated noise (b) with magnitude and phase spectral characteristics from [13] (Fig. 5)

The **auto correlation** of a complex spectral sequence can be defined as

$$C_F(\Delta k) = \mathrm{E}[X^*(k)X(k+\Delta k)] = \mathrm{E}[|X^*(k)||X(k+\Delta k)|\mathrm{e}^{\mathrm{i}(\phi(k+\Delta k)-\phi(k))}], \quad (14.4)$$

where $|X(k)|$ denotes the magnitude and $\phi(k)$ shows the phase of the complex spectral sequence of interest. By discarding the magnitude component, the auto correlation of the phase components can be defined by

$$C_F(\Delta k) = \mathrm{E}[\mathrm{e}^{\mathrm{i}(\phi(k+\Delta k)-\phi(k))}] \quad (14.5)$$

$$= \mathrm{E}[\mathrm{e}^{\mathrm{i}\Delta\phi(k,\Delta k)}] \quad (14.6)$$

$$= C_{F_r}(\Delta k) + \mathrm{i}C_{F_i}(\Delta k). \quad (14.7)$$

Thus, by taking the magnitude of the phase correlation above,

$$C_{F_r}(\Delta k) = \mathrm{E}[\cos\Delta\phi(k,\Delta k)] \quad (14.8)$$

$$C_{F_i}(\Delta k) = \mathrm{E}[\sin\Delta\phi(k,\Delta k)] \quad (14.9)$$

$$|C_F(\Delta k)| = \sqrt{C_{F_r}^2(\Delta k) + C_{F_i}^2(\Delta k)} \quad (14.10)$$

can be derived.

By applying the **phase correlation** analysis to the spectrum of the magnitude signal shown in Fig. 14.5, the envelope frequency can be observed as shown by

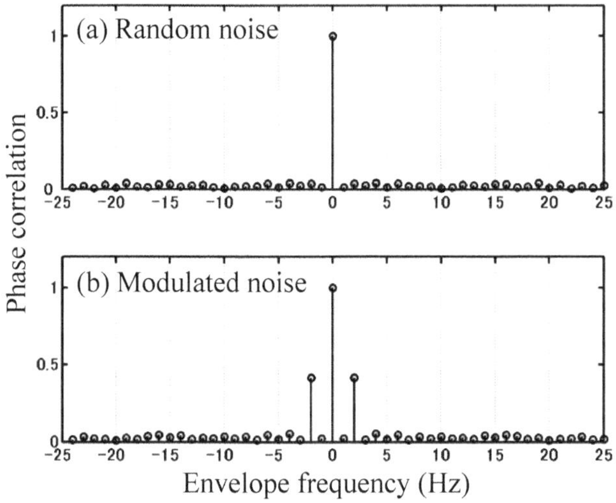

Fig. 14.6 Phase spectral auto-correlation analysis for signals shown in Fig. 14.5 from [13] Fig.6

Fig. 14.6. Here, the horizontal axis corresponds to the frequency shift for the phase correlation analysis Δk. Actually, Fig. 14.6(b) indicates the envelope frequency of 2 Hz that corresponds to that for the modulated signal in Fig. 14.5(b). Only for the modulated noise case (Fig. 14.6(b)) can the modulation frequency be estimated from phase information alone.

Figure 14.7 is an example of a hybrid signal for the modulated signal shown in Fig. 14.5 following the procedure synthesizing the PSS hybrid signal. The original envelope that was illustrated in Fig. 14.5 is partly preserved even when reconstructing the signal on the basis of the phase spectrum only. The frequency bin in the

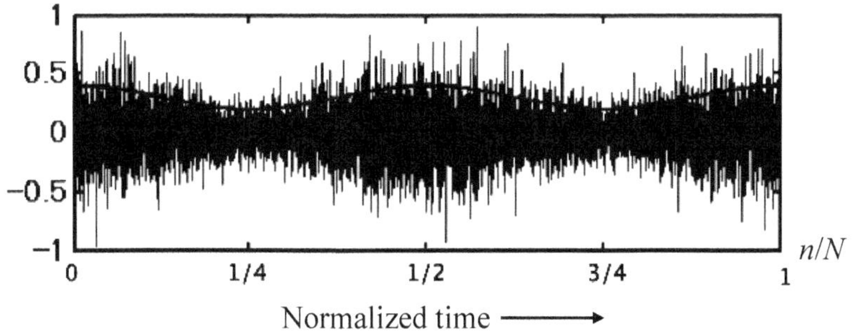

Fig. 14.7 Reconstruction of modulated noise of Fig. 14.6(b), using corresponding phase spectrum and random magnitude spectrum from [13] (Fig.7)

phase spectrum resulting from the DFT should be narrow enough to reflect the envelope frequency in the phase correlation sequence. The frame length used in the DFT should be longer than the period of the envelope modulation of interest because the frequency bin is determined by the inverse of the frame length.

14.1.5 Envelope Recovery from Phase for Short Time Window

As Figs. 14.4 (b)-(e) indicated, the frame length should be shorter than the period of the center frequency of interest for recovery of the narrow-band envelopes from the phase. For the experimental results, the limit of the shortest window length is 1/16 ms (i.e.,the sampling interval), corresponding to a single-point DFT. The single-point DFT can be defined as

$$X(k)|_{k=0} = \sum_{n=0}^{0} x(n)\mathrm{e}^{-\mathrm{i}\frac{2\pi k}{N}n}\Big|_{k=0} = x(0) \tag{14.11}$$

for a signal $x(n)$. The result of a single-point DFT is each sample itself, and the phase is the sign of the sample, $x(n)$. Thus, the phase information of a single-point DFT keeps the **zero crossings** of the original signal[79], if the sampling frequency is adequate. As previously described in subsection 13.7.2, the density of zeros for a waveform can be estimated by the power spectrum for the entire waveform. It can be assumed, therefore, that short-term distribution of zero crossings might be mostly expressed by the short-term power spectrum that reproduces the narrow-band envelopes of the original speech waveform. The recovery of narrow-band envelopes basically might be due to this assumption.

Keeping the zero crossings of a waveform is the same as applying infinite **peak clipping** to the signal waveform, which also preserves the zero-crossing information while losing all amplitude information. Figure 14.8 shows an example of modulated sinusoidal waveforms (Figs.(a)-(c)) and its clipped version. Spectral records for the envelope (Fig.8(d)), its carrier (Fig.8(e)), and the modulated signal (Fig. 8(f)) are represented by the line-spectral characteristics. Here, the solid lines and solid circles show the original ones, while the dotted lines and open circles indicate the infinitely clipped ones.

The spectral structure of the modulated signal can be expressed as the **convolution of the spectral sequences** for the envelopes and the carrier such that

$$\mathrm{FT}[y(n)] = \mathrm{FT}[w(n)] * \mathrm{FT}[x(n)] \tag{14.12}$$

where

$$y(n) = w(n)x(n). \tag{14.13}$$

If only the zero-crossing property is preserved with magnitude of unity (discarding the envelope of the modulated signal), the convolved spectral-structure is expanded, including its higher harmonics.

The modulation property, such as the **temporal envelope**, can be recovered by applying appropriate filtering, as shown in Fig. 14.9, or despite that the higher harmonics are not contained in the original modulated signal[13]. Figure 14.9 (a) is a close-up of Fig. 14.8 (f). For a bandwidth of denoted by (i) in Figs. 14.9 and 14.10, representing sub band analysis in different frequency region, the waveforms shown in (b) of both figures are obtained. Here, the broken line represents the original envelope. However, if the bandwidth increases according to the examples illustrated by (c) or (d) in both figures, the original envelope is no longer recovered. This explains why the original envelope can be recovered from zero-crossing information when applying sub band filtering, provided that bandwidth is adapted to the modulation frequency of interest. This may also explain why envelope recovery from phase spectra for very short window is poorer at high frequencies (Fig. 14.4: higher frequencies are associated with broader absolute bandwidth.) It is well known that infinite peak clipping version of a speech waveform, keeping zero-crossing information, is almost perfectly intelligible. Recovery of the envelope from the carrier-like information, such as zero-crossing information, may partly explain why zero-crossing speech could be intelligible.

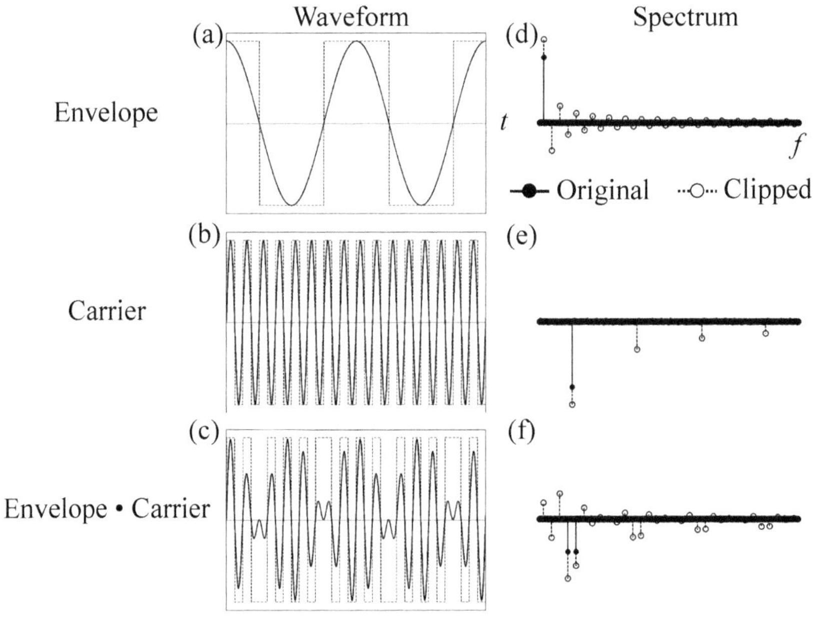

Fig. 14.8 Spectrum of infinitely clipped version of modulated sinusoidal signal from [13] (Fig.10)

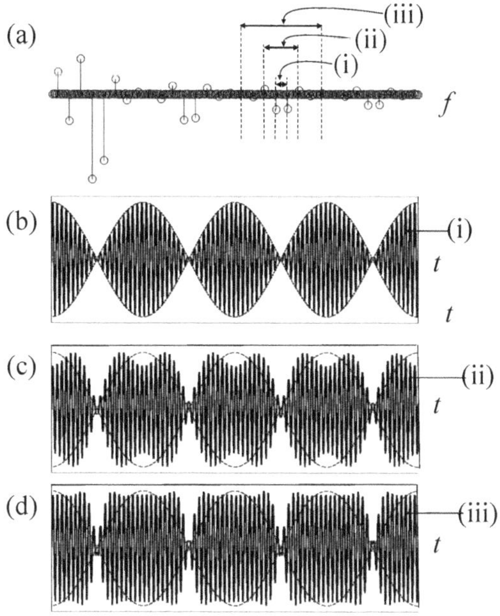

Fig. 14.9 Sinusoidal envelope recovery in base band from the clipped wave as shown in Fig. 14.8 after applying sub band filtering with different bandwidth, indicated by (i)-(iii) from[13] (Fig.11)

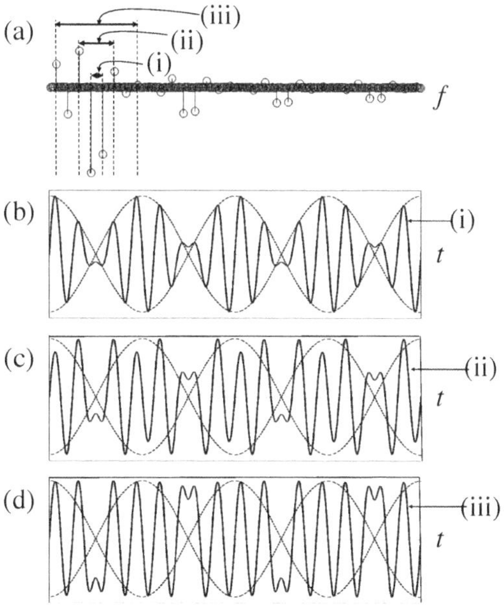

Fig. 14.10 Sinusoidal envelope recovery in higher frequency band from the clipped wave after applying sub band filtering [117]

14.1.6 Relationship between Magnitude and Phase Spectral Components

Subsection 14.1.2 described that the imaginary (real) part of the spectrum can be derived from the real (imaginary) part for the real and causal sequences. Namely, those two components of the spectrum are not independent of each other for the real and causal sequences. However, the magnitude and phase spectral components are not always derived each other even for a real and causal sequence. It is possible to get the phase (magnitude) spectral components from the magnitude (phase) spectral information only for the **minimum-phase sequence**. Recall the schematic for decomposition shown by Fig. 13.42. Such decomposition can be performed using **cepstral sequences**.

Suppose a real sequence $x(n)$ and its Fourier transform $X(k) = |X(k)|e^{i\phi(k)}$. The cepstrum of the real sequence is defined as

$$C_{ep}(k) = \ln X(k) = \ln|X(k)| + i\phi(k) \tag{14.14}$$

$$c_{ep}(n) = \text{IDFT}[C_{ep}(k)]. \tag{14.15}$$

Note that the cepstral sequence $c_{ep}(n)$ of a real sequence $x(n)$ is also a real sequence. Recall that the singularities of a causal sequence are located within the **unit circle** on the z-plane; thus the cepstral sequence must be **causal** for a **minimum-phase sequence**. Figure 14.11 shows an example of the decomposition of the minimum-phase and all-pass components using the cepstral sequences. Note that the minimum-phase cesptrum is real and causal, and its even and odd components correspond respectively to the magnitude- and phase cepstrum for the minimum-phase sequence.

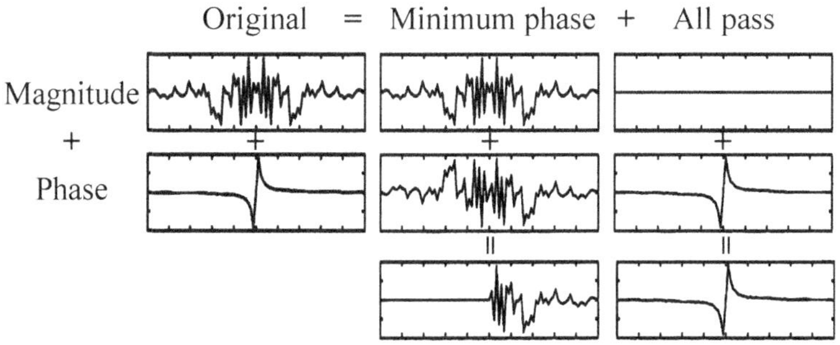

Fig. 14.11 Cepstral decomposition of real sequence into minimum-phase and all-pass components [69]

On the other hand, the all-pass component has only the phase cepstrum, which is an odd sequence. Therefore, when decomposing the cepstrum of a minimum-phase sequence into the even and odd sequences, the even sequence yields the magnitude spectrum and the odd component yields the phase spectrum. Consequently, the phase spectrum (magnitude spectrum) can be derived from the magnitude spectrum (phase spectrum) for a minimum-phase sequence, because the even and odd sequences can be derived each other for a causal sequence.

14.2 Speech Signal Reconstruction by Spectral Peak Selection

Speech material can be represented by magnitude spectral information subject to adequate analysis/synthesis window lengths, such as 32-128ms. The magnitude spectrum of sound such as speech can be basically decomposed into the **fundamental frequency** and its **harmonics**. Such a spectral harmonic structure of sound more or less can be characterized by the dominant spectral peaks, which represent the resonance frequencies for the organ that produces the sound. This section describes how a signal is analyzed and reconstructed by spectral peak selection.

14.2.1 Representation of Sinusoidal Signal Using Spectral Peak Selection of Discrete Fourier Transform(DFT)

Discrete Fourier transformation (DFT) is a mathematical tool of signal representation using the fundamental and its harmonics. However, note that the fundamental that DFT defines is determined independent of the signal signature but by the length of the window used for DFT. This indicates that the fundamental of a signal to be observed cannot be estimated by DFT, except for when the signal length L taken for observation(window length) is $L = pT$, where T denotes the fundamental period of the signal to be estimated and p is an integer. Therefore, spectral analysis of a signal seems almost impossible by DFT from a practical point of view, if it is true[118][119].

Spectral interpolation, however, would be an inevitable tool for signal analysis by DFT. The signal's **original spectrum**, which is called the **true spectrum**, of a finite-length (finite window length) record can be estimated from the **spectral peak selection** from the interpolated spectra[120]. Here, the original spectrum refers to the spectrum of a virtual signal that could be obtained if a record of infinite length could be taken. Namely, the relationship between the original and observed spectra is expressed by the following formula:

$$X_0(k) * W(k) = X(k), \tag{14.16}$$

where $X_0(k)$ denotes the original spectrum for the original signal $x_0(n)$, $W(k)$ shows the spectrum of the window function used for the DFT, $X(k)$ is the observed spectrum, and $x(n) = x_0(n) \cdot w(n)$. Suppose a sinusoidal signal of the analytic signal form

$$x_0(n) = \mathrm{e}^{\mathrm{i}\Omega_0 n}. \tag{14.17}$$

By applying a **rectangular window function** of the length N to the signal above and taking the Fourier transform,

$$\begin{aligned} X(\mathrm{e}^{-\mathrm{I}'\mathrm{d}\Omega}) &= \frac{1}{N}\sum_{n=0}^{N-1} \mathrm{e}^{\mathrm{i}\Omega_0 n}\mathrm{e}^{-\mathrm{i}\Omega n} = X_0(\mathrm{e}^{-\mathrm{i}\Omega}) * W(\mathrm{e}^{-\mathrm{i}\Omega}) \\ &= \frac{1}{N}\frac{1-\mathrm{e}^{-\mathrm{i}(\Omega-\Omega_0)N}}{1-\mathrm{e}^{-\mathrm{i}(\Omega-\Omega_0)}} \qquad (\Omega_0 \neq \Omega) \\ &\quad 1 \qquad\qquad\qquad\qquad\quad (\Omega_0 = \Omega), \end{aligned}$$

where

$$X_0(\mathrm{e}^{-\mathrm{i}\Omega}) = \delta(\Omega - \Omega_0) \qquad 0 < \Omega < 2\pi \tag{14.18}$$

$$W(\mathrm{e}^{-\mathrm{i}\Omega}) = \frac{1}{N}\sum_{n=0}^{N-1} \mathrm{e}^{-\mathrm{i}\Omega n} = \frac{1}{N}\frac{1-\mathrm{e}^{-\mathrm{i}\Omega N}}{1-\mathrm{e}^{-\mathrm{i}\Omega}} \tag{14.19}$$

$$X_0(\mathrm{e}^{-\mathrm{i}\Omega}) * W(\mathrm{e}^{-\mathrm{i}\Omega}) = \int_0^{2\pi} \delta(\Omega' - \Omega_0)\frac{1}{N}\frac{1-\mathrm{e}^{-\mathrm{i}(\Omega-\Omega')n}}{1-\mathrm{e}^{-\mathrm{i}(\Omega-\Omega')}} d\Omega'. \tag{14.20}$$

Figure 14.12 illustrates examples of the power spectra calculated from the Fourier transform $X(\mathrm{e}^{-\mathrm{i}\Omega})$. Spectral components at the frequencies other than $\Omega = \Omega_0$ can be seen in the Fourier transform, despite that the original signal is a sinusoid with a single frequency. If the original angular frequency is $\Omega_0 = 2\pi k/N$ $(0 < k \leq N-1)$ where k and N are integers, then the Fourier transform observed at $\Omega = 2\pi l/N$ for an integer l $(0 < l \leq N-1)$ becomes

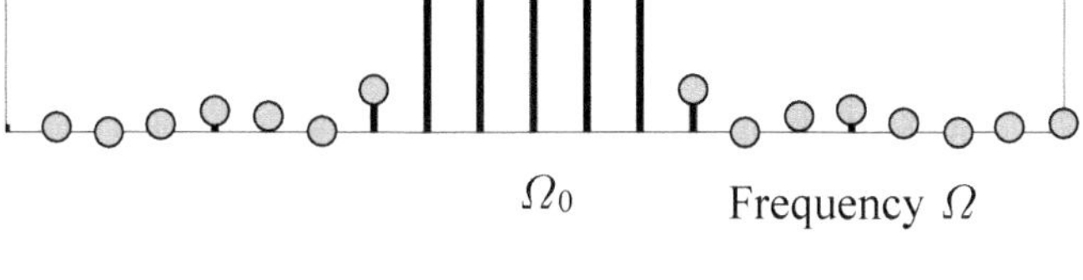

Fig. 14.12 Windowed sinusoidal spectrum (Power spectrum)

$$X(e^{-i\Omega})|_{\Omega=\frac{2\pi}{N}l} = \begin{cases} 1 & l = k \\ 0 & l \neq k. \end{cases} \tag{14.21}$$

That is, the original spectrum for the single sinusoid can be seen. The results suggest that the original spectrum, i.e., the true spectrum, of a finite-length record can be estimated from the **spectral peak** in the interpolated spectra from the DFT of the windowed signal[120].

Suppose that a target signal is expressed in analytic form as

$$x(n) = Ae^{i\frac{2\pi}{N}(k_0+\Delta k)n}, \tag{14.22}$$

where A denotes the complex magnitude including the initial phase, k_0, p, q are integers, and $\Delta k = q/p$. **Spectral interpolation** can be performed by taking DFT of the signal after making the original window length longer so that the original signal record is followed by the newly added zeros. By taking the DFT of the signal for which the record length is increased to M by appending $M-N$ zeros, the interpolated DFT becomes

$$\begin{aligned} X(k) &= \frac{1}{N}\sum_{n=0}^{N-1} x(n)e^{-i\frac{2\pi k}{M}n} \\ &= \frac{1}{N}\sum_{n=0}^{N-1} x(n)e^{i2\pi n\frac{pk_0+q}{N\cdot p}-i2\pi n\frac{k}{M}} \\ &= \frac{1}{N}\sum_{n=0}^{N-1} x(n)e^{-i\frac{2\pi n}{M}(k-(pk_0+q))} \end{aligned} \tag{14.23}$$

where $M = N\cdot p$. The original spectrum can be seen at the spectral peak where $k = pk_0 + q$. Figure 14.13 demonstrates the process where the spectral peak, which denotes the original spectrum, can be estimated by **spectral peak selection** from the interpolated DFT spectra[120].

A signal that is composed of sinusoidal components can also be represented by spectral peak selection. Assume a target signal that is expressed in the analytic form as

$$x(n) = \sum_{k=1}^{K} A(k)e^{i2\pi\nu(k)n} + \varepsilon_K(n), \tag{14.24}$$

where $A(k)$, $\nu(k)$ denote the k-th sinusoidal component's complex magnitude and frequency, respectively, K is the number of dominant sinusoidal components, and ε_K denotes the residual component such as external noise. Figure 14.14 is an example showing that the original spectrum, i.e., the true spectrum, of a finite-length record can be estimated from the spectral peak in the interpolated spectra. Several spectral peaks corresponding to the dominant sinusoidal components are seen in the figure.

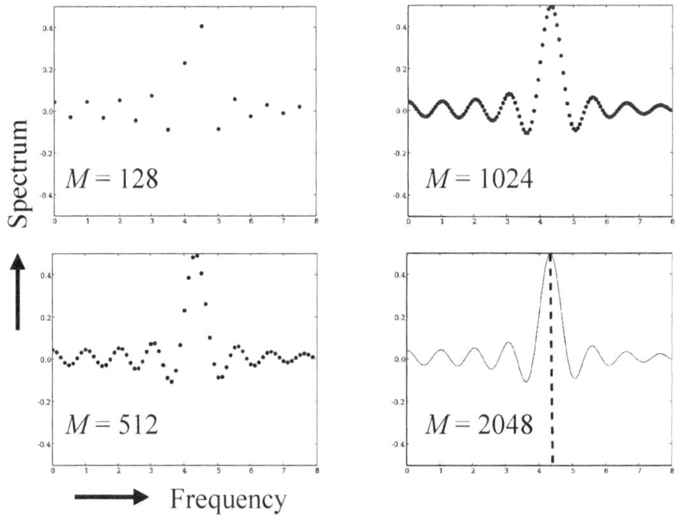

Fig. 14.13 Interpolated ($N \rightarrow M$) spectral records for windowed sinusoid (N=64)

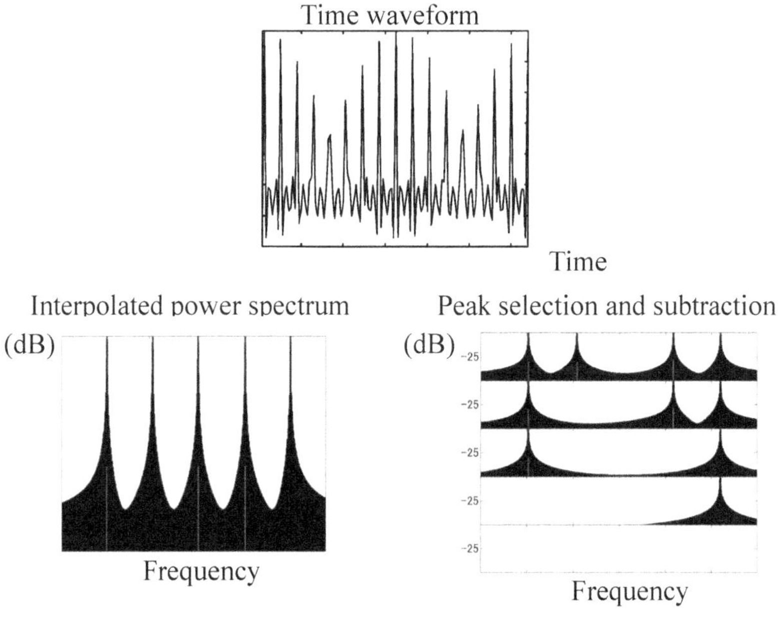

Fig. 14.14 Spectral peak selection and subtraction for compound signal of five sinusoids from [34] (Fig.7.3)

The following procedure is used for spectral peak selection.

Step 1: Take the M-point DFT of the signal in the analytic form after padding $M-N$ zeros so that

$$X(k) = \frac{1}{N}\sum_{n=0}^{N-1} x(n)\mathrm{e}^{-\mathrm{i}\frac{2\pi k}{M}n}, \tag{14.25}$$

where the length of a record is N. This DFT indicates the dominant frequency-spectrum estimate at the spectral peak that corresponds to the maximum power spectrum frequency.

Step 2: Select the maximum component in the power spectrum record obtained in Step 1 as $X(K_p)$ for making $|X(k_p)|^2$ maximum.

Step 3: Subtract the maximum component from the original signal so that

$$e(n) = x(n) - X(k_p)\mathrm{e}^{\mathrm{i}\frac{2\pi}{M}k_p n} \qquad n = 0, 1, ..., N-1 \tag{14.26}$$

and set

$$x(n) \leftarrow e(n) \qquad n = 0, 1, ..., N-1. \tag{14.27}$$

This subtraction that is performed in the signal length N excluding the newly added zeros is crucial for signal representation without the leakage spectrum due to the truncating window. However, it is also possible to perform the subtraction process in the frequency domain by subtracting the spectrum of the windowed sinusoid.

Step 4: Repeat steps 1 to 3 until

$$\sum_{n=0}^{N-1} |e(n)|^2 < E, \tag{14.28}$$

where E is the allowable error. Figure 14.14 showed the results obtained after five repetitions of steps 1 to 3 for the signal waveform plotted in Fig. 14.14 (a) where only the real part was shown for the complex signal. Note that the windowed spectrum including the leakage spectrum is removed from around every spectral peak by each subtraction step.

Speech waveforms can also be represented by the spectral peak selection[121]. Figure 14.15 is an example of speech waveform reconstruction by using only the spectral peak components. The rectangular window length for DFT analysis is 512 samples corresponding to 32 ms, and a triangular window was used for reconstruction. Each frame starts with the last 256 data points of the previous frame to avoid discontinuities between successive frames. The envelope of the entire waveform can be mostly reconstructed on a frame-by-frame basis by selecting the maximal spectral component every frame, subject to an adequate frame length.

Note, however, the waveform reconstructed by selecting only the maximal component is not intelligible any longer. As described in the previous subsection, it is necessary to recover the **narrow-band envelopes** to obtain an intelligible speech

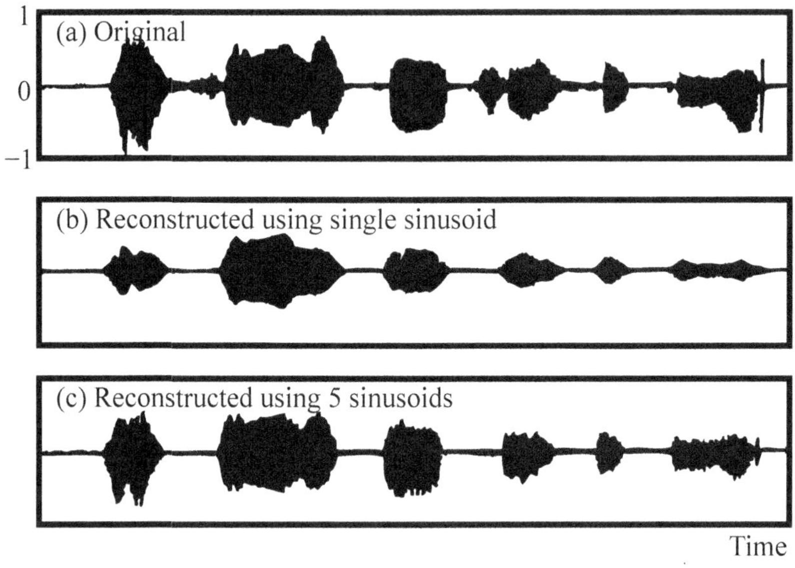

Fig. 14.15 Original and reconstructed waveforms from [121] (Fig.2)

waveform, e.g., every 1/4 octave bands, instead of the envelope of the entire waveform[11][12][13]. Namely, the reconstructed waveform only looks intelligible, it does not sound intelligible any more. In contrast, the waveform shown in Fig. 14.15 (c), which was reconstructed using five maximal components in every frame, looks similar to the waveform of Fig. 14.15 (b), but surprisingly it sounds almost perfectly intelligible. The narrow-band envelopes could be resynthesized by using five maximal components at least every frame.

14.2.2 Fundamental Frequency Analysis by Spectral Peak Selection

A waveform of sound such as speech could be represented by spectral peak selection as described in the previous subsection. The signal properties of sound are mostly characterized by the periodic structure constructed by the fundamental and its harmonics. In particular, the **fundamental frequency** is the most significant signal signature of sound; however, it is not yet simple to extract the fundamental frequency from the signal. Actually, the temporally changing pitch, which is an effect of the fundamental frequency, could be heard from the reconstructed waveform by the maximal spectral selection as shown in Fig. 14.15 (b). Such temporal tonal change characterizes sound, even speech, although it is not intelligible any more. This type of temporal change might be due to the change of the dominant energy components of a signal that mostly correspond to the fundamental frequencies on the frame-by-frame basis.

However, note that pitch can be created even if such a fundamental component is not contained in any frame. Speech through the telephone network might be a good example of sound that creates pitch for listeners without the fundamental components because of the frequency-band limitation in the low frequencies of communication channels. This phenomenon, **pitch sensation without the fundamental**, is called pitch under missing of the fundamental. From a perceptual point of view, auto-correlation analysis of the narrow-band envelopes has been performed[122]. The average of the ACFs for the narrow-band envelopes over a wide frequency range indicates the fundamental period or frequency, even if the fundamental is removed from the signal of the target signal.

The **correlation and convolution** sequences are similar to each other. Suppose sequences of $a(n)$ and $b(n)$. The **generating functions** can be defined as

$$A(X) = \sum_n a(n)X^n \tag{14.29}$$

$$B(X) = \sum_n b(n)X^n \tag{14.30}$$

for the two sequences. The **convolved sequence** $c_v(n)$ between $a(n)$ and $b(n)$ can be generated by the generating function of $C_v(X)$;

$$C_v(X) = A(X)B(X) = \sum_n c_v(n)X^n. \tag{14.31}$$

Here, the convolved sequence can be written as

$$c_v(n) = \sum_m a(m)b(n-m) = \sum_m a(n-m)b(m). \tag{14.32}$$

Similar to in the convolved sequence, define the generating function of the sequence $a(n)$ as

$$A^*(X) = \sum_n a(n)X^{-n}. \tag{14.33}$$

The **cross-correlation** sequence $c_f(n)$ between the two sequences is generated by the generating function such that

$$C_f(X) = A^*(X)B(X) = \sum_n c_f(n)X^n. \tag{14.34}$$

Here, the cross-correlation sequence can be written as

$$c_f(n) = \sum_m a(m)b(n+m). \tag{14.35}$$

Similar to the cross-correlation sequence, the **auto correlation** sequence $c_f(n)$ can be generated as

$$C_f(X) = A^*(X)A(X) = \sum_n c_f(n)X^n. \tag{14.36}$$

Suppose that the sequence $a(n)$ is composed of unit pulses, the entries of which are all unities or zeros. The auto-correlation sequence represents the histogram of spacings between successive unities. Therefore, by taking the spacing that is most probable, the period of the sequence can be estimated. This explains why the auto-correlation analysis has been taken as a tool for estimating the fundamental period of a signal of the target.

The auto-correlation analysis works for estimation of the period independent of the condition of whether the fundamental component is missing or not. Note, however, the results of the auto-correlation analysis depend on the power spectral properties of the signal in addition to the fundamental component. This can be understood by recalling the **relationship between the auto-correlation sequence and signal power spectrum**.

Again suppose a signal of $x(n)$ whose power spectrum is $|X(\mathrm{e}^{-\mathrm{i}\Omega})|^2$. The auto-correlation sequence can be written as

$$\left.\sum_n c_f(n)z^{-n}\right|_{z=\mathrm{e}^{\mathrm{i}\Omega}} = \left.X^*(z^{-1})X(z^{-1})\right|_{z=\mathrm{e}^{\mathrm{i}\Omega}} \tag{14.37}$$

or

$$|X(\mathrm{e}^{-\mathrm{i}\Omega})|^2 = \sum_n c_f(n)\mathrm{e}^{-\mathrm{i}\Omega n} \tag{14.38}$$

by substituting $X = z^{-1} = \mathrm{e}^{-\mathrm{i}\Omega}$ for X in the generating function. The formulation above states that the auto-correlation sequence and the signal power spectrum are converted by the Fourier transformation to each other. This means that the fundamental frequency could not be estimated independent of the power spectral properties of the signal of the target following the auto-correlation sequence. It might be quite naturally understood, however, that the fundamental frequencies might be the same, even if the sound is made by different musical instruments. Here, the difference of the musical instruments could be represented by the difference of the power spectral properties with the same fundamental frequencies.

The periodic structure that is composed of the fundamental and its harmonics reflects the resonant mechanism of the sound production process. Therefore, the resonant frequencies including higher harmonics are important signatures that characterize the sound, as are the power spectral properties. Such resonant frequencies can be estimated by peak-spectral selection. Figure 14.16 outlines a method of estimating the fundamental and its harmonics by using spectral-peak selection and auto-correlation analysis in the frequency domain instead of the time domain[123].

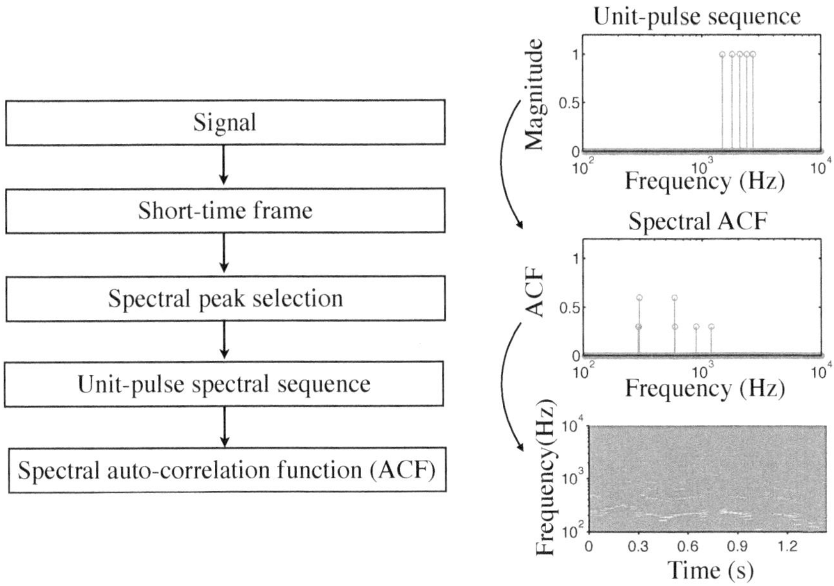

Fig. 14.16 Method of estimating fundamental frequency and its harmonics by spectral-peak selection and auto-correlation analysis in frequency domain from [123] (Fig.4)(Fig.5)

A spectral sequence composed of unit pulses is obtained by selection of the spectral peaks removing the magnitude information. If the auto-correlation analysis is carried out on the unit spectral sequence along the frequency axis, a histogram of frequency spacing for the dominant peaks can be derived.

Figure 14.17 shows examples of the fundamental and its harmonics extracted following the procedure described in Fig. 14.16 under the condition of missing the fundamental[123]. The fundamental frequencies and the harmonics can be estimated by the histograms, which are a result of the auto-correlation analysis for the selected spectral-pulse sequence in the frequency domain, even under the condition of missing the fundamental. All of the figures were obtained on the frame-by-frame basis every 30 ms and using 6 spectral peaks selected in every frame. Figure 14.17(a) and (b) shows the results of a piano tone A4 that clearly indicate the fundamental with the harmonics. Similarly Fig. 14.17 (c) and (d) are the results for a piano tone of A4 and Cis5. Note that frequencies can be seen even lower by 1 octave (or even 2 octaves) than those for the A4. These low frequency components correspond to mathematical estimation of the fundamental periods for the compound sound. However, the fundamental-frequency-related pitch sensation should be defined by the most probable frequency in the frame average shown in Fig. 14.17 (b) or (d) rather than the mathematical estimates.

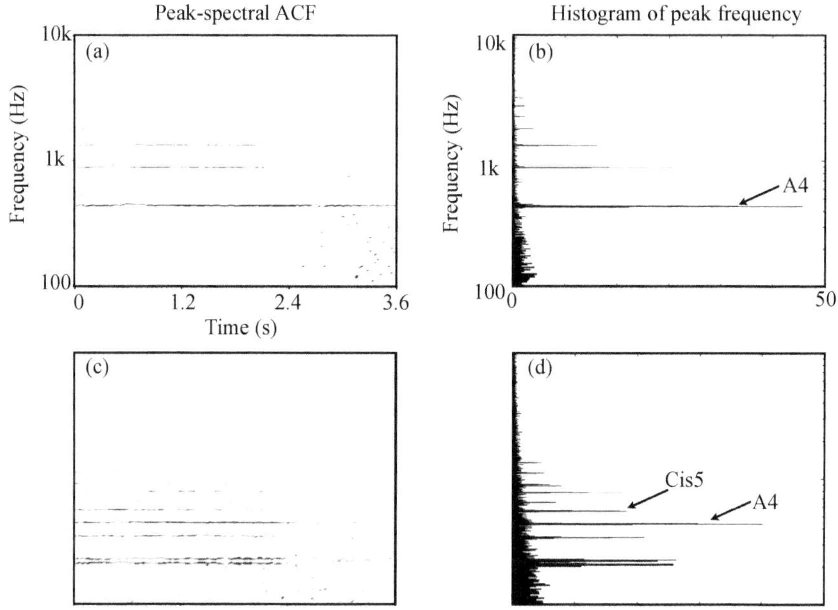

Fig. 14.17 Fundamental frequency and its harmonics analysis by auto correlation of selected spectral peaks; left column: frame-based auto-correlation analysis, right column: average for frame-based results from [123] (Fig.6) (Fig.7)

14.2.3 Speech Representation of Envelope Modulation Modeling

As described in subsection 14.1, intelligible speech can be synthesized by using magnitude spectral components subject to adequate frame lengths. This could be explained by how closely the narrow-band envelopes are resynthesized to those of the original speech waveform. According to this, a speech waveform can be also described based on the envelope-modulation scheme from a signal theoretic point of view[124][125]. Figure 14.18 is a schematic for envelope modulation modeling of speech signals[126]. A speech signal $x(n)$ can be expressed as superposition of modulated sub band signals such that

$$x(n) = \sum_k A_k(n) \cos \phi_k(n), \tag{14.39}$$

where $A_k(n)$ and $\phi_k(n)$ denote the **envelope** (**instantaneous magnitude**) and the **instantaneous phase** for the k-th sub band, respectively. A typical bandwidth for the sub band is 1/4-octave. If the original sub band envelopes could be preserved, intelligible speech could be obtained even by using the narrow-band noise or the sinusoid with the single center-frequency as a carrier signal for each sub band. Note that sub band processing is crucial to obtain an intelligible speech waveform. If only

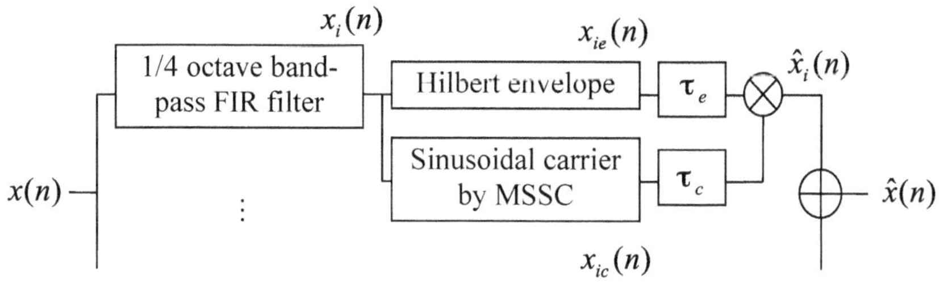

Fig. 14.18 Envelope modulation modeling of speech from [126] (Fig.1)

a few band signals whose bandwidths are too wide were taken, no intelligible speech could be synthesized any longer.

Figure 14.19 shows samples of instantaneous phase analysis of original speech by subtracting central-frequency components from the phase records. The instantaneous phase could be approximated fairly well by using the central-frequency components, but fluctuating components were still included. Brass-like tones or musical noise might be generated if such fluctuations are ignored, although almost perfectly intelligible speech could be synthesized by superposition of the envelope-modulated sinusoidal signals of the central frequencies for corresponding sub bands. In contrast, Figure 14.20 shows examples of instantaneous phase records of speech synthesized using frame-wise sinusoidal carrier in each sub band. The frequencies for

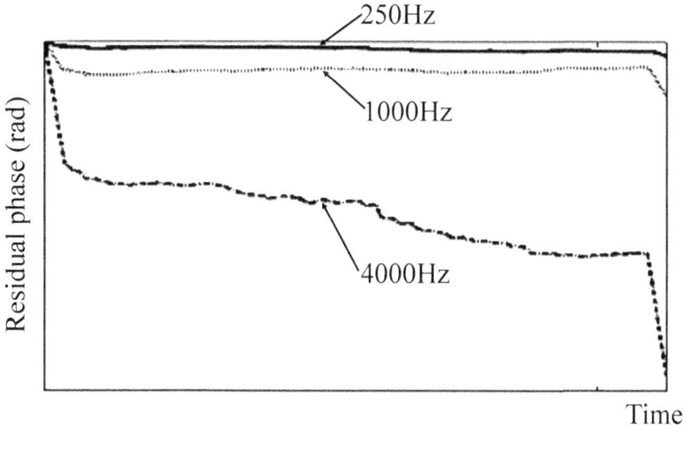

Fig. 14.19 Residual phase obtained by subtracting central frequency components from original phase records (solid: center frequency is 250Hz, dotted: 1000Hz, dash-dot: 4000Hz from [125] (Fig.4)

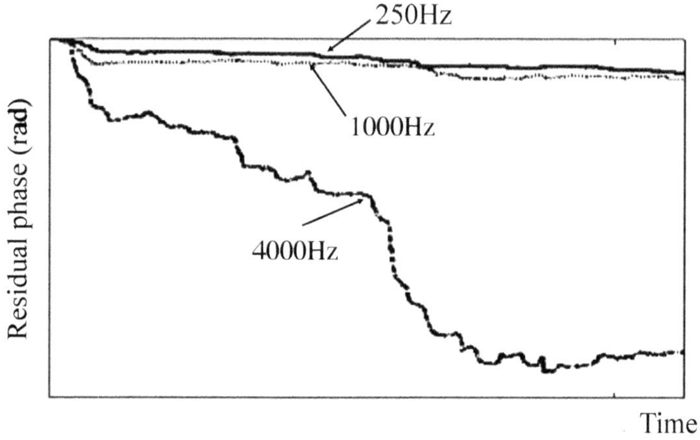

Fig. 14.20 Residual phase for frame-wise sinusoidal carriers after subtracting central frequency components (solid: center frequency is 250Hz, dotted: 1000Hz, dash-dot: 4000Hz from [125] (Fig.5)

the frame-based sub band carriers were estimated from the greatest magnitude spectral components in each frequency band. Namely, the carrier in each sub band was reconstructed by the **maximal-peak-spectral selection** on the frame-by-frame basis. Intelligible speech with speaker's voice quality preserved could be represented through superposition of the envelope-modulated sinusoidal signals with the greatest magnitude in each frequency band. The frame-wise sinusoidal carriers might be good candidates for representing the temporally-changing narrow-band carriers.

14.3 Clustered Line-Spectral Modelling (CLSM) of Sound

As described in the preceding subsection, a signal can be expressed by peak-spectral selection. However, when the target signal is composed of two or more spectral components around the peaks, it is difficult to describe by peak-spectral selection. This is because the spectral peaks, even in the interpolated spectral records, may no longer correspond to the true spectra because of the spectral leakage overlap.

Speech intelligibility is closely related to the envelopes of sub band speech signals, such as 1/4 octave-band speech signals. The envelope of a narrow-band signal can generally be expressed as a form of slowly varying amplitude modulation. This modulation characteristic in the time waveform can be represented by clustered spectral components around spectral peaks in the frequency domain. **Clustered line-spectral modeling (CLSM)** will be described in this section.

14.3.1 Formulation of CLSM

CLSM is based on the **least square error** solution[2] on the frequency domain[120]. Suppose that a compound signal is composed of two sinusoids whose frequencies are closely located to each other such that

$$x(n) = A_1 e^{i\frac{2\pi}{N}(k_0+\Delta k_1)n} + A_2 e^{i\frac{2\pi}{N}(k_0+\Delta k_2)n} = x_1(n) + x_2(n) \tag{14.40}$$

where k_0 is a positive integer, $\Delta k_1 = q_1/p$, $\Delta k_2 = q_2/p$, and p,q_1,q_2 are positive integers, respectively. Figure 14.21 shows waveforms that are composed of two sinusoidal signals as stated above. When the observation window length is long enough to separately observe the two spectral peaks, those peaks are seen as shown in Fig. 14.21. In contrast, if the window length becomes too short, those spectral peaks cannot be separated any more as shown in Fig. 14.22.

By taking the Fourier transform for the compound signal stated above, the transforms can be written as

$$X(k_1) = X_1(k_1) + W(k_1 - k_2)X_2(k_2) \tag{14.41}$$

$$X(k_2) = W(k_2 - k_1)X_1(k_1) + X_2(k_2) \tag{14.42}$$

at two sufficiently interpolated frequency bins in the discrete form, where

$$X(k_1) = \frac{1}{N}\sum_{n=0}^{N-1} x(n)e^{-i\frac{2\pi k}{M}n}\Bigg|_{k=k_1} = X(k)|_{k=k_1} \tag{14.43}$$

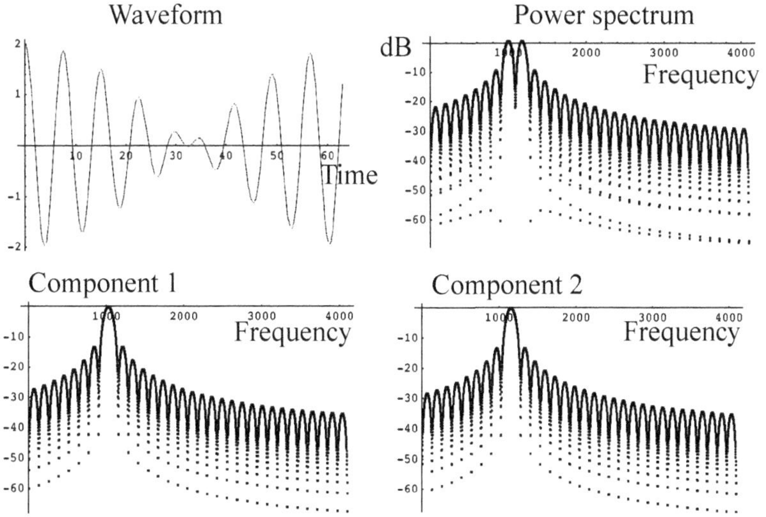

Fig. 14.21 For long-time window that is long enough to separate peaks

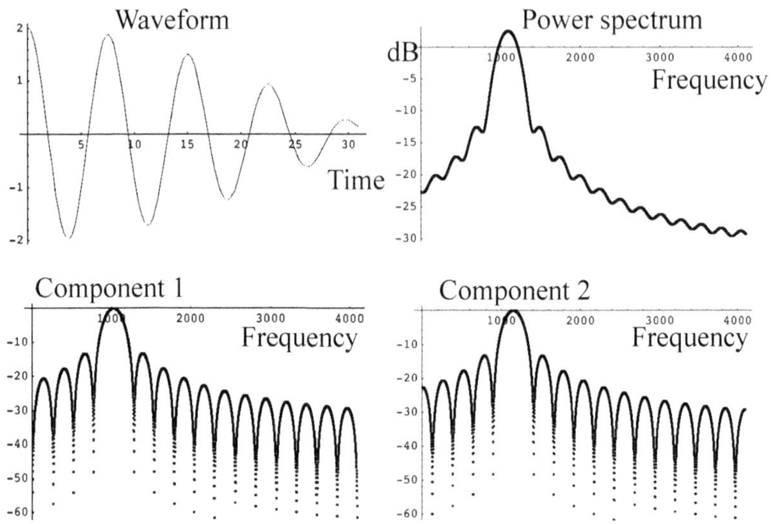

Fig. 14.22 Spectral properties for two truncated sinusoids with frequencies close to each other when window length is too short to separate.

$$X(k_2) = X(k)|_{k=k_2} \tag{14.44}$$

$$W(k_2 - k_1) = \frac{1}{N} \sum_{n=0}^{N-1} w(n) e^{-i\frac{2\pi k}{M}n} \Big|_{k=k_2-k_1} \tag{14.45}$$

$$M = N \cdot p, \tag{14.46}$$

and $w(n)$ shows the windowing function applied to the target signal. This spectral representation can be interpreted graphically as shown in Fig. 14.23. Namely, if the frequencies of the two sinusoids could be assumed k_1 and k_2, then the spectral expression above states that the spectrum of the compound signal is made by the overlapped leakage spectral components. Consequently, if the spectrum record is observed for at least two frequency bins, the spectral magnitude and phase for the two components can be estimated by solving the simultaneous equations on the frequency domain because the leakage spectral characteristics are determined according to the windowing function[120].

Suppose a signal with a record of length N and its interpolated spectrum analyzed by M-point DFT after zero padding to the original record. Assume that the signal is composed of K clustered sinusoidal components around the peak $k = k_p$ such that

$$x(n) = \sum_{k=1}^{K} A(k) e^{i2\pi \nu(k) n} + \varepsilon_K(n), \tag{14.47}$$

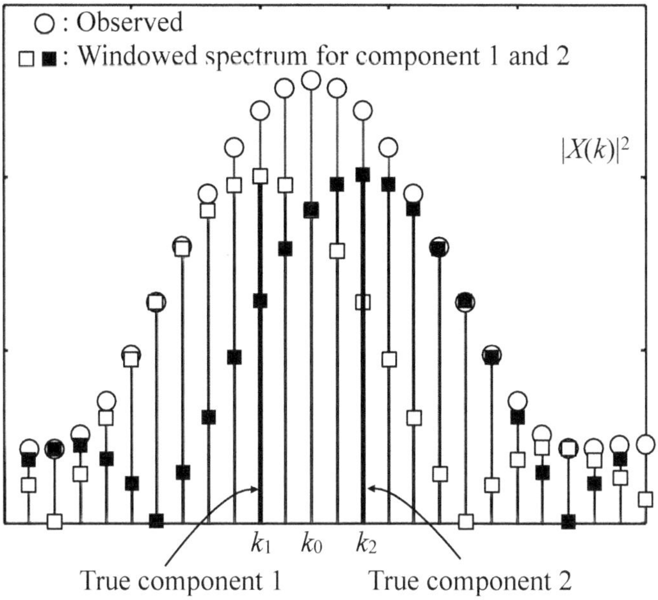

Fig. 14.23 Overlap of leakage spectra of two clustered sinusoids truncated by rectangular window

where $A(k)$, and $\nu(k)$ denote the complex magnitude and frequency for the k-th frequency bin, respectively. K is the number of components that are clustered around the peak $k = k_p$, and $\varepsilon_K(n)$ is the residual noise or modeling error. Try to represent the signal by P sinusoidal components clustered between $k = k_{p-m}$ and $k = k_{p-m+P-1}$. The P sets of components can be estimated with the **least square error (LSE) criterion** using a set of linear equations for spectrum observations at L frequency points between k_{p-l} and $k_{p-l+L-1}$ as follows[120]:

$$\mathbf{x}_{observe} = W\mathbf{x}_{signal}, \tag{14.48}$$

where

$$\begin{pmatrix} X(k_{p-l}) \\ \vdots \\ X(k_{p-l+L-1}) \end{pmatrix} = \mathbf{x}_{observe} \tag{14.49}$$

denotes the spectrum observed at L frequency points, and

$$\begin{pmatrix} X_s(k_{p-m}) \\ \vdots \\ X_s(k_{p-m+P-1}) \end{pmatrix} = \mathbf{x}_{signal} \tag{14.50}$$

denotes the P spectral components for the signal where $L > P$, $l > m$, and

$$m = \begin{cases} \frac{P-1}{2} & P:\text{odd} \\ \frac{P}{2} & P:\text{even} \end{cases} \qquad \begin{matrix} l = \frac{L-1}{2} & L:\text{odd} \\ l = \frac{L}{2} & L:\text{even}. \end{matrix} \tag{14.51}$$

The matrix W is given by

$$W = \begin{pmatrix} W_{NM}(k_{p-l} - k_{p-m}) & \cdots & W_{NM}(k_{p-l} - k_{p-m+P-1}) \\ \vdots & \ddots & \vdots \\ W_{NM}(k_{p-l+L-1} - k_{p-m}) & \cdots & W_{NM}(k_{p-l+L-1} - k_{p-m+P-1}), \end{pmatrix} \tag{14.52}$$

where

$$W_{NM}(q) = \frac{1}{N} \sum_{n=0}^{N-1} w(n) \mathrm{e}^{-\mathrm{i}\frac{2\pi kn}{M}} \Bigg|_{k=q} \tag{14.53}$$

for the window function $w(n)$. The spectral components of the signal can be estimated by finding LSE solutions such as

$$\hat{\mathbf{x}}_{signal} = (W^{\mathrm{T}}W)^{-1}W^{\mathrm{T}}\mathbf{x}_{observe}. \tag{14.54}$$

14.3.2 LSE Solution of Simultaneous Equations

As described in the preceding subsection, CLSM obtains the spectral information of a signal as the LSE solution of a set of simultaneous equations on the frequency plane. This subsection formulates the **LSE solution of simultaneous equations**[2].

Suppose a set of simultaneous linear equations such that

$$A\mathbf{x} = \mathbf{b}. \tag{14.55}$$

The equation above has a set of unique solutions when the matrix A is a square matrix with independent column vectors. When the matrix A is rectangular (N rows and M columns) and $N < M$, solutions are available but not unique. This is because the number of equations N is smaller than that of the unknown solutions[2][3]. In contrast, if $N > M$, no solutions are available, but **LSE solutions** are obtained instead[2].

When $N > M$, namely the number of equations is larger than that of the unknown solutions, the LSE solution $\hat{\mathbf{x}}$ minimizes the squared error that is defined by

$$|\mathbf{e}|^2 = |\mathbf{b} - A\hat{\mathbf{x}}|^2, \tag{14.56}$$

where $|\mathbf{e}|^2$ denotes the **square norm** of a vector. In other words, the LSE solution $\hat{\mathbf{x}}$ solves the equation

$$A\hat{\mathbf{x}} = \hat{\mathbf{b}}, \tag{14.57}$$

where $\mathbf{b} = \hat{\mathbf{b}} + \mathbf{e}$, instead of the equation $A\mathbf{x} = \mathbf{b}$.

The linear equation $A\mathbf{x} = \mathbf{b}$ can be written as a **linear combination of the column vectors v** of the matrix A:

$$x_1\mathbf{v}_1 + x_2\mathbf{v}_2 + \ldots + x_M\mathbf{v}_M = \mathbf{b}, \tag{14.58}$$

where

$$A = (\mathbf{v}_1 \quad \mathbf{v}_2 \cdots \mathbf{v}_M) \tag{14.59}$$
$$\mathbf{x} = (\mathbf{x}_1 \quad \mathbf{x}_2 \cdots \mathbf{x}_M)^{\mathrm{T}} \tag{14.60}$$
$$\mathbf{b} = (\mathbf{b}_1 \quad \mathbf{b}_2 \cdots \mathbf{b}_N)^{\mathrm{T}}. \tag{14.61}$$

When the vector **b** is a vector in the column space, the combination coefficients vector **x** gives the solution vector for the simultaneous equation. On the other hand, when the vector **b** is not located in the column space, it cannot be expressed as a linear combination of the column vectors. Figure 14.24 shows the orthogonal projection vector $\hat{\mathbf{b}}$ of the vector **b** on the column space[46]. This projection vector meets the LSE criterion. Namely, the squared norm of the error becomes minimum. The LSE solution vector **x** satisfies the linear equation $A\hat{\mathbf{x}} = \hat{\mathbf{b}}$.

The **orthogonal relationship** $\mathbf{e} \perp \hat{\mathbf{b}}$ can be rewritten, using the **inner product**[3], as

$$\hat{\mathbf{b}}^{\mathrm{T}}\mathbf{e} = 0, \tag{14.62}$$

or equivalently

$$\mathbf{v}_1^{\mathrm{T}}\mathbf{e} = \mathbf{v}_2^{\mathrm{T}}\mathbf{e} = \ldots \mathbf{v}_M^{\mathrm{T}}\mathbf{e} = 0 \tag{14.63}$$

where $\mathbf{e} = \mathbf{b} - \hat{\mathbf{b}}$ and $\mathbf{x}^{\mathrm{T}}$ denotes taking the transpose of **x**. By substituting the error vector **e** into the equations above,

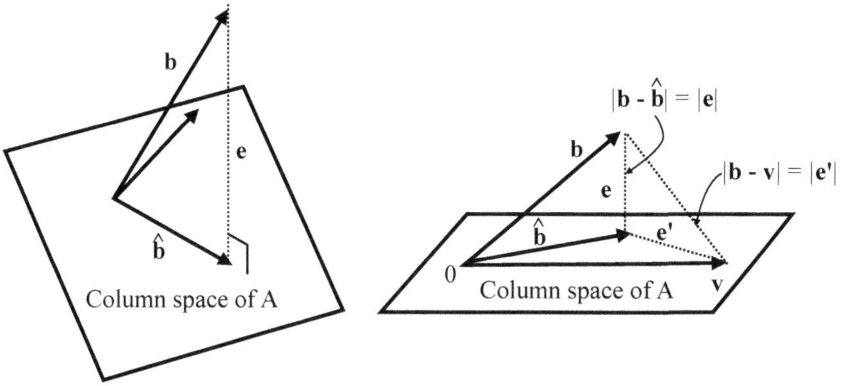

Fig. 14.24 Orthogonal projection onto column space: (left) orthogonal projection vector; (right) orthogonal projection vector and least square error criterion

a set of linear equations

$$A^{\mathrm{T}}\hat{\mathbf{b}} = A^{\mathrm{T}}\mathbf{b} \tag{14.64}$$

is obtained where $N > M$. Thus, recalling the equation $A\hat{\mathbf{x}} = \hat{\mathbf{b}}$,

$$A^{\mathrm{T}}A\hat{\mathbf{x}} = A^{\mathrm{T}}\mathbf{b} \tag{14.65}$$

is obtained where the matrix $A^{\mathrm{T}}A$ is a square and symmetric matrix. When the square matrix is not singular, the LSE solution can be obtained as

$$\hat{\mathbf{x}} = (A^{\mathrm{T}}A)^{-1}A^{\mathrm{T}}\mathbf{b}. \tag{14.66}$$

Finding the linear regression line is a typical example of getting the LSE solutions[46].

14.3.3 CLSM Examples

Figure 14.25 is an example of signal analysis based on CLSM, where $A_k = 1$, K=5, N=512, M=4096, and the signal-to-noise-ratio is 20dB[120]. The waveform is re-

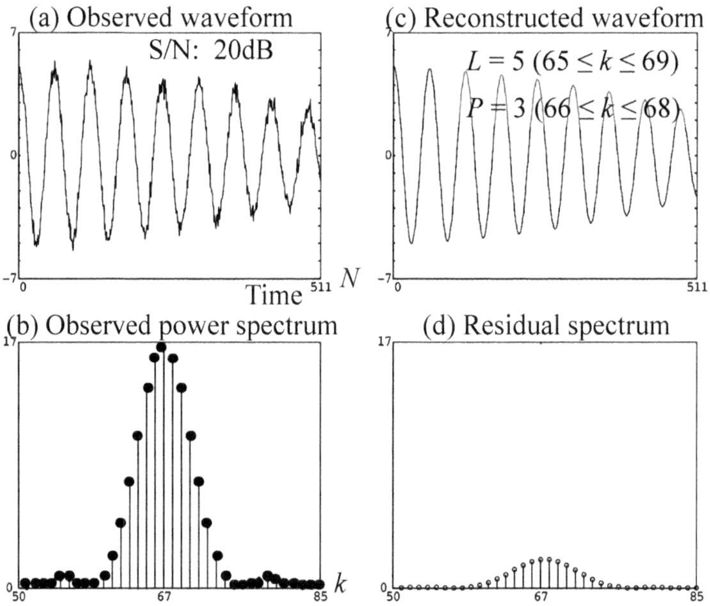

Fig. 14.25 CLSM example for compound signal $x(n) = \sum_{k=1}^{5} \mathrm{e}^{\mathrm{i}2\pi\nu_k n} \cdot w(n)$, $w(n) = 1$ $(0 \le n \le N-1)$, $\nu_k = (8 + k/8)/512$ from [120] (Fig.2)

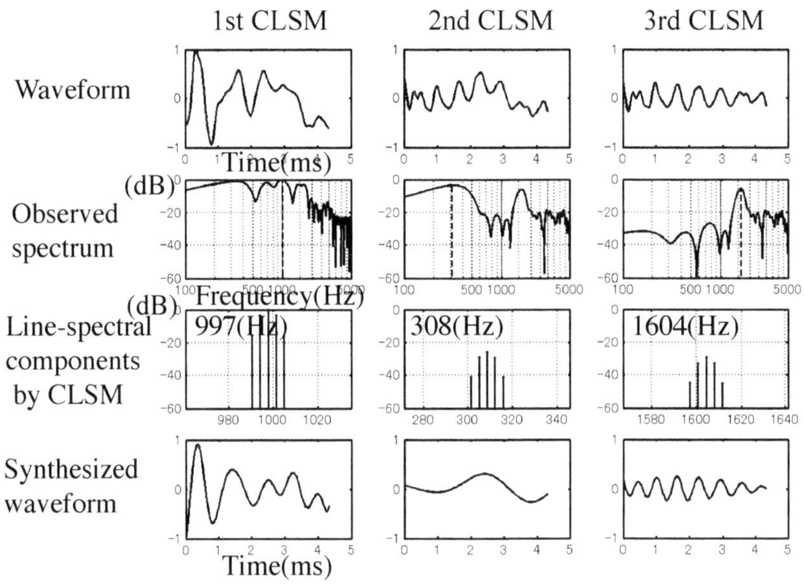

Fig. 14.26 CLSM representation of spoken vowel 'a' from [127] (Fig.3)

constructed based on CLSM, where L=5 between $k = k_p - 2 = 65$ and $k = k_p + 2 = 69$ for $P = 3$ sinusoidal components between $k = k_p - 1 = 66$ and $k = k_p + 1 = 68$. The CLSM approach can be repeated to represent multiple dominant spectral peaks.

Figure 14.26 is an example for the spoken vowel 'a'[127]. The top row in the figure shows the waveform to be analyzed, the middle row displays the power spectral components, the third one presents the line-spectral components (magnitude) extracted by CLSM where $P = 5$ and $L = 7$, and the bottom one illustrates the synthesized waveforms by CLSM. The top figure in the left column is a cycle of the waveform for the vowel 'a', and the second top figure shows that the dominant spectral peak is around 1kHz. By applying the CLSM approach to the dominant peak, the line spectral components shown by the third figure in the left column were obtained. Consequently, as shown by the bottom one, the synthesized waveform resembles the envelope of the entire waveform of the top figure.

The top figure in the center column in the figure shows the residual component that was left after the first CLSM at around 1kHz. It can be seen that the spectral components were lost at around 1kHz, as shown in the second top figure of the center column. By applying the second CLSM approach to the second dominant peak around 300 Hz, the synthesized waveform as shown by the bottom figure in the center column was obtained. Consequently, the second-residual was obtained as plotted in the top figure of the right column where the third dominant spectral peak can be seen because the first and second ones were already removed.

The right column displays the third trial of the CLSM approach to the second residual as shown in the top figure. As seen in the second figure in the column, the third dominant peak is located at around 1600 Hz. As shown in the column, this dominant component can be represented by CLSM, and thus only a few residual components are left.

CLSM can be applied to transient signals such as impulse-response type records. Suppose a decaying sinusoidal waveform representing the resonant impulse response of a single-degree-of-freedom system. Figure 14.27 is an example of the CLSM approach to the decaying signal[127]. The top graph shows the waveform, and the second one presents the power spectrum of the waveform as well as that for the synthesized signal by CLSM. The decaying envelope can be represented by 5 clustered spectral components, as shown in the bottom figure. Note that the number of sinusoidal components constructing the dominant spectral peak in a target signal is unknown. The required number of components P for CLSM can be estimated practically, however, by repeating the CLSM process for the dominant peaks so that the residual energy of the signal becomes as small as possible in the practical sense.

Figure 14.28 is an example of the CLSM approach to vibration[128]. Figure (a) shows a vibrating waveform for a piano string that was previously described in subsection 11.5.1. Figure (b) presents the power spectral properties of the string vibration (acceleration) that represent the fundamental and its harmonics. By repeating the CLSM approach to the dominant spectral components where $P = 5$ $L = 7$, the synthesized waveform can be made as shown in Fig. (c) with its power spectrum as shown in Fig.(d). Again it can be seen that the decaying characteristic of the target signal can be represented by the CLSM approach with a residual part (as shown in Figs. (e) and (f)) left.

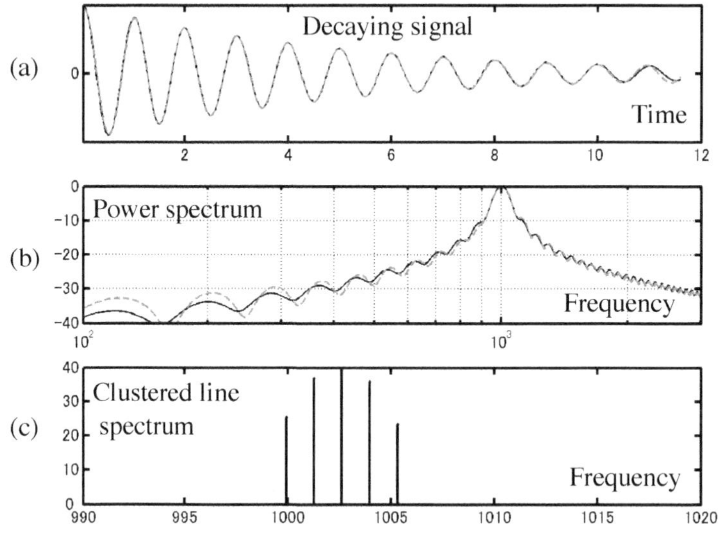

Fig. 14.27 CLSM analysis for decaying signal from [127] (Fig.1)

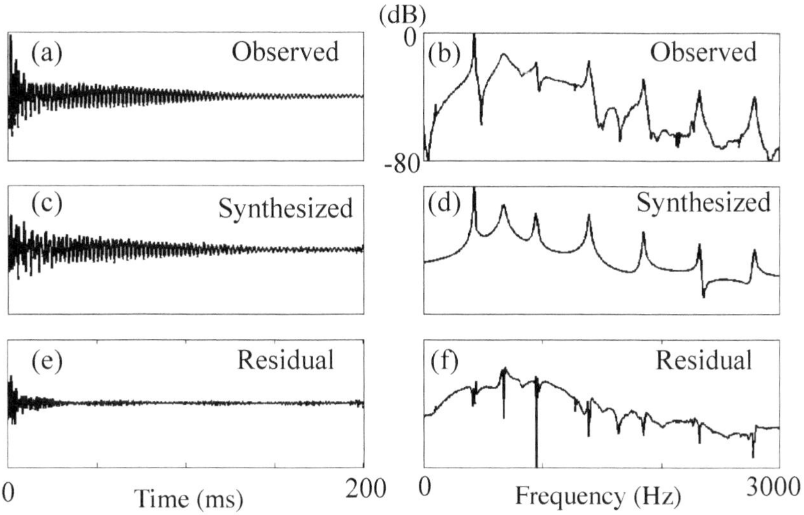

Fig. 14.28 CLSM analysis for string vibration of piano;(a) observed waveform sampled by 6 kHz, (b) interpolated power spectra of (a), (c) synthesized by CLSM ($P = 5$, $L = 7$), (d) interpolated power spectra of (c), (e) residual (a)-(c), (f) residual power spectra for (e) [128]

The CLSM, in principle, is a method for representing the dominant spectral components that are produced by the resonant mechanism of the target signal. Therefore, the CLSM approach might not be a good way to represent the transient portion of signals such as the initial parts of impulse response records. Indeed, many residuals are left in the initial portions of the impulsive records as shown in Fig. 14.28.

The transient portions might be important for signal analysis, in particular, for musical sound analysis, as are the spectral harmonic analysis in the frequency domain. A transient signal with a brief record length could be characterized by the **zeros** rather than the **poles** due to the resonance frequencies. This could be understood by the fact that the occurrence of zeros is sensitive to the residual or remainder signals as described in subsection 13.4.2. Signal representation in the time domain will be described in the following sections.

14.4 Clustered Time-Sequence Modelling (CTSM)

14.4.1 Correspondence between Time and Frequency Regions

The Fourier transformation defines a signal or function from both sides of the time and frequency domains. Figures 14.29 and 14.30 illustrate a single pulse in the time domain and its spectral components. Namely, the Fourier transform of a single pulse is a sinusoidal function in the frequency plane in the complex form. Therefore, if there are two pulses in the time domain, its Fourier transform is the modulated complex sinusoidal sequence as shown in Fig. 14.31. Namely, the zeros are produced in

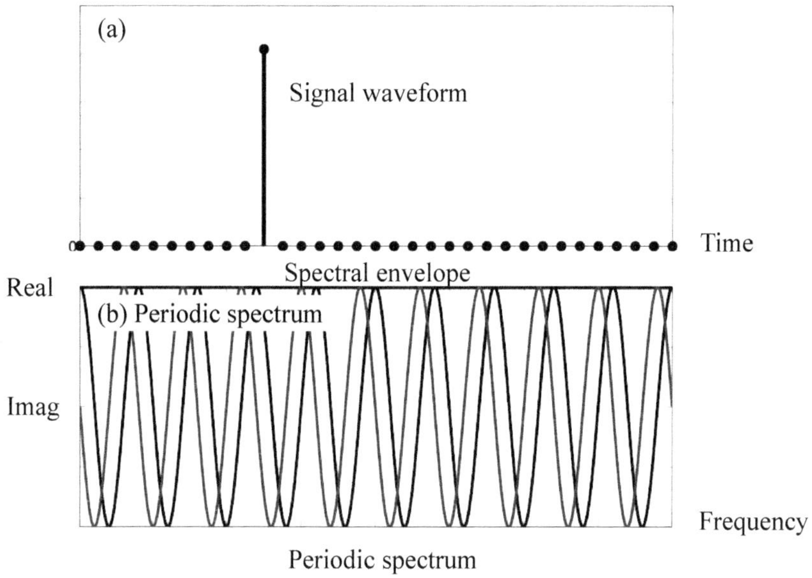

Fig. 14.29 Unit pulse (a) and its periodic spectrum with flat envelope(b)

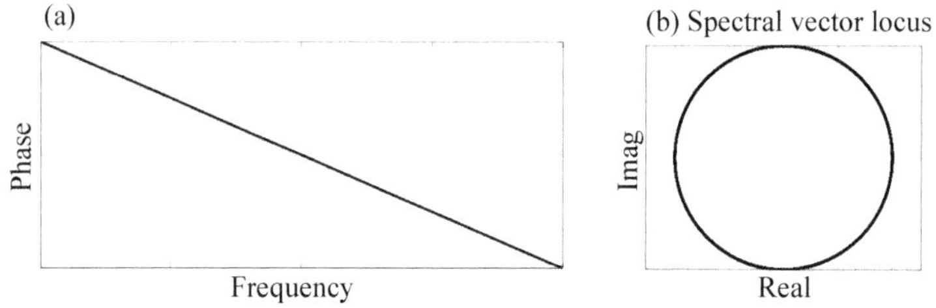

Fig. 14.30 Phase spectrum of unit single impulse (a) and its vector locus (b)

the frequency domain. These zeros can be interpreted as the spectral troughs due to the early echo in terms of room acoustics. In this example, the zeros are minimum phase as shown in Fig. 14.31 (c).

In contrast, suppose that there is a single line-spectral component in the frequency domain as shown in Fig. 14.32 and 14.33. By taking the inverse Fourier transform for the single line-spectral component, the analytic or complex sinusoidal signal is obtained in the time domain as illustrated in Fig. 14.32. Similar to the previous example, suppose that two line-spectral components are located in the frequency plane as displayed in Fig. 14.34. By taking the inverse Fourier transform for the two line-spectral components, the modulated time waveform can be obtained as shown in Fig. 14.34 (b). The **zeros** can be defined in the time region, and those zeros are interpreted as **minimum phase** in the **complex time domain** as shown in Figure (c).

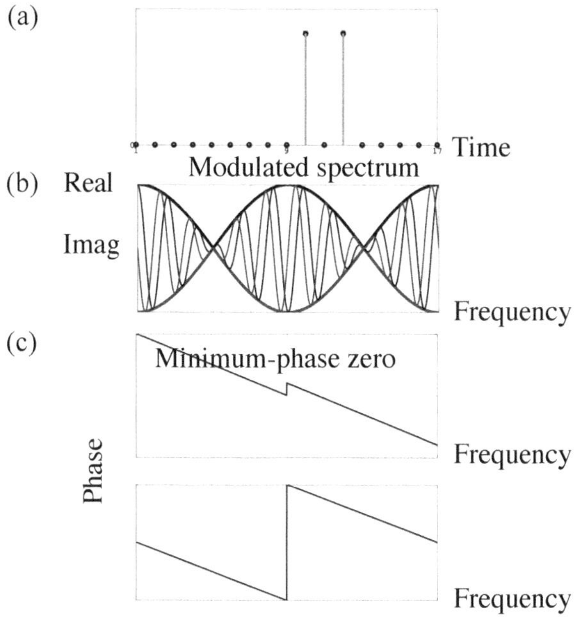

Fig. 14.31 Pair of unit pulse (a) and its modulated spectrum (b) and phase spectra (c)

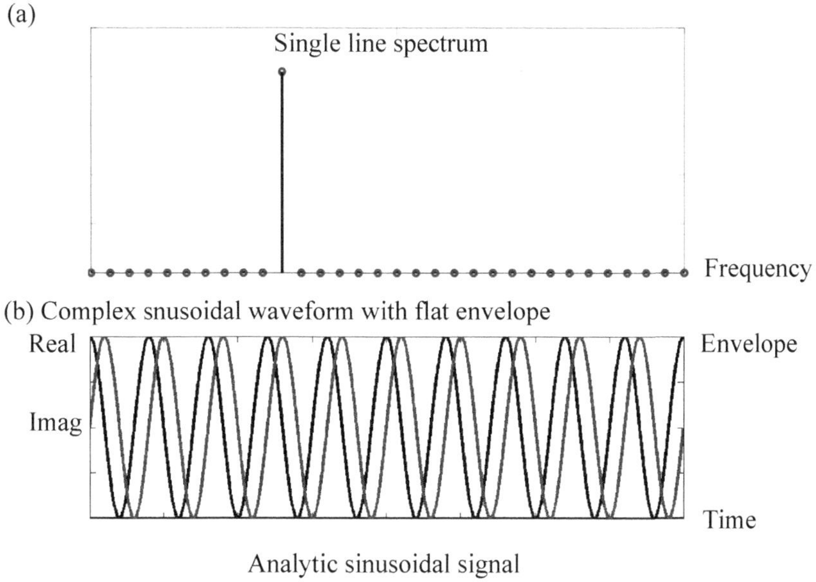

Fig. 14.32 Single line-spectral component (a) and complex (analytic) time waveform (b)

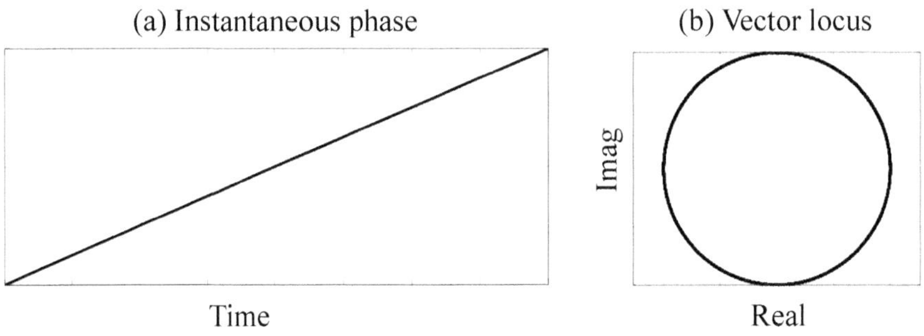

Fig. 14.33 Instantaneous phase (a) and vector locus (b) for analytic sinusoid

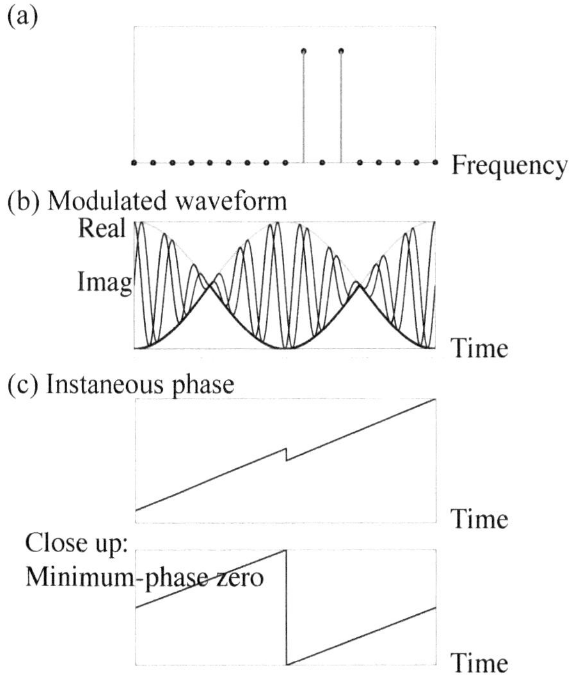

Fig. 14.34 Pair of two line-spectral components (a) and its modulated waveform (b) with instantaneous phase (c)

This can be understood by seeing the sign changes in the real and imaginary parts of the carrier signal as shown in Figs. 14.35 and 14.36. These results illustrated by Figs. 14.29-14.36 clearly show the **complementarity of time and frequency** as conjugate variables.

The phase change due to the **minimum-phase zeros on the complex time domain** can be represented by the positive **instantaneous frequency** in terms of **analytic signal** representation. The **non-minimum-phase zeros** can also be produced in the **complex time domain** as well as in the frequency region. Suppose that there are two line-spectral components as shown in Fig. 14.37 (a). The inverse Fourier transform is illustrated in Fig. 14.37 (b). It looks similar to the curves plotted in Fig. 14.34, the sign changes in the carrier are different as shown in Fig. 14.38. Namely, the zeros are interpreted as the non-minimum-phase zeros as illustrated in Fig. 14.39 and 14.40. Such **non-minimum-phase zeros** correspond to the **negative instantaneous frequencies**.

It might be interesting to see the example as presented in Figs. 14.41, where the three line-spectral components are arranged in the frequency plane. If this symmetric arrangement of pulse-like components is available in the time region, the **linear-phase** characteristic can be obtained. The same thing happens in the time domain, if the inverse Fourier transform is taken for the line-spectral sequence, the inverse Fourier transform shows the actual **linear phase** in the **complex time domain**. The **zeros** might be located as **symmetric pairs** with respect to the **real time axis** in the complex time domain. Consequently, the phase effects due to the symmetrically located zeros cancel each other, and thus the linear phase can be seen.

As described above, both the signal and spectrum can be represented by the complex variables. Namely, the magnitude and phase are applied to the spectrum, and similarly the envelope and instantaneous phase are assigned to the complex signals. Recall that the magnitude or phase spectral components can be converted to each other for the minimum-phase signals. The same thing is possible for the minimum-phase complex-time signals. In other words, the envelope and carrier part can be converted to each other if the complex time signal is minimum phase. However, sound, such as speech, mostly seems to be non-minimum phase in the complex-time domain from experimental studies[129].

Filtering with the filtered impulse response and windowing with the windowed spectral function make a corresponding pair between the time and frequency planes. The effect of filtering in the frequency domain can be seen by the smearing of signals in the time domain, and by the smearing of spectral records due to windowing in the time domain. Actually, the CLSM approach is based on the spectral leakage as a result of time windowing of the signal. The same type of approach might be able to represent a narrow-band signal according to the signal smearing by the filtering in the frequency plane. The name **clustered time-sequence modeling (CTSM)** can be given to this time-signal analysis inspired by CLSM. It will be described in the next subsection.

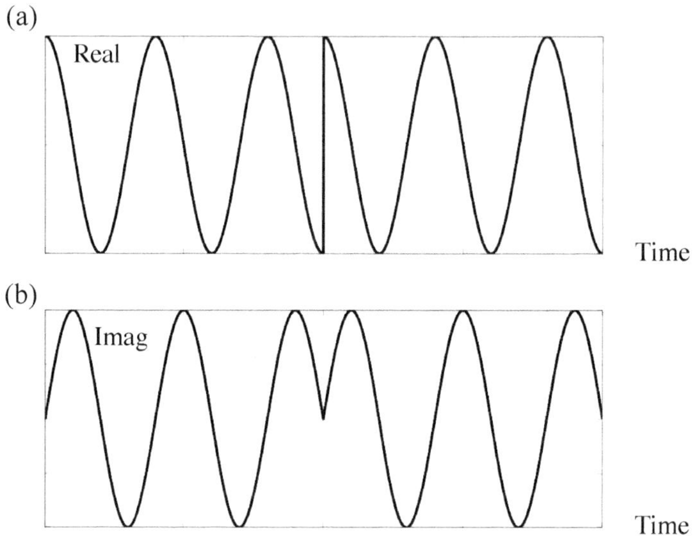

Fig. 14.35 Real (a) and imaginary (b) parts for minimum-phase carrier

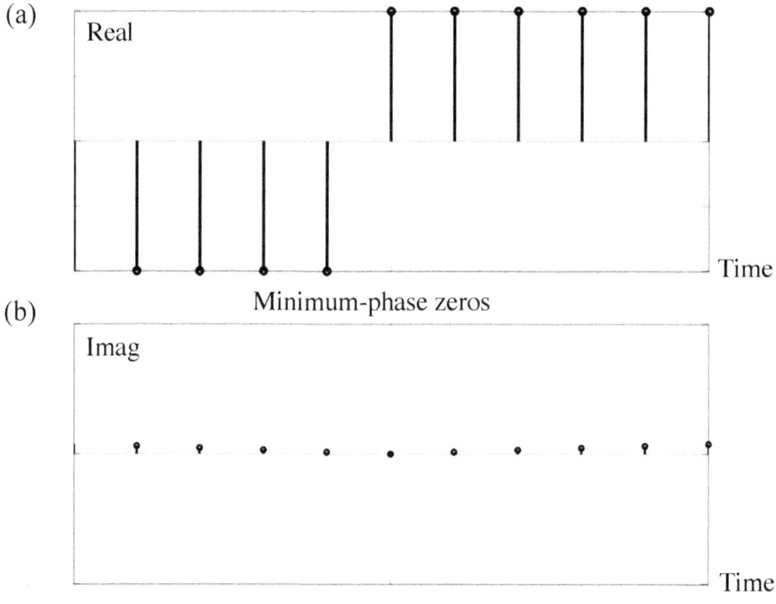

Fig. 14.36 Sign change of real (a) and imaginary (b) part for minimum-phase carrier

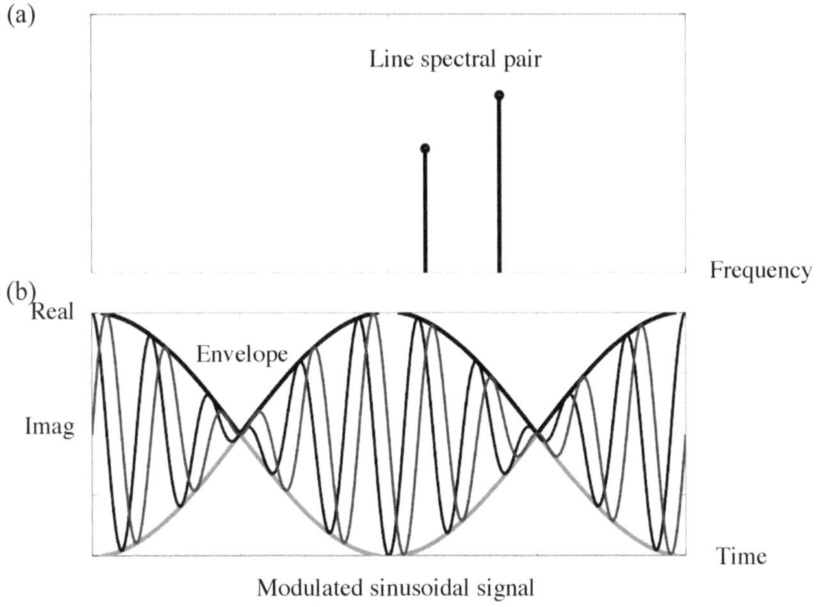

Fig. 14.37 Asymmetric line-spectral pair (a) and modulated (analytic) sinusoidal signal (b)

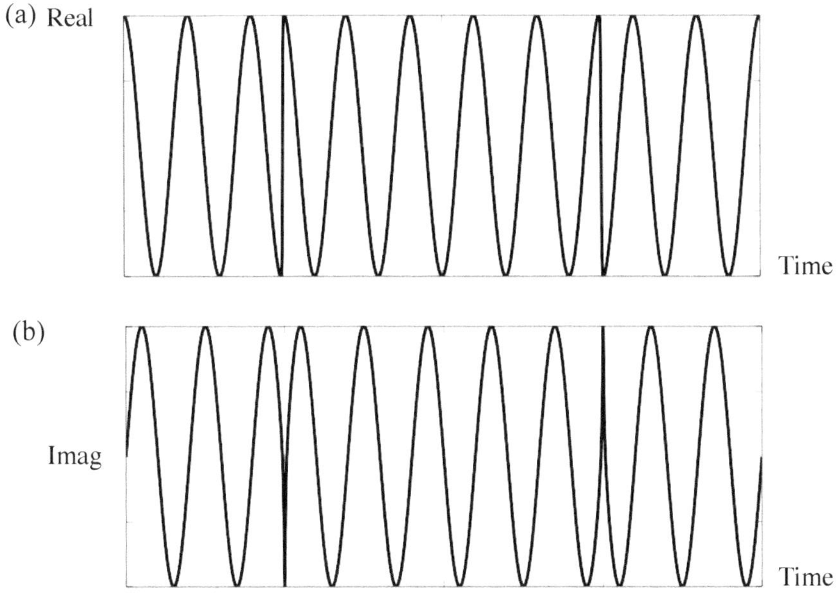

Fig. 14.38 Real (a) and imaginary (b) parts for non-minimum phase carrier

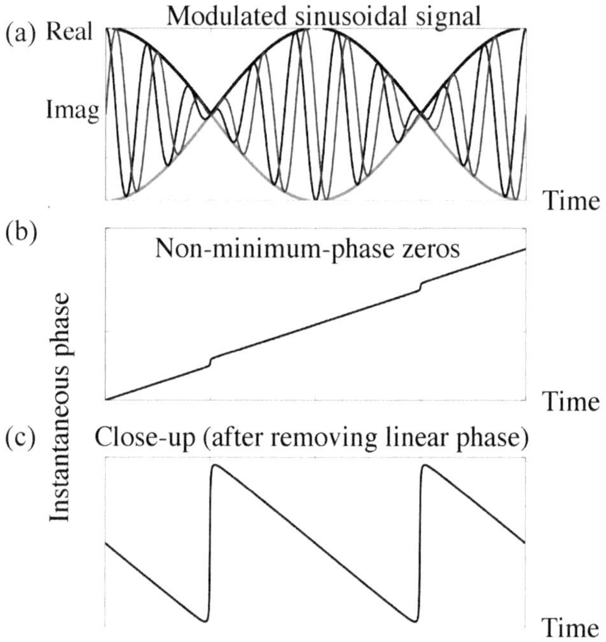

Fig. 14.39 Modulated (analytic) waveform (a) and non-minimum phase instantaneous phase with (b) and without (c) linear-phase component

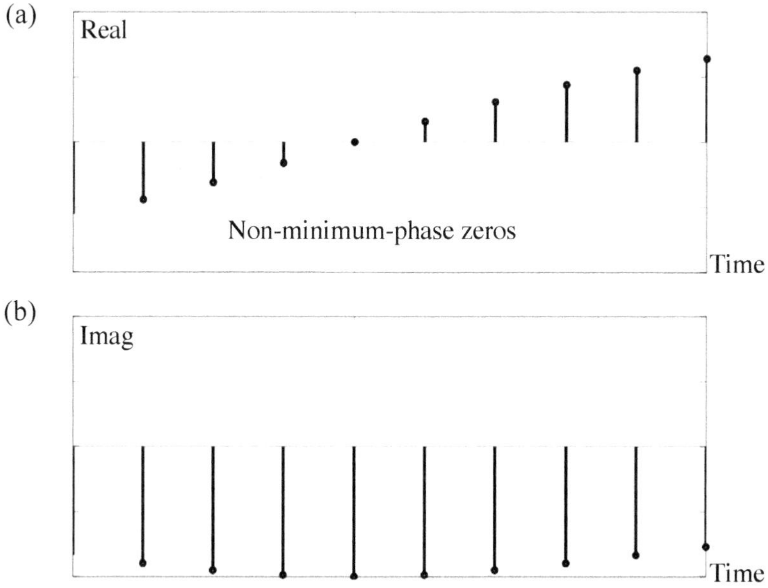

Fig. 14.40 Sign change of real (a) and imaginary (b) parts for non-minimum phase carrier

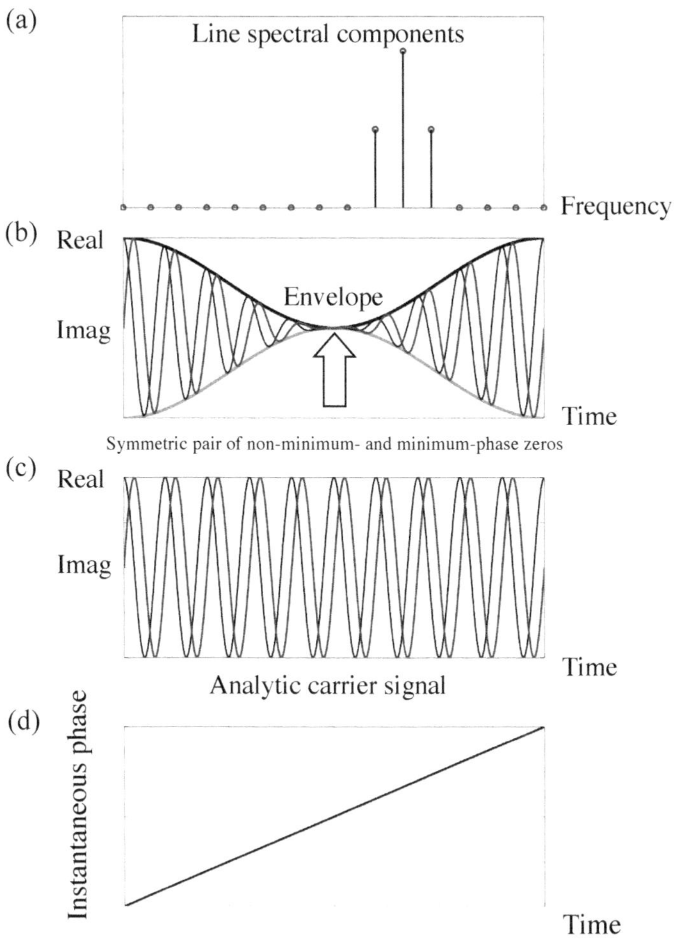

Fig. 14.41 Three symmetric sinusoidal components (a) and amplitude-modulated analytic waveform (b), its carrier (c), and instantaneous phase (d)

14.4.2 Formulation of CTSM

CTSM is a method for representing a transient signal in a short time period as an output signal from a narrow-band filter to an input signal composed of a clustered time sequence. Therefore, CTSM is formulated in the time region based on the same type of principle that formulates the CLSM in the frequency plane.

According to the correspondence between the time and frequency domains described in previous, the spectral leakage due to windowing the target signal can be interpreted as the impulse response due to the narrow-band filtering. Namely, the window length used for CLSM in the time region corresponds to the bandwidth of the filtering for CTSM. Thus, the overlap of the leakage spectra that is the basis of the CLSM represents superposition of the impulse response records in the time domain for the CTSM[130].

Figure 14.42 explains the CTSM approach graphically. Assume that a signal is written as superposition of the impulse responses of filtering:

$$x(n) = \sum_{m=1}^{M} a(m)h(n-l_m), \tag{14.67}$$

where l_m denotes the time-shift for the impulse response $h(n)$. By taking L points of observation around the signal peak ($L > M$), the clustered time series $a(m)$ can be obtained as the LSE solutions for L simultaneous equations similar to the CLSM approach but in the time domain. Figure 14.42 (a) shows a sample of a time waveform, Fig. (b) is the impulse response for the narrow-band filtering, Fig. (c) is the solution with respect to the dominant peak expressed as the clustered time series, Fig. (d) shows the synthesized response by narrow-band filtering the solution of the sequence, and Fig. (e) is the residual signal. Figure 14.43 illustrates the close-up of superposition of the impulse responses that correspond to the overlap of the leakage spectra for CLSM in the frequency domain. By applying again the CTSM approach to the residual signal, the second dominant peak can be characterized by the second clustered time sequence. By repeating the process so that the residual signal becomes sufficiently small, the transient signal can be represented by CTSM as shown by Fig. 14.44.

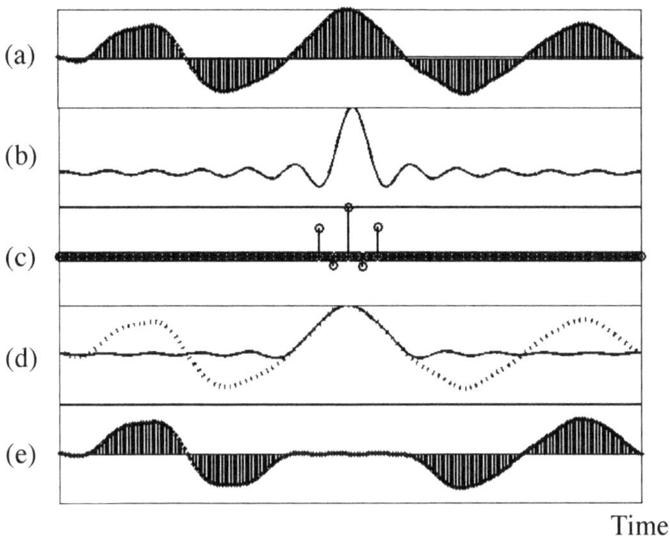

Fig. 14.42 Schematic for CTSM; (a)sample of time waveform, (b)impulse response of narrow-band filter, (c)CTSM solution vector as clustered time series, (d) synthesized response, (e) residual error [128]

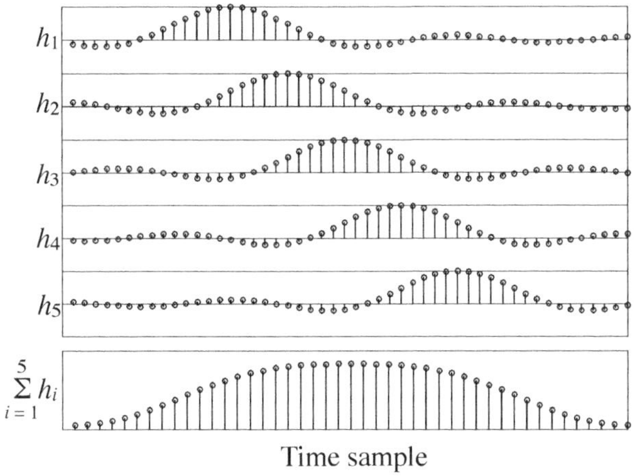

Fig. 14.43 Superposition of impulse response records for CTSM[128]

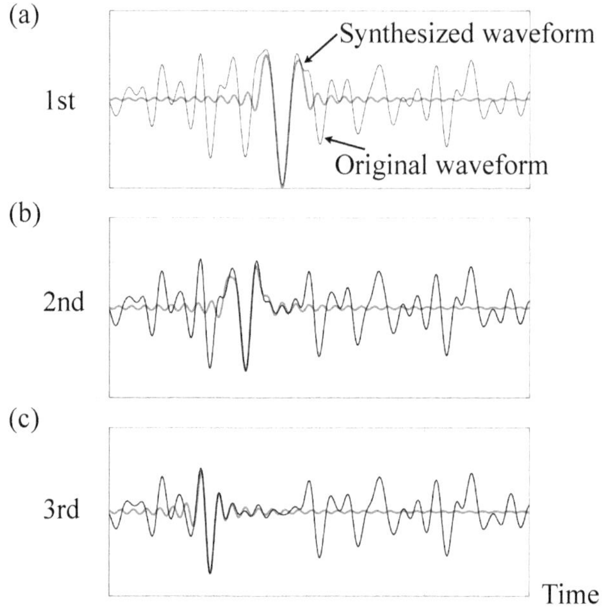

Fig. 14.44 Repeated CTSM[128]

14.4.3 CTSM Example

Figure 14.45 is an example of the CTSM approach to the initial transient portion of the vibration that was displayed in Fig. 14.28 (a). As mentioned in subsection 14.3.3, the CLSM approach might not be good for analyzing such a transient portion of a signal. Figure 14.45(a) displays the initial portion between 0 and 10 ms of the vibration record. Figure (b) is a close-up of the first dominant peak of the signal sampled by 48kHz. Figure (c) displays the impulse response for narrow-band filtering (lower than 3kHz). Recall that narrow-band filtering corresponds here to the time-windowing for CLSM. That is, the narrow-band filtering should be interpreted as over sampling rather than filtering, corresponding to the time-windowed signal being expressed by interpolated spectral components obtained by DFT with zero-padding in the CLSM approach. Figure (d) presents the clustered time sequence that produces the first dominant peak by the filtering. This solution was obtained by assuming five pulses and observing the waveform at seven points around the peak. Figure (e) shows the synthesized waveform and the original one. By repeating this process on the residual that is defined by subtraction of the synthesized waveform from the original one, the residual shown by Fig. (f) is left after 20 repetitions.

In contrast to the fact that resonant responses could be characterized by clustered-line spectral components including the envelopes of the waveforms obtained by CLSM, the clustered time series might be informative for representing transient signal portions in a short term period.

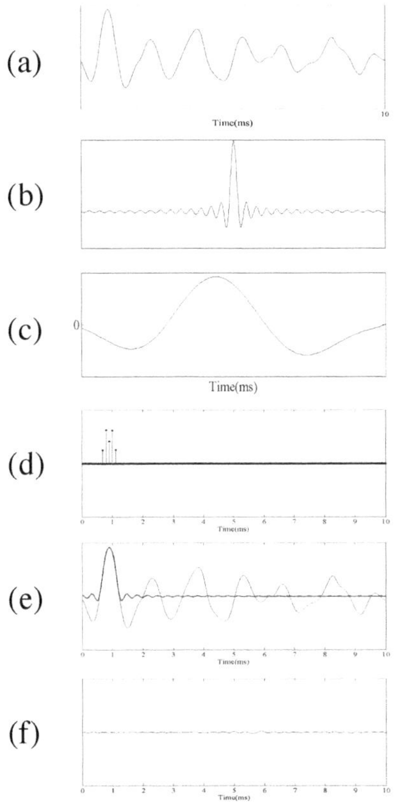

Fig. 14.45 CTSM example for string vibration analysis (a) initial portion of string vibration record (0-10ms), (b) impulse response of low-pass filtering lower than 3 kHz, (c) close-up of first dominant peak sampled by every 1/48 ms, (d) clustered time sequence obtained for first dominant peak (b), (e) synthesized waveform (thick) with original one (thin), (f) residual left after 20 repetitions [128]

Chapter 15
Poles, Zeros, and Path Information

Poles and zeros characterize the transfer function as described in previous chapters. The transfer function, which can be expressed by the Green function from a physical point of view, represents path information between the source and receiver positions in the space. This chapter describes the poles and zeros from the point of view of estimation of the path information.

15.1 Single Echo in Space

15.1.1 Auto-correlation Sequence and Power Spectrum

A simple example of the transfer function that can be characterized by the zeros is shown by the impulse response composed of a single echo. The transfer function can be written as

$$H(z^{-1}) = 1 - \mu^M z^{-M}, \tag{15.1}$$

assuming an M-point of delay where μ represents the reflection coefficient of sound. The zeros can be given by the solution of

$$H(z^{-1}) = 0. \tag{15.2}$$

Namely,

$$z_{0_k} = |\mu| e^{i\frac{2\pi}{M}k} \quad (k = 0, 1, 2, \cdots, M-1). \tag{15.3}$$

The path information conveyed can be seen in the zeros of the transfer function.

However, the path information can also be observed in the power spectrum of the transfer function or the auto-correlation sequence. The power spectrum can be written as

$$|H(e^{-i\Omega})|^2 = \sum_{n=0}^{1} c_f(Mn)\cos(M\Omega n), \tag{15.4}$$

M. Tohyama: *Sound and Signals*, Signals and Communication Technology, pp. 351–375.
springerlink.com

where $c_f(Mn)$ denotes the **auto-correlation sequence** in the causal sequence form. That is, the auto-correlation sequence can be derived from

$$(1-\mu^M z^{-M})(1-\mu^{*M} z^M) = 1+|\mu|^{2M}+2\mu^M\cos(M\Omega) \tag{15.5}$$
$$= \sum_{n=0}^{1} c_f(Mn)\cos(M\Omega n),$$

assuming that μ is a real number.

Figure 15.1 illustrates a sample of the auto-correlation sequence and its spectrum. The path information can also be seen here in the spectral peaks that are interlaced with the troughs (zeros).

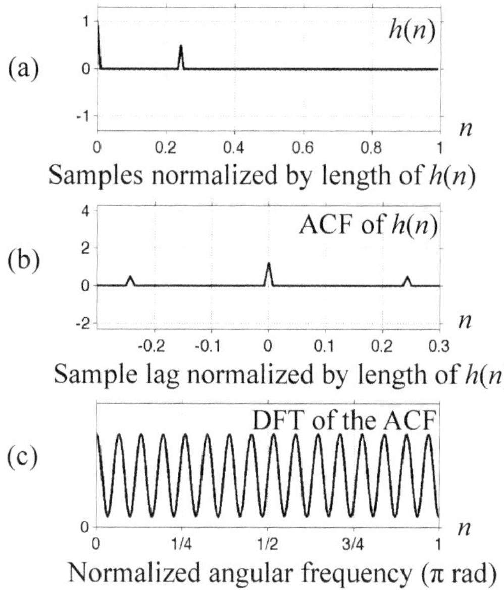

Fig. 15.1 Sample of impulse response represented by single echo (a), auto-correlation sequence (b) and its power spectrum (c)

15.1.2 Estimation of Path Information

Path information cannot be estimated by using the auto-correlation sequence or the power spectral record without phase information in general. However, the power spectrum information itself could be important. The power spectrum can be obtained by using the auto-correlation sequence which can be formulated by convolution of the auto-correlation sequences between the source and transfer function. Namely, the auto-correlation sequence $c_{f_{ob}}(n)$ of the received signal is expressed as

$$c_{f_{ob}}(n) = c_{f_s}(n) * c_{f_{path}}(n), \tag{15.6}$$

by using **convolution** where $c_{f_s}(n)$ and $c_{f_{path}}(n)$ denote the auto-correlation sequences of the source and the transfer function, respectively.

The relation above states that the path information conveyed by the power spectral record can be estimated subject to the auto-correlation sequences $c_{f_s}(n)$ being close to the **delta sequence**. This fact indicates it is possible to estimate the power spectral path information from the observed signal only, if the source could assumed to be **white noise**.

15.2 Resonant Path Information

15.2.1 Single-Pole Transfer Function

The single-pole transfer function that represents a single-degree of freedom system conveys important path information as well as the single echo systems do. Consider the transfer function

$$H(z^{-1}) = \frac{1}{1-\mu^M z^{-M}}, \tag{15.7}$$

where

$$\mu = |\mu| e^{i\Omega_p}. \tag{15.8}$$

The impulse response that follows the transfer function above can be written as

$$h(n) = \mu^{Mn} = |\mu|^{Mn} e^{iM\Omega_p n} \tag{15.9}$$

in the analytic signal form. Note here that the impulse response can be approximated by a sinusoidal wave with a slowly decaying envelope as long as the damping effect is not too significant.

Assume that the source signal can be approximated by using the white noise. The source waveform is random, but the observation signal can be regarded as a superposition of the sinusoidal waveforms as far as observing the signal in the frequency band around the spectral peak corresponding to the resonance frequency $M\Omega_p$. It indicates that the path information represented by the resonant frequency can be estimated by using the spectral-peak selection or CLSM, even if the observation window length is short subject to the response signal being observed after the system reaches the steady state.

The fact that the resonant frequency can be estimated even by using a short window length is crucial for the path-information estimation[131]. The observation signal can be assumed to be a random signal under non-deterministic excitation of a source. Therefore, the estimates obtained from the observation records are also random variables, and thus taking the ensemble average is inevitable to get a stable estimate. It is possible to get the average over the entire response record, if the random samples are taken by using short intervals (window length).

15.2.2 Estimation of Resonant Frequencies for Path Information

Figure 15.2 is an example of estimation of the path information[117]. Peak-spectral frequencies are important estimates rather than the magnitude response under random source excitation. This is because the resonant frequencies might be more robust to random fluctuations of the excitation than the spectral magnitude response [131]. Figure 15.2 illustrates the impulse response, its power spectrum, and statistical frequencies for the estimate of the resonant frequency with the highest spectral peak selected in every short frame from the interpolated spectral record [131][117]. The results indicate that the resonant frequencies are quite likely to be selected from the spectral record, if the window length (W in the figure) is longer than L, i.e., the inverse of frequency separation of the two resonant spectral peaks. Here spectral interpolation was performed by DFT after zero adding to the record. Note that the spectral troughs can also be estimated by using the statistical frequencies for the highest peak as shown by Fig. 15.2 (right). The spectral peaks due to the resonant frequencies are most likely to be selected, while the troughs are the most unlikely to be selected.

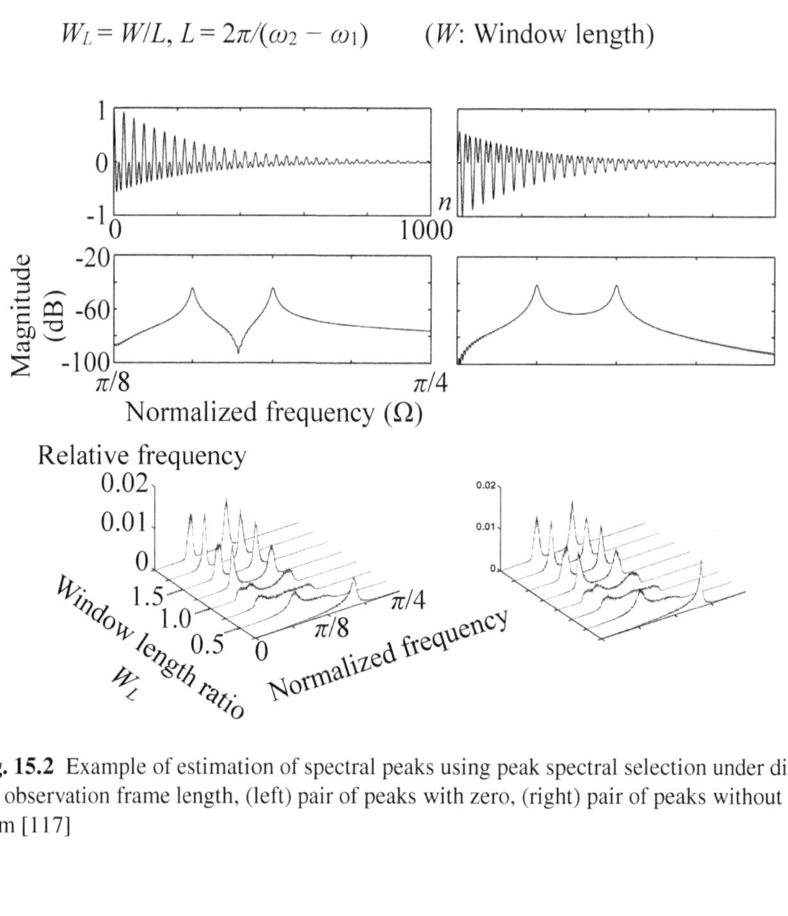

Fig. 15.2 Example of estimation of spectral peaks using peak spectral selection under different observation frame length, (left) pair of peaks with zero, (right) pair of peaks without zero from [117]

Recall the relationship of the auto-correlation sequences between the source and the transfer function for a path given by Eq. 15.6. Even if $c_{f_s}(n)$ could be assumed to be a delta function, $c_{f_{ob}}(n)$ has a record length equal to that for $c_{f_{path}}(n)$. Therefore, if a shorter frame length is taken for observation than that for $c_{f_{path}}(n)$, the estimated auto-correlation sequence $\hat{c}_{f_{path}}(n)$ could be different from $c_{f_{path}}(n)$. This is because the estimated one is periodic with the period of the frame length, which must be different from the original one. This fact indicates that the entire auto-correlation sequence cannot be estimated with its whole spectrum; only the decimated (or sampled) power-spectral record can be estimated. Note again that the spectral records can be only estimated around the dominant spectral peaks, which represent the significant characteristics for the path information[120][131].

Figure 15.2 actually confirms that the power spectral dominant peaks can be estimated using the interpolated spectral records for the truncated auto-correlation sequences. However, the overlapped spectral responses, due to closely-located spectral peaks, cannot be separately estimated using the spectral peak selection due to the loss of spectral resolution when the window length is short. The window length required to separately estimate the spectral peaks is given by the distance $\Delta\nu$ between the two frequencies of the peaks, i.e., the window length independent of the record length of the impulse response for the transfer function.

15.3 Combination of Poles and Zeros

15.3.1 Inverse Filtering

Suppose that $y(n)$ is an output signal of a linear system for an input signal $x(n)$. The output signal can be written as

$$y(n) = x(n) * h(n) \tag{15.10}$$

by using the **convolution** between the input signal and the impulse response $h(n)$. It is generally called **filtering** to get the output signal from the input signal through the linear system. In contrast, estimating the input signal from the output $y(n)$ is called **inverse filtering**.

Inverse filtering, in principle, is possible for a linear system, subject to the impulse response and the entire signal record of $y(n)$ being perfectly known without errors. This is because the input signal can be obtained as the solutions for the linear equation, such as

$$H\mathbf{x} = \mathbf{y}, \tag{15.11}$$

where H denotes the matrix composed of the time-shifted impulse response records, $\mathbf{y}$ shows the vector of observation signal records, and $\mathbf{x}$ is the vector to be solved.

Consider an example where the impulse response is composed of

$$h(0), h(1), \cdots, h(N-1), \tag{15.12}$$

and the entire output signal is listed as

$$y(0), y(1), \cdots, y(M-1). \tag{15.13}$$

The output signal record can be written as

$$y(0) = x(0)h(0) \tag{15.14}$$
$$y(1) = x(1)h(0) + x(0)h(1) \tag{15.15}$$
$$y(2) = x(2)h(0) + x(1)h(1) + x(0)h(2) \tag{15.16}$$
$$\vdots$$
$$y(n) = x(n)h(0) + x(n-1)h(1) + \cdots + x(n-(N-1))h(N-1) \tag{15.17}$$
$$\vdots$$
$$y(L-1) = x(L-1)h(0) + x(L-2)h(1) + \cdots + x(L-1-(N-1))h(N-1) \tag{15.18}$$
$$\vdots$$
$$y(M-2) = x(M-2-(N-2))h(N-2) + x(M-2-(N-1))h(N-1) \tag{15.19}$$
$$y(M-1) = x(M-1-(N-1))h(N-1), \tag{15.20}$$

where $M = N + L - 1, L > N$, and L is the record length of the input signal. The expression above indicates that the input signal $x(n)$ can be recursively obtained from the output signal $y(n)$, if the impulse response is known. However, this is a tricky and virtual process because it is quite unlikely to get the entire waveform of the output signal without errors, even if the impulse response is known. A different approach is necessary to the inverse filtering that is an important tool for source signal analysis by reducing the effects of the path information on the source signal signatures[29].

The effect of the zeros on the path characteristics can be cancelled or equalized by inverse filtering. The zeros can be interpreted as the frequencies of the source components that are not propagated to the observation point along the path of interest. Inverse filtering is a fundamental tool for reducing the loss of source information during traveling of sound on the path to the observation position.

Consider the transfer function

$$H(z^{-1}) = \frac{N(z^{-1})}{D(z^{-1})} = \frac{1 - a^M z^{-M}}{1 - a^M z^{-M}} = 1 \quad (N(z^{-1}) = D(z^{-1})). \tag{15.21}$$

This indicates that the zeros of the numerator $N(z^{-1})$ were cancelled by the zeros of the denominator $D(z^{-1})$ (namely, the poles of the transfer function subject to $N(z^{-1}) = D(z^{-1})$). This type of cancellation is called inverse filtering, where $\frac{1}{D(z^{-1})}$ is called the **inverse filter** for $N(z^{-1}) = D(z^{-1})$. Inverse filtering, however, is possible only for $|a| < 1$ because the response due to the poles could be **non-causal** for $|a| > 1$. If inverse filtering is possible, the source information that is lost by the sound-traveling path could be recovered from the observation response after being obtained through the path.

When the source signal characteristics are **minimum phase**, namely all the zeros of the source spectral components are located within the unit circle on the frequency domain (z-plane), the inverse filtering is always possible independent of the conditions of the path[132]. Suppose that the source spectral record is written as

$$X(k) = X_{min}(k)X_{ap}(k), \tag{15.22}$$

where $X_{min}(k)$ and $X_{ap}(k)$ correspond to the minimum phase and all-pass components, respectively. Similarly, by assuming that the transfer function for the path can be written as

$$H(k) = H_{min}(k)H_{ap}(k), \tag{15.23}$$

the spectral record observed through the path is given by

$$Y(k) = X(k)H(k) = Y_{min}(k)Y_{ap}(k) \tag{15.24}$$

$$Y_{min}(k) = X_{min}(k)H_{min}(k). \tag{15.25}$$

Consequently, the minimum-phase components of the source signal can be recovered according to

$$X_{min}(k) = Y_{min}(k)/H_{min}(k). \tag{15.26}$$

This outcome indicates that the minimum-phase component of the source signal can be estimated through inverse filtering of the path-transfer function only using its minimum-phase part independent of the all-pass component of the path[132].

Figure 15.3 illustrates the schematic of the patterns of poles and zeros. If a source waveform does not have any non-minimum-phase zeros as shown by Fig. 15.3 (a) and the non-minimum-phase shift is always due to the path-transfer function, then the source waveform can be recovered using only the inverse filter for the minimum-phase part as stated above. However, note here that the equation

$$Y(z^{-1}) = X(z^{-1})H(z^{-1}) \tag{15.27}$$

always holds between the input and output signals of a linear system whose transfer function is defined as $H(z^{-1})$. Therefore, if the path information could be perfectly known without error, inverse filtering is possible even for the non-minimum-phase path information. This is because all the non-minimum-phase zeros due to the path-transfer function will, ideally, be cancelled as shown in Fig. 15.3(b). Consequently, there are no residual unstable poles left in the inverse filter, and thus the source waveform can be recovered. However, this is a tricky only virtual process for inverse filtering, and it is quite unlikely in practical situations. This is quite similar to the fact that inverse filtering could be interpreted as solving the set of linear equations mentioned in the beginning of this subsection.

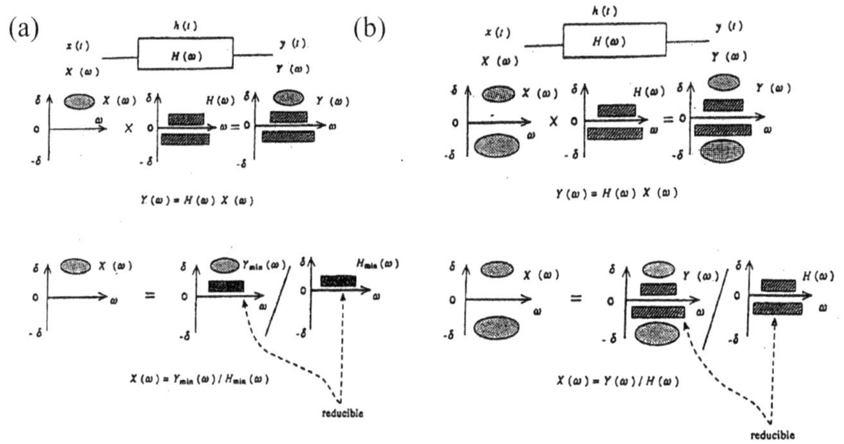

Fig. 15.3 Patterns of poles and zeros in input/output signals. (a) Minimum-phase input signal $x(t)$. (b) Non-minimum-phase input signal $x(t)$. $Y_{min}(\omega)$: minimum phase component of $Y(\omega)$; $H_{min}(\omega)$: minimum phase component of $H(\omega)$ from [132] (Fig.2)

Source waveform recovery is an important issue for machinery noise diagnostics from practical and engineering view points[29]. Figure 15.4 illustrates an experimental schematic for minimum-phase source-waveform recovery in reverberant space[132]. A series of measurements concerning the reverberant response to a train of pulse waveforms of source signals ($\nu_c = 500$Hz) were taken in an 86(m^3) room. Figure 15.5(b) shows the recovered waveform from the response data (Fig. 15.5(a)) observed at M1. A pulse-like source waveform is extracted from the reverberant response.

However, inverse filtering is quite sensitive to the fluctuations of the path-transfer function[132]. Figure 15.6 shows examples of waveforms at M2a and M2b recovered by the inverse filter for M2 using an exponential time window and smoothing average in the frequency domain of the transfer function. **Exponential time windowing** on the minimum-phase component[132] and taking a smoothing average in the frequency domain[133] might be a possible way to getting a robust recovery process for unpredictable changes of the path information. This is partly because the density of the zeros closely located to the frequency axis (or inside of the observation frequency circle in z-plane) decreases by the exponential windowing or the smoothing average following the statistical properties of the transfer functions in reverberant space.

Figure 15.7 plots the magnitude and phase response for a sample of minimum-phase component, with and without exponential windowing. Windowing with a negative exponent moves the zero along the radial direction farther from the unit circle, whereas with a positive exponent it moves them closer to the unit circle. Plot A was modified with the negative exponent, plot B was obtained with the positive one. Another application of the exponential-time windowing on the cepstral domain including the all-phase component will be described in subsection 15.5.2.

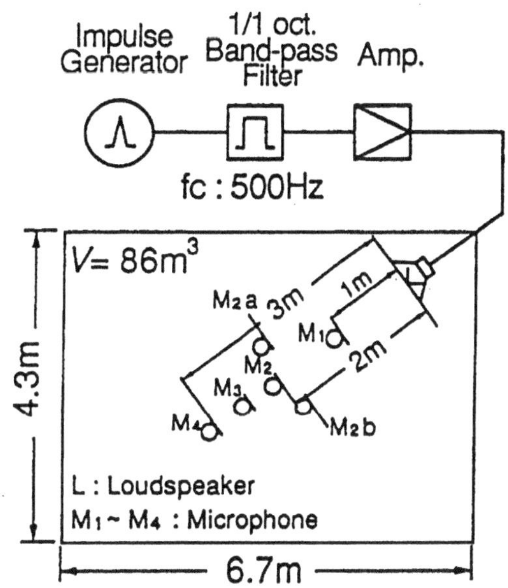

Fig. 15.4 Experimental arrangement for source waveform recovery in reverberant space from [132] (Fig.3)

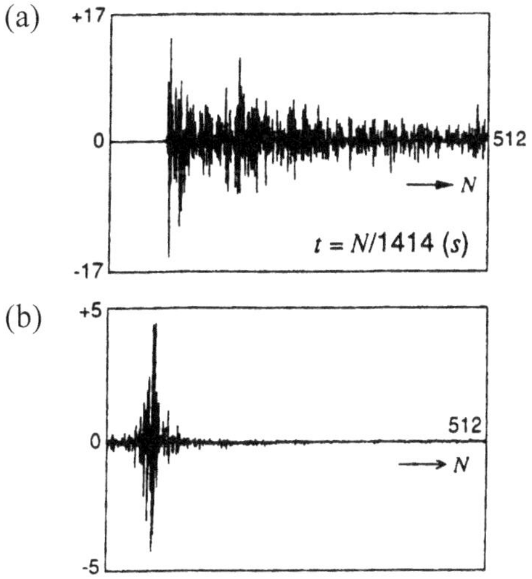

Fig. 15.5 Reverberant response (a) and recovered source waveform (b) from [132] (Fig.9)

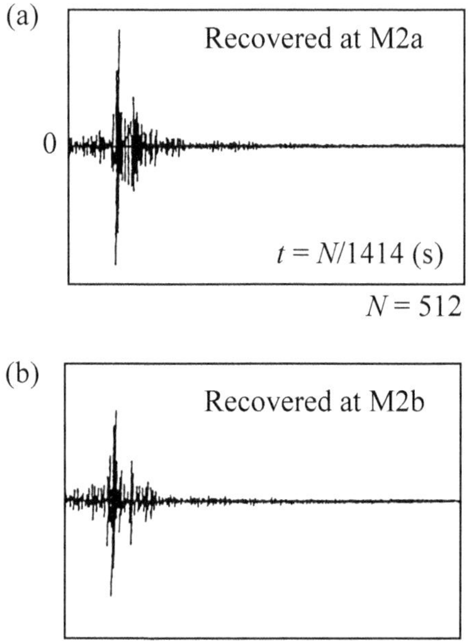

Fig. 15.6 Recovery at M2a and M2b by inverse filter for transfer function at M2 where smoothing average (within 60 Hz) and exponential time window is applied from [133] (Fig.10)

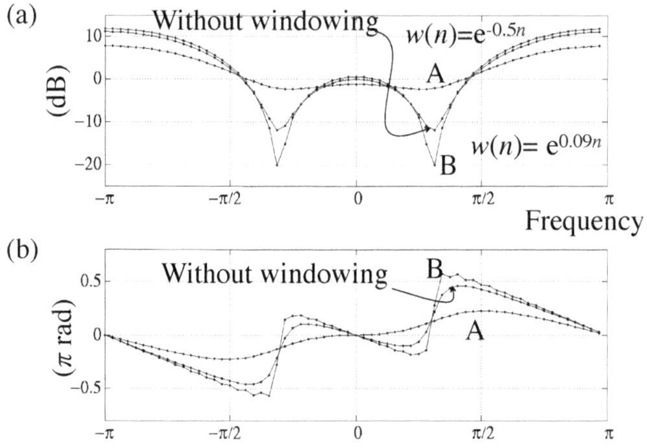

Fig. 15.7 Magnitude (a) and phase (b) modification of minimum-phase component using exponential time windowing for minimum-phase component from [113] (Fig.4)

15.3.2 Transfer Function of Closely Located Pair of Pole and Zero

As described in the previous subsection, inverse filtering uses the transfer function of which poles and zeros cancel each other, i.e., those locations coincide with each other. However, if the pole and zero are very closely located to each other, different effects on the path information may be possible.

Consider the transfer function

$$H(z^{-1}) = \frac{1 - bz^{-M}}{1 - az^{-M}}, \tag{15.28}$$

where $a = |a|e^{i\Omega_0}$, $b = |b|e^{i\Omega_0}$, and $|a| \simeq |b|$. If $|b|$ is slightly greater the $|a|$ as shown in Fig. 15.8(a), then a very deep and steep trough can be seen in the magnitude frequency response. This type of system is called a **notch filter** that is usable to sharply suppress the source spectral components around particular frequencies. In contrast, if $|a|$ is slightly larger than $|b|$, very steep spectral peaks can be seen (Fig. 15.8 (b)). This type of filter is usable to sharply reinforce the signal components around particular frequencies.

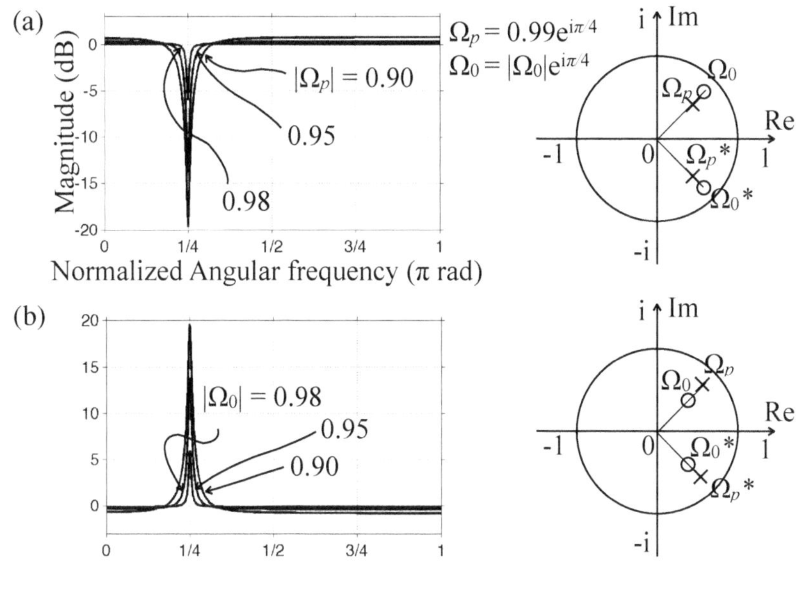

Fig. 15.8 Magnitude frequency response due to pair of pole and zero, (a): notch filter, (b): presence filter from [69]

15.4 Creation and Control of Path Information

Perception of sound is highly sensitive to path information in addition to source characteristics. Reverberant sound in a concert hall is necessary to reproduce musical sound, as if a listener might listen to the sound in the concert hall. A direction from which sound comes to a listener is also important path information for reproduction of the sound. **Sound image projection** is a fundamental audio-engineering tool for sound reproduction[46]. Again inverse filtering is a key issue for creating path information that is necessary to control the virtual sound source in the reproduced field.

15.4.1 Simultaneous Equations for Sound Image Control

Figure 15.9 illustrates a schematic of the 2-channel sound reproduction system. Suppose that $S_L(z^{-1})$ and $S_R(z^{-1})$ represent the z-transform for the binaural signals that were recorded at a listener's ear positions in an original field. The hypothesis of the virtual sound-image projection is that if the binaural signal stated above, $S_L(z^{-1})$ and $S_R(z^{-1})$, could be reproduced at the listener's ears even in a field different from the original one, no difference could be perceived by the listener between the originally recorded and reproduced ones.

For that purpose, four filters are necessary as shown in the figure. In addition, the transfer function of those filters must be solutions for the equations such that

$$X_{LL}(z^{-1})H_{LL}(z^{-1}) + X_{LR}(z^{-1})H_{RL}(z^{-1}) = 1 \tag{15.29}$$

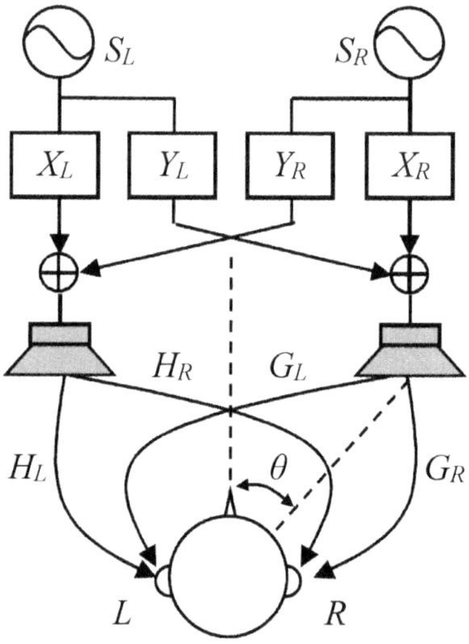

Fig. 15.9 Sound image projection system from [135] (Fig.3)

$$X_{LL}(z^{-1})H_{LR}(z^{-1}) + X_{LR}(z^{-1})H_{RR}(z^{-1}) = 0 \tag{15.30}$$

for X_{LL} and X_{LR}, and similarly

$$X_{RL}(z^{-1})H_{LL}(z^{-1}) + X_{RR}(z^{-1})H_{RL}(z^{-1}) = 0 \tag{15.31}$$
$$X_{RL}(z^{-1})H_{LR}(z^{-1}) + X_{RR}(z^{-1})H_{RR}(z^{-1}) = 1 \tag{15.32}$$

for X_{RL} and X_{RR}. Here $H(z^{-1})$ indicates the **head-related transfer functions**, respectively; namely, $H_{LL}(z^{-1})$ denotes the transfer function from the loudspeaker L to the left ear entrance of the listener in the reproduced field. Other $H(z^{-1})$ similarly indicates the corresponding transfer functions[134].

The solutions of the equation above, however, contain the process of inverse filtering. Namely, the transfer function of the filters $X_{LL}(z^{-1})$ and $X_{RR}(z^{-1})$ are written as

$$X_{LL}(z^{-1}) = \frac{H_{RR}(z^{-1})}{D(z^{-1})} = \frac{H_{RR}(z^{-1})}{D_{min}(z^{-1})D_{ap}(z^{-1})} \tag{15.33}$$
$$X_{LR}(z^{-1}) = \frac{-H_{LR}(z^{-1})}{D(z^{-1})} = \frac{-H_{LR}(z^{-1})}{D_{min}(z^{-1})D_{ap}(z^{-1})} \tag{15.34}$$
$$\begin{aligned} D(z^{-1}) &= H_{LL}(z^{-1})H_{RR}(z^{-1}) - H_{LR}(z^{-1})H_{RL}(z^{-1}) \\ &= D_{min}(z^{-1})D_{ap}(z^{-1}) \end{aligned} \tag{15.35}$$

for example. In general, the denominator $D(z^{-1})$ is non-minimum phase. Therefore, inverse filtering is impossible. Rewrite the simultaneous equations by substituting the solutions, such that

$$\frac{H_{RR}(z^{-1})}{D(z^{-1})}H_{LL}(z^{-1}) + \frac{-H_{LR}(z^{-1})}{D(z^{-1})}H_{RL}(z^{-1}) = 1 \tag{15.36}$$
$$\frac{H_{RR}(z^{-1})}{D(z^{-1})}H_{LR}(z^{-1}) + \frac{-H_{LR}(z^{-1})}{D(z^{-1})}H_{RR}(z^{-1}) = 0. \tag{15.37}$$

By multiplying $D_{ap}(z^{-1})$ to both sides,

$$\frac{H_{RR}(z^{-1})}{D_{min}(z^{-1})}H_{LL}(z^{-1}) + \frac{-H_{LR}(z^{-1})}{D_{min}(z^{-1})}H_{RL}(z^{-1}) = D_{ap}(z^{-1}) \tag{15.38}$$
$$\frac{H_{RR}(z^{-1})}{D_{min}(z^{-1})}H_{LR}(z^{-1}) + \frac{-H_{LR}(z^{-1})}{D_{min}(z^{-1})}H_{RR}(z^{-1}) = 0 \tag{15.39}$$

are obtained. This outcome indicates that inverse filtering by taking only the minimum-phase component makes it possible to control the binaural difference of the magnitude spectral properties, but the all-pass component is left uncontrolled [135].

15.4.2 Stabilization of Recursive Path

Figure 15.10 is a schematic of sound reinforcement in a closed space. **Howling** of sound is a difficult issue to settle for such a closed system. Howling can be interpreted as a phenomenon that occurs when unstable poles are created in the transfer function of the closed loop. By assuming the gain of the amplifier to be constant A, the transfer function of the closed loop can be written as

$$G(z^{-1}) = \frac{AH(z^{-1})}{1 - AH(z^{-1})}, \quad H(z^{-1}) = H_{min}(z^{-1})H_{ap}(z^{-1}), \tag{15.40}$$

where $H(z^{-1})$ denotes the transfer function between the loudspeaker and the microphone. When the zeros of the denominator (poles) moves across the unit circle in the frequency domain to the outside, the howling occurs. In general, as the gain increases, howling is likely, i.e., the zeros approach the unit circle in the frequency plane.

In principle, if it were possible to remove the unstable poles from the transfer function of the closed loop, howling will not happen. For that purpose, if the next condition

$$|AH(z^{-1})|_{z=e^{i\Omega}} < 1 \tag{15.41}$$

holds for all the frequencies of interest, the impulse response of the closed-loop transfer function will converge. However, the open-loop transfer function, namely, the room transfer function in the example of Fig. 15.10 , has a lot of poles and zeros due to the reverberation in general. Therefore, it seems difficult to control the system so that the condition above might be satisfied for a wide range of frequencies.

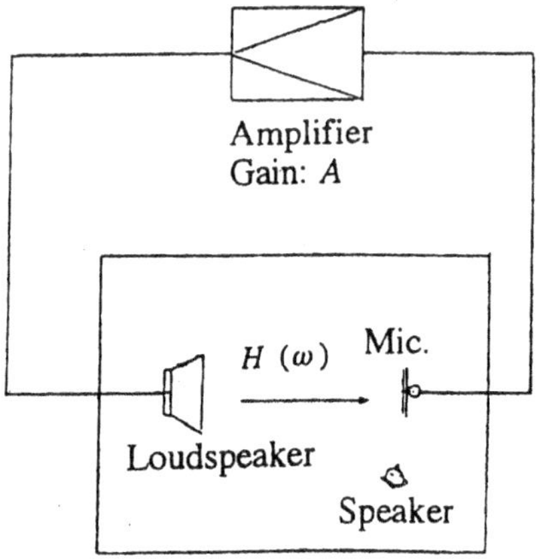

Fig. 15.10 Schematic of public address system with feedback loop in closed space

Equalization of the magnitude response of the open loop of the transfer function is a possibility to get a stable reinforcement system. Figure 15.11 is a schematic of the open-loop equalization using minimum-phase inverse filtering[136]. The spectral peaks become noticeable when the amplifier gain A increases, and the levels of those peaks are not uniform because they reflect the magnitude response of the open-loop frequency characteristics. Consequently, the impulse responses eventually diverge. In contrast, the equalization makes that the impulse response records converge, despite gain A increasing. In addition, the levels of the spectral peaks are mostly uniform.

The equalization effect can be interpreted using the closed-loop transfer function such that

$$G_{min}(z^{-1}) = \frac{H_{ap}(z^{-1})}{1 - AH_{ap}(z^{-1})} \tag{15.42}$$

for the equalized system using the **minimum-phase inverse**. Here, the condition

$$|AH_{ap}(e^{-i\Omega})| < 1 \tag{15.43}$$

is more easily satisfied than that given by Eq. 15.41 without equalization because $|H_{ap}(e^{-i\Omega})| = 1$.

The closed-loop impulse response has a shorter record length than that without equalization. This result also indicates that **coloration**, which is change of sound quality due to echoes of the closed loop, might be reduced. Figure 15.12 shows the **reverberation decay curves** of the closed-loop impulse responses[136]. It can be seen that reverberation effects are decreased by the equalization. The open-loop magnitude frequency response with that for the minimum-phase inverse filter. From a practical view point, a smoothing or averaging process, however, is necessary to work the minimum-phase equalization effectively because of the fluctuations of this open-loop transfer function[136].

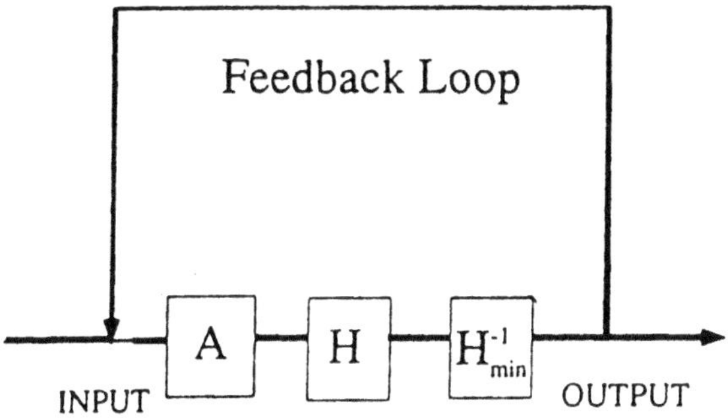

Fig. 15.11 Block diagram of feedback system with minimum-phase inverse filter from [136] (Fig.2)

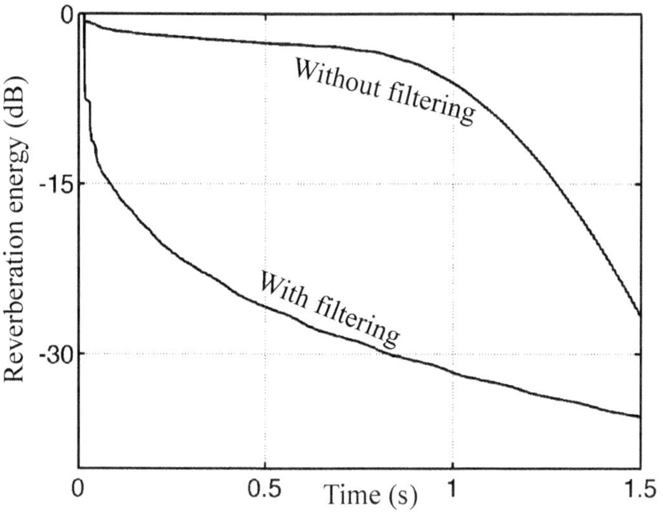

Fig. 15.12 Reverberation decay curves of closed loop transfer functions [69]

15.4.3 Detection of Howling Frequency

Time-frequency analysis plays a fundamental role in the detection of spectral distortion. The time-dependent frequency distortion of sound through a feedback loop is commonly called **spectral coloration** because of the system (hidden) resonance [137]. Even if the spectral coloration can be only slightly perceived by listening, the free-oscillation components of the resonance could be main factor in howling (i.e., unstable poles of the transfer function). Therefore spectral-coloration analysis is significant for predicting the howling frequency of a closed loop and realizing a stable acoustic system. **Cumulative spectral analysis (CSA)**[138] or **cumulative harmonic analysis (CHA)**[137] is a way to determine quickly what the principal resonant frequency of a public-address system is before it starts howling.

CSA is formulated by means of Fourier transform of an impulse response record using a unit-step time-window function. It performs a time-frequency analysis of the transient response of a linear system such as a loudspeaker corresponding to a tone-burst input signal. The spectral accumulation effect can be emphasized by substituting a spectral accumulation function for the unit-step time-window function of CSA. This substituted formula is called cumulative harmonic analysis (CHA)[137]. The spectral accumulation process, such as a growing spectral peak inherent signals, might be effectively displayed by CSA or CHA.

A feedback loop, in principle, yields periodic signals. The harmonic structure of a periodic signal is the result of the superposition of repeated spectral records with a fixed phase lag. That is, as the periodic-signal length becomes longer, spectral peaks grow due to the in-phase accumulation of the harmonic sinusoidal components corresponding to the period. However, if the superposition makes the resultant signal unstable beyond the steep but stable spectral peaks, then the system that produces the superposition starts howling. Howling can be interpreted as a change from the spectral periodicity of signals to the spectral selectivity dominated by only a few components.

Suppose a signal $x(n)$ and a spectral accumulation function $w(n)$. CHA of $x(n)$ is defined by

$$\mathrm{CHA}(n, \mathrm{e}^{-\mathrm{i}\Omega}) = \sum_{m=0}^{n} w(m)x(m)\mathrm{e}^{-\mathrm{i}m\Omega}. \tag{15.44}$$

Introducing an example of simple spectral accumulation function such as

$$w(n) = n + 1 \quad (n \geq 0) \tag{15.45}$$

into $w(n)$,

$$\begin{aligned} \mathrm{CHA}(n, z^{-1}) &= \sum_{m=0}^{n} (m+1)x(m)z^{-m} \\ &= 1x(0)z^{-0} + 2x(1)z^{-1} + 3x(2)z^{-2} + \cdots + (n+1)x(n)z^{-n} \end{aligned} \tag{15.46}$$

is derived where $z = \mathrm{e}^{\mathrm{i}\Omega}$. The effect of the transfer function pole on the frequency characteristics can be emphasized by CHA. Suppose that the sequence

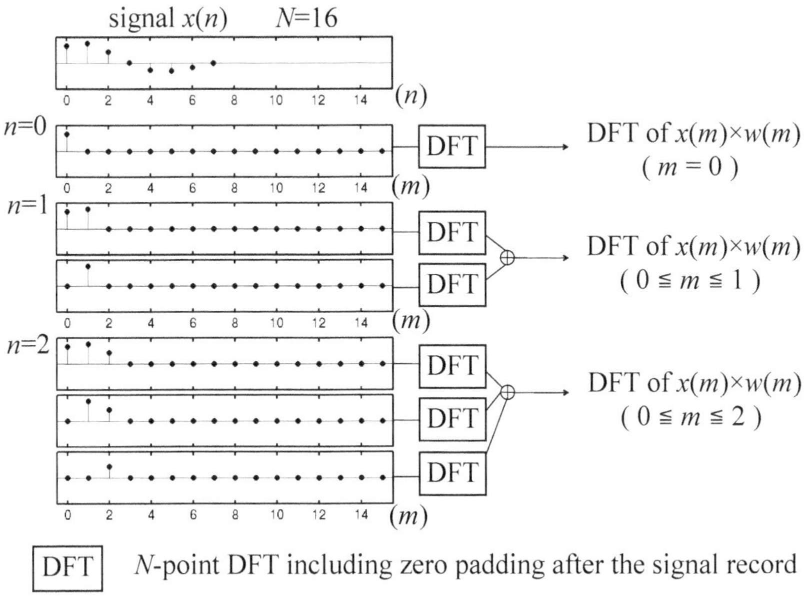

Fig. 15.13 Schematic of cumulative harmonic analysis (CHA) using numerical example where $w(n) = n + 1$ from [137] (Fig.1)

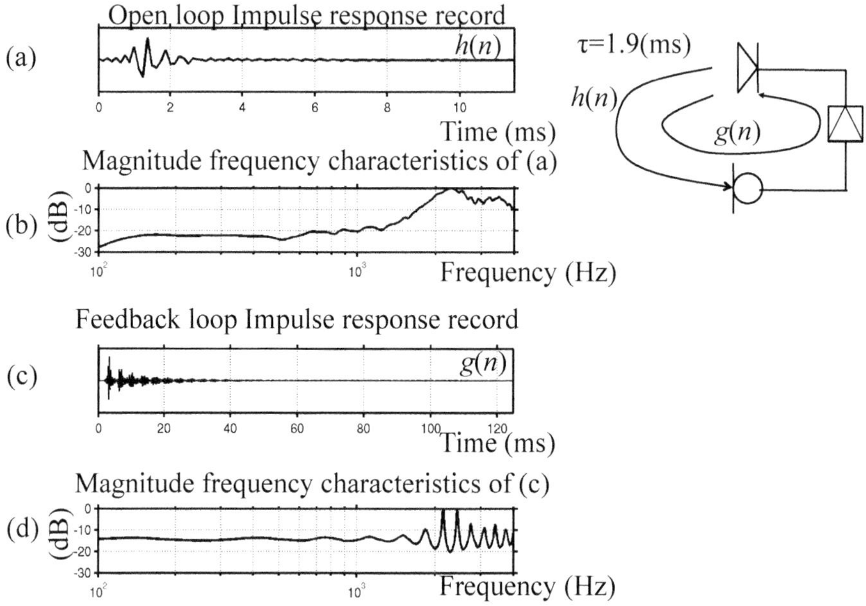

Fig. 15.14 Impulse responses and frequency characteristics, (a) open-loop impulse response with its frequency characteristic (b), (c) closed-loop impulse response including feedback with its spectrum (d) from [137] (Fig.3)

$$h(n) = a^n \quad (n = 0, 1, 2, \cdots) \quad 0 < |a| < 1. \tag{15.47}$$

If a limit of $H(z^{-1})$ is taken such as

$$\begin{aligned} \lim_{n\to\infty}\mathrm{CHA}(n, z^{-1}) &= \sum_{n=0}^{\infty}(n+1)a^n z^{-n} \\ &= \frac{1}{(1-az^{-1})^2} \end{aligned} \tag{15.48}$$

then CHA increases the order of the pole. Figure 15.13 is a schematic of CHA with the accumulation function $w(n)$ given by Eq. 15.45[137]. The time-window function $w(n)$ can be interpreted as a spectral accumulation function.

To estimate the howling frequency (hidden resonance) and avoid unstable amplification, it is useful to visualize under stable conditions, prior to howling, how the frequency components of the input signals, including feedback, can be narrowed down to a single element. If it is possible to test a system including a feedback loop, then observation of the reverberation sound might be a good way to diagnose the system, i.e., the path information independent of the source-signal properties. Figure 15.14 illustrates a system with a stable feedback loop. Figure 15.14 (a) shows an open-loop impulse response with its magnitude frequency response (b) from the loudspeaker to the microphone. Similarly, Figs. 15.14 (c) and (d) show

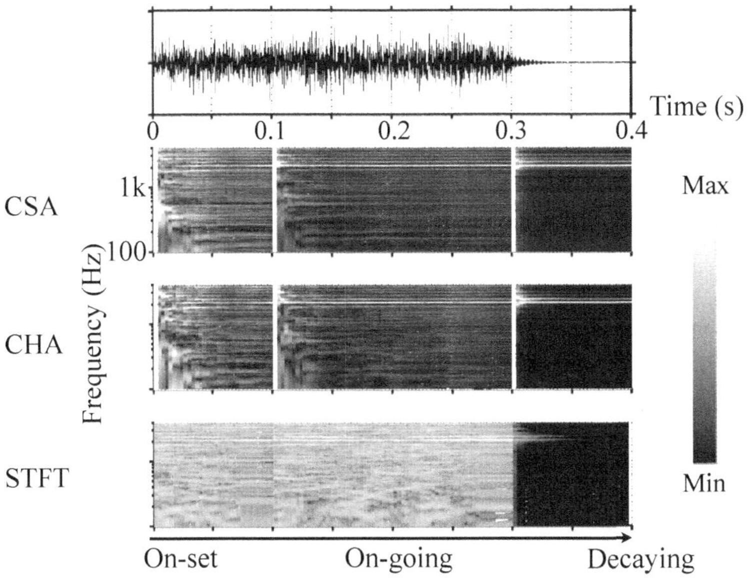

Fig. 15.15 Time and frequency analysis (CSA, CHA, and Spectrogram(STFT)) in initial portion (left: On-set), steady state (center: On-going) and reverberation (right: Decaying) responses to broad-band random noise for feedback system from [34] (Fig.9.3)

the response and its spectrum including the stable feedback loop. Figure 15.15 shows time-frequency analysis of initial state responses (left), steady state responses (center) and reverberation responses (right) to a random noise signal for stable feedback system.

It is desirable to predict the howling frequency under in situ conditions for a system including a feedback loop. However, the left or center column implies that estimating the howling frequency might be difficult under stable-loop conditions even by CSA or CHA. In contrast, a clear indication about the howling frequency can be obtained from the reverberation sound. The decaying portion of the signal in principle are composed of free oscillations of a linear system of interest. Therefore there are clear differences in the responses between the reverberation sound and the stationary signals.

15.5 Control and Modification of Reverberant Sound Path

15.5.1 Control of Reverberant Sound Path

Reverberant sound mostly reduces the intelligibility of speech. **Speech intelligibility** is highly sensitive to the **energy ratio** of the direct and reverberant sound defined in subsection 11.4.3. The intelligibility increases as the ratio increases. It is possible to increase the ratio by arranging multiple sound sources in addition to decreasing the distance from the source[139]

Suppose that sound sources are located on a circle with a radius r from the listening position. The number of uncorrelated points on the circle can be estimated as

$$N_{u_c} \simeq \frac{2\pi r}{c/2\nu} \tag{15.49}$$

according to the spatial correlation properties mentioned in subsection 12.3.1, where ν is the center frequency of the narrow-band sources and c is the sound speed in the air. Similarly, on the spherical surface, such a number is given by

$$N_{u_s} \simeq \frac{4\pi r^2}{c^2/4\nu^2}. \tag{15.50}$$

The energy ratio at the central position, namely, the listening position, can be estimated as

$$D_{R_c} = \frac{A\nu}{4rc} = K_c \tag{15.51}$$

$$D_{R_s} = \frac{A\nu^2}{c^2} = K_s. \tag{15.52}$$

Note that the ratio for the multiple sources on the spherical surface is constant independent of the distance r.

According to the outcome of the energy ratio, which states the independence of the distance, the minimum of the ratio can be found for the sources on the spherical surface. By assuming that

$$S = 4\pi R^2, \tag{15.53}$$

the ratio denoted by K_s above can be rewritten as

$$K_s = N_{S_{max}} K_{S_{min}} = N_{S_{max}} \frac{-\ln(1-\alpha)}{4}, \tag{15.54}$$

where $N_{S_{max}}$ gives the number of uncorrelated sources arranged on the whole room surface. The minimum of the ratio becomes

$$K_{S_{min}} = \frac{-\ln(1-\alpha)}{4}, \tag{15.55}$$

which states the sound absorption coefficient is a representative number of the room acoustics rather than the reverberation time. Even if the reverberation time becomes longer, reverberation sound may not be noticeable when the room volume is large.

In contrast, even if the reverberation time is short, the reverberation sound might be harmful to speech intelligibility for listeners when the room volume is small.

Figure 15.16 illustrates examples of sound source arrangements on coaxial surfaces[139]. By assuming that appropriate time delay is given to the sources on the inner surface so that all the direct sound might reach the central position simultaneously, the energy ratio is given by

$$D_{R_{dc}} = \frac{(1+a)^2}{1+a^2 b} K_c = K_{dc} \tag{15.56}$$

at the center, where W_{X_1} and W_{X_2} are the sound power outputs of the inner and outer sources,

$$b = r_2/r_1, and \quad a = \sqrt{W_{X_2}/W_{X_1}}. \tag{15.57}$$

This outcome indicates that the ratio takes its maximum by controlling the sound power outputs of inner and outer sources. Such a maximum becomes

$$K_{dc_{max}} = (1+\frac{1}{b})K_c \tag{15.58}$$

for the sources on the coaxial circles by setting the sources as

$$a = \sqrt{W_{X_2}/W_{X_1}} = 1/b. \tag{15.59}$$

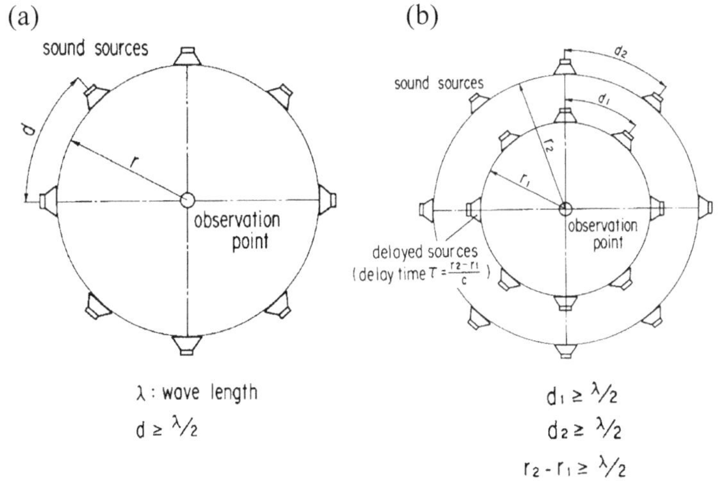

Fig. 15.16 Sound source arrangement on circle (a) and coaxial double circle (b) [139] (Fig. 1)(Fig.2)

Similarly, the maximum is obtained as

$$K_{ds_{max}} = 2K_s \tag{15.60}$$

for the spherical source arrangement when $a = 1/b$.

Note that the sound power output of the outer sources must be smaller than that for inner sources to get the maximum ratio. It can be found that the energy ratio becomes two times greater for the coaxially double spherical surfaces than that for the sources on the single surface independent of the radius ratio between the two layers. However, such a increase cannot be expected for the coaxially double circles of sound source arrangements. An example of loudspeaker arrays can be seen in the article [139], where speech intelligibility could be improved even in reverberant space.

15.5.2 Modification of Path Information Using Time Windowing

Spectral smoothing, including phase, is significant for obtain stable inverse filtering effects on the room reverberation transfer functions. In general, smoothing the frequency characteristics shortens the reverberation time, whereas increasing the range between the peaks and dips in the frequency characteristics renders longer reverberations. This subsection extends the concept of complex spectral smoothing to reverberation-time control for reverberators or room acoustic simulators[113]. Artificial reverberation control that does not change the locations of spectral peaks and dips at the original frequencies is desired independent of the reverberation time because of maintaining the original reverberation sound characteristics.

Suppose an impulse response of a finite record length. Figure 15.17 (a) shows a sample impulse response in reverberant space, (b) is the magnitude, and (c) gives the phase response. Although the impulse response has a finite record length, dips and many peaks in the magnitude response can be seen in addition to abrupt phase changes. Steep magnitude peaks will become less steep, as damping increases. On the contrary, the dips depend on the location of the zeros, which cannot be estimated at present in practical situations. Therefore, the phase characteristics are not fully predictable because the reverberation conditions vary. This might only be possible using the static model, which assumes that the density of zeros decreases following a Cauchy distribution in the complex frequency plane[109]. This model, however, creates isolated dips in the magnitude response under the short reverberation condition, because the zero density becomes sparse as the reverberation time decreases.

It is important to preserve the locations of the peaks and dips in the magnitude response, independent of the reverberation conditions. The transfer function can be decomposed into minimum-phase and all-pass components as shown in Fig. 13.42. In the illustration only non-minimum phase zeros for the original transfer function are shown. If there is a minimum-phase zero, then it remains at the original location in the minimum-phase part. The poles and zeros for the all-pass part make symmetric pairs with respect to the unit circle. The locations of the all-pass poles are the same as those for the minimum-phase zeros in Fig. (b), and similarly the all-pass

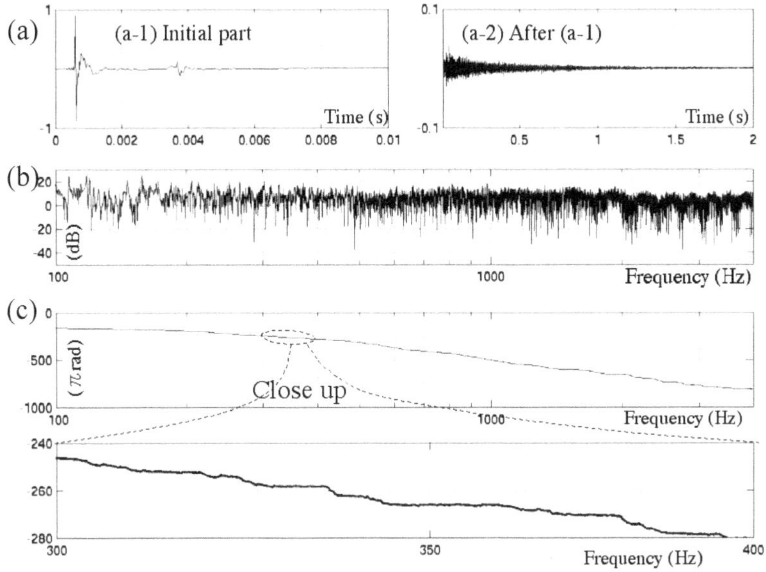

Fig. 15.17 (a)Example of impulse response record in reverberant space, (b)magnitude response, (c)phase response with close-up view from [113] (Fig.2)

zeros are located at the original non-minimum phase positions in Fig.(a). The magnitude and phase can be controlled by relocating the poles and zeros in Figs. (b) and (c).

Reverberation control, with respect to the magnitude-response, can be interpreted as exponential time-windowing the minimum-phase component as illustrated in Fig. 15.7. This is because exponential windowing is equivalent putting the whole figure of the poles and zeros close to (apart from) the unit circle on the z-plane along the radial direction. The magnitude and phase of the minimum-phase response and also the all-pass phase is important for reverberation sound rendering. As shown in Fig. 13.42, the poles and zeros for the all-pass component were composed of symmetric pairs of poles and zeros, with respect to the unit circle. If the minimum-phase zeros moved radially apart from the unit circle after exponential windowing, then the all-pass poles have to be relocated at just the same locations as those of the minimum-phase zeros. On the other hand the all-pass zeros have to move outside (apart from the unit circle) to the symmetric positions with respect to the unit circle.

Reverberation control, moving the poles and zeros of the all-pass response to correspond with the exponential windowing for the minimum-phase part can be done in the cepstral domain. Figure 15.18 (a) plots the all-pass impulse response in Fig. 13.42 (c) and (b) shows its cepstrum[113]. Figure (c) is a newly constructed response from the causal part of the cepstrum in Fig. (b). The causal cepstral components are created by the minimum-phase zeros or the poles. Thus this time-sequence in Fig.(c) could be interpreted as the response due to the all-pass poles. Figure

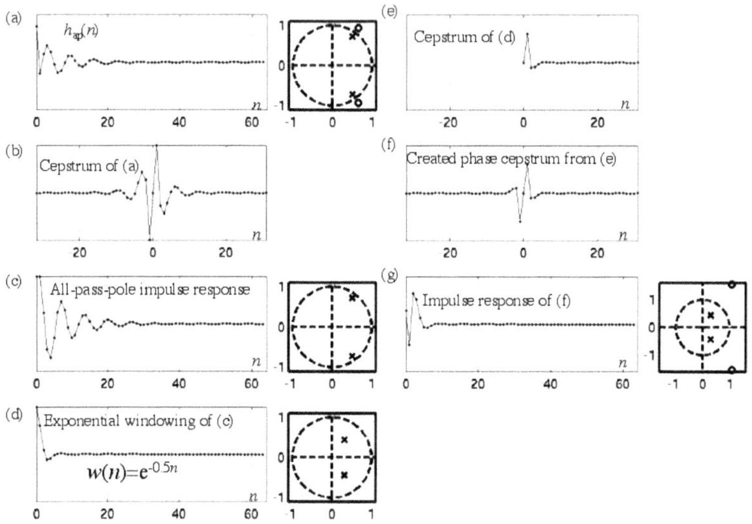

Fig. 15.18 All-pass phase modification, pole-zero movement using cepstral decomposition from [113] (Fig.5)

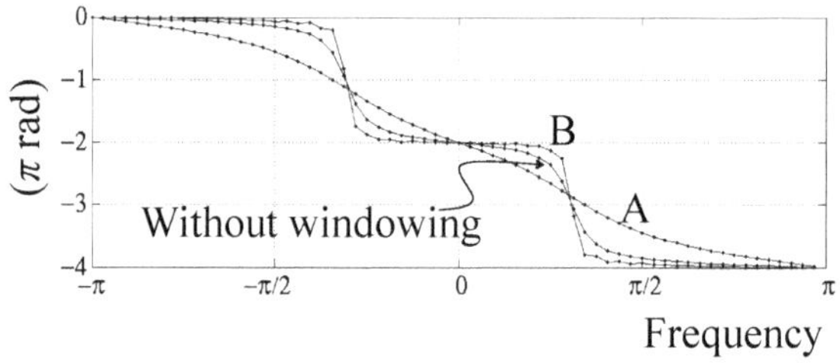

Fig. 15.19 All-pass phase modification corresponding to minimum-phase magnitude of Fig. 15.7, A(B): exponential windowing with negative (positive) exponent [113] (Fig.6)

(d) shows the modified pole-response obtained by exponentially-windowing the sequence in Fig. (c), and (e) shows its cepstrum. Thus the all-pass phase cepstrum is obtained as shown in Fig. (f) by adding its non-causal part into the causal cepstrum shown in Fig. (e), because the phase cepstrum is an odd sequence. The all-pass impulse response can be constructed from the all-pass phase cepstrum in Fig. (g), because the all-pass components have no magnitude cepstrum. Figure 15.19 plots an all-pass phase control, corresponding to the minimum-phase magnitude shown in Fig. 15.7. Exponential windowing in the minimum-phase component is performed in the time domain, while the windowing is applied to the all-pass part in the cepstral domain. Abrupt phase changes cannot be seen in plot A, where windowing with a negative exponent is applied. However, steeper steps in the phase changes are obtained for a positive exponent, as shown in plot B.

Figure 15.20 shows an example of reverberation control[113]. Plot O in Figs. (a) and (b) denotes the original impulse response record and reverberation energy-decay curve, respectively. Here, $h(n)$ denotes the impulse response record with a length of N. Plot A(B) illustrates a modified example using exponential windowing with a negative(positive) exponent. The change in the reverberation decay curve can be observed. Figures (c) and (d) are close-up for the magnitude and all-pass phase responses, respectively. Similarly the effects of reverberation control can be confirmed, as shown in plots A and B. Reverberation control is possible in a global range of reverberation time. Smoothing effects are obtained for both magnitude and phase responses without pole/zero modeling but keeping the global trend (frequency locations of peaks and dips) of the responses.

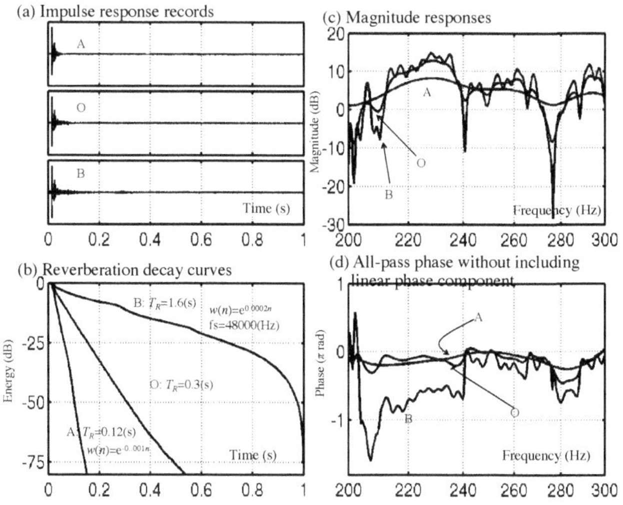

Fig. 15.20 Magnitude and phase modification using exponential windowing with a negative (A) and positive (B) exponent for the impulse response shown by O from [113] (Fig.7)

Erratum: Oscillation and Resonance

M. Tohyama: *Sound and Signals,* Signals and Communication Technology, pp. 9–30.
springerlink.com

DOI 10.1007/978-3-642-20122-6_16

In the original version, error has occurred in the following pages:

page		error	correct
p.26	eqn.(2.63)	$= (1 + \frac{m_o}{2}\cos\frac{\Delta\omega}{2}t)\sin\omega_c t$	$= (1 + m_o\cos\frac{\Delta\omega}{2}t)\sin\omega_c t$
	eqn.(2.66)	$A = \sqrt{1 + \frac{m_o^{\ 2}}{4}\cos^2\frac{\Delta\omega}{2}t}$	$A = \sqrt{1 + m_o^{\ 2}\cos^2\frac{\Delta\omega}{2}t}$
	eqn.(2.67)	$\phi(t) = \tan^{-1}\left(\frac{m_o}{2}\cos\frac{\Delta\omega}{2}t\right)$	$\phi(t) = \tan^{-1}\left(m_o\cos\frac{\Delta\omega}{2}t\right)$
	eqn.(2.70)	$\frac{d\phi(t)}{dt} = -\frac{1}{1 + \frac{m_o^{\ 2}}{4}\cos^2\frac{\Delta\omega}{2}t}\frac{m_o\Delta\omega}{4}\sin\frac{\Delta\omega}{2}t$ $\cong -\frac{m_o\Delta\omega}{4}\sin\frac{\Delta\omega}{2}t$	$\frac{d\phi(t)}{dt} = -\frac{1}{1 + m_o^{\ 2}\cos^2\frac{\Delta\omega}{2}t}\frac{m_o\Delta\omega}{2}\sin\frac{\Delta\omega}{2}t$ $\cong -\frac{m_o\Delta\omega}{2}\sin\frac{\Delta\omega}{2}t$
p.27	Fig.2.8	$m_o =$	$2m_o =$
	Fig.2.9	$m_o =$	$2m_o =$
p.28	Fig.2.10	$m_o =$	$2m_o =$

The original online version for this chapter can be found at
http://dx.doi.org/10.1007/978-3-642-20122-6_2

Erratum: Representation of Sound in Time and Frequency Plane

M. Tohyama: *Sound and Signals*, Signals and Communication Technology, pp. 305–349.
springerlink.com

DOI 10.1007/978-3-642-20122-6_16

In the original version, error has occurred in the following page:

Page	error	correct
p.315 Fig.14.9 (title)	… in base band from … … by (i)-(iii) from[13] (Fig.11)	… in high frequency band from … … by (i)-(iii) [117]
Fig.14.10 (title)	… in higher frequency band from … … filtering [117]	… in base band from … … filtering from[13] (Fig.11)

The original online version for this chapter can be found at
http://dx.doi.org/10.1007/978-3-642-20122-6_14

Erratum: Sound and Signals

M. Tohyama: *Sound and Signals,* Signals and Communication Technology, pp.1–412.
springerlink.com

DOI 10.1007/978-3-642-20122-6_16

In the original version, error has occurred in the following page:

page	error	correct
p.381 ref.[59]	...and the Definition of Reproduced Sound 41(3), 222–224 (1978)	...and the Definition of Reproduced Sound. Acustica 41(3), 222–224(1978)

The original online version for this chapter can be found at
http://dx.doi.org/10.1007/978-3-642-20122-6

Appendix

Resonance Frequency for Displacement and Velocity Resonance

Recall expressions of the magnitude and phase resonanses for the displacement such that

$$A(\omega) = \frac{F_X/M}{\sqrt{({\omega_0}^2 - \omega^2)^2 + k^2\omega^2}} \tag{1}$$

$$\tan\phi(\omega) = \frac{k\omega}{{\omega_0}^2 - \omega^2}, \quad k = R/M = 2\delta_0, \tag{2}$$

for a simple resonator under the sinusoidal external force with magnitude F_X. The magnitude response takes its maximum at the frequency of the external force such that

$$\omega_M = \sqrt{{\omega_0}^2 - \frac{k^2}{2}}, \tag{3}$$

where the denominator becomes the minimum. Note that the resonance frequency is slightly lower than the eigenfrequency ω_0 that is the frequency of the free oscillation. Consequently, the terms resonance and eigenfrequencies should be distinguished[5].

As stated above, the resonance frequency becomes lower as the damping factor increases; however, the phases take $-\pi/2$ at the eigenfrequency independent of the damping factor as shown in Fig. 2.3 in the main text. The phase of the displacement decreases to $-\pi/2$ from the initial phase position 0 through $-\pi/2$.

The resonance frequency, however, differs for the velocity response. The magnitude and phase for the velocity response are given by

$$B(\omega) = \omega A(\omega) = \frac{F_X}{\sqrt{(\frac{K}{\omega} - \omega M)^2 + R^2}} \tag{4}$$

$$\tan\Phi(\omega) = \frac{k\omega}{\omega_0^2 - \omega^2} + \frac{\pi}{2}. \tag{5}$$

The magnitude for the velocity response takes its maximum at the frequency $\omega = \omega_0$ that is equal to the eigenfrequency independent of the damping conditions. Namely, the eigenfrequency is uniquely determined for the vibrating system; however, the resonance frequency depends on the response to be observed. The phase for the velocity response decreases from $\pi/2$ at the initial phase position to $-\pi/2$ through 0 at the resonance frequency, i.e., the eigenfrequency.

Figure 1 shows the magnitude and phase responses for the displacement, velocity, and acceleration. The differences of the magnitude and phase responses are due to the variables to be observed. Only the velocity is in-phase with the external force at the resonance frequency. Note again the resonance frequencies are not unique in general but depend on the quantity to be observed and the damping conditions.

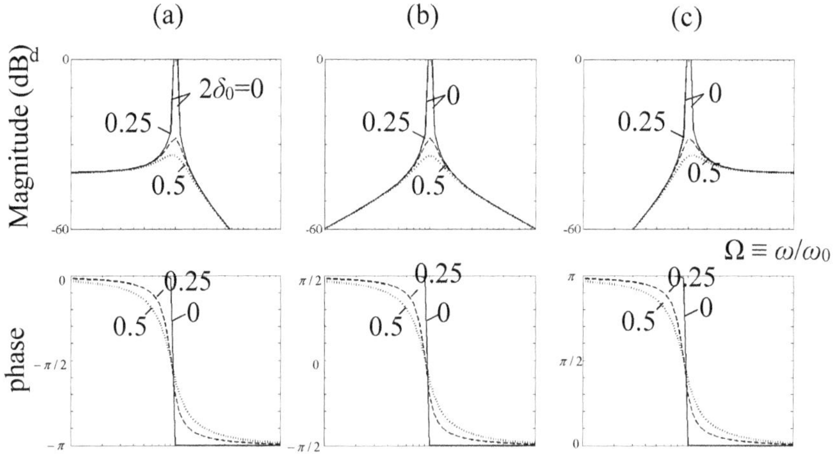

Fig. 1 Magnitude and phase responses for simple oscillator, (a) displacement, (b) velocity, and (c) acceleration

References

[1] Lang, S.: A First Course in Calculus, 5th edn. Springer, Heidelberg (1986)
[2] Lay, D.C.: Linear Algebra and Its Applications. Addison-Wesley Publishing Company, Reading (1994)
[3] Lang, S.: Linear Algebra. Springer, Heidelberg (1987)
[4] Itow, T.: Onkyo Kogaku Genron (Principles of Acoustics). Corona Publishing Co., Ltd, Tokyo (1967) (in Japanese)
[5] Blauert, J., Xiang, N.: Acoustics for Engineers. Springer, Heidelberg (2008)
[6] Hayasaka, T.: Denki Onkyogaku (Electro Acoustics), Iwanami Shoten, Tokyo, Japan (1979) (in Japanese)
[7] Rayleigh, J.W.S.B.: The Theory of Sound. Dover Publication, New York (1945)
[8] Rossing, T.D., Fletcher, N.H.: Principles of Vibration and Sound. Springer, Heidelberg (1995)
[9] Tohyama, M.: Oto no Butsuri (Physics of sound). Corona Publishing Companry (2010) (in Japanese)
[10] Hartmann, W.M.: Signals, Sound, and Sensation. Springer, Heidelberg (1998)
[11] Houtgast, T., Steeneken, H., Plomp, R.: Predicting Speech Intelligibility in Rooms from the Modulation Transfer Function. I. General Room Acoustics, Acustica 46, 60–72 (1980)
[12] Drullman, R.: Temporal Envelope and Fine Structure Cues for Speech Intelligibility. J. Acoust. Soc. Am. 97, 585–592 (1995)
[13] Kazama, M., et al.: On the Significance of Phase in the Short Term Fourier Spectrum for Speech Intelligibility. J. Acoust. Soc. Am. 127(3), 1432–1439 (2010)
[14] Hirata, Y., Kawai, T.: On the Isolation of Solid-borne Noise from a Water Pump. Report of architectural acoustics research meeting, AA 78-20, Acoust. Soc. Japan (1978) (in Japanese)
[15] Born, M.: Atomic Physics. Blackie and Son Limited (1951)
[16] Hara, Y.: Private communication (2009)
[17] Mori, T.: Gendai no Koten Kaiseki (Modern Classical Mathamatics), Chikuma Shobo (2006) (in Japanese)
[18] Sakakibara, S.: Quick Reference to Mathematica. Kyoritsu Publishing Company (2000) (in Japanese)
[19] Hayasaka, T.: Gakki no Kagaku (Science of Musical Instruments). Corona Publishing Company (1992) (in Japanese)
[20] Weinreich, G.: Coulpled Piano Strings. J. Acoust. Soc. Am. 62(6), 1474–1484 (1977)

[21] Norton, M.P.: Fundamentals of Noise and Vibration Analysis for Engineers. Cambridge University Press, Cambridge (1996)
[22] Weinreich, G.: Physics of Piano Strings, Gabriel Weinreich, Science (March 1979)
[23] Lyon, R.H.: Progressive Phase Trend in Multi-degree-of-freedom Systems. J. Acoust. Soc. Am. 73(4), 1223–1228 (1983)
[24] Lyon, R.H.: Range and Frequency Dependence of Transfer Function Phase. J. Acoust. Soc. Am. 76(5), 1433–1437 (1984)
[25] Rossing, T.D., Fletcher, N.H.: The Physics of Musical Instruments. Springer, Heidelberg (1988)
[26] Hannah, L.: A Brief Histroy of the Speed of Sound. Acoustics Bulletin 32(4), 28–31 (2007)
[27] Harris, C.M.: Effects of Humidity on the Velocity of Sound in Air. J. Acoust. Soc. Am. 49, 890 (1971)
[28] Moore, B.C.J.: An Introduction to the Psychology of Hearing, 4th edn. Academic Press, London (1997)
[29] Lyon, R.H.: Machinery Noise and Diagnostics. ButterWorth, London (1987)
[30] Tohyama, M., Suzuki, H., Ando, Y.: The Nature and Technology of Acoustic Space. Academic Press, London (1995)
[31] Tohyama, M., Itow, T.: Theoretical Interpretation of the Mass Law Based on the Wave Theory. Acustica 30(1), 1–11 (1974)
[32] Feynman, R.P., Leighton, R.B., Sands, M.L.: The Feynman Lectures on Physics II (1965)
[33] Taylor, C.: Exploring Music, IOP Publishing Ltd. (1992)
[34] Tohyama, M.: Signal Analysis and Acoustics. Springer, Japan (2007) (in Japanese)
[35] Hekt, E.: Optics. Pearson Education, Inc., London (2002)
[36] Courant, R., Robbins, H.: What is Mathematics? An Elementary Approach to Ideas and Methods, 2nd edn. Oxford University Press, Inc., Oxford (1969)
[37] Suzuki, A., Tohyama, M.: Active Sound-power Control in a Reverberant Space. NTT-R&D 38(8), 931–938 (1989)
[38] Eisner, E.: Complete Solutions of the "Webster" Horn Equation. J. Acoust. Soc. Am. 41(4), 1126–1146 (1967)
[39] Mast, T.D., Yu, F.: Simplified Expansions for Radiation from a Baffled Circular Piston. J. Acoust. Soc. Am. 118(6), 3457–3464 (2006)
[40] Hayasaka, T.: Onkyokogaku Nyumon(Introduction to Engineering Acoustics), Nikkan Kogyo Shinbun-sha, Tokyo, Japan (1978) (in Japanese)
[41] Sugiyama, K.: Private communication (2005)
[42] Tohyama, M.: Radiation Fields far from Point or Ring Source on a Rigid Cylindrical Baffle. Acustica 60(3), 230–235 (1986)
[43] Hayasaka, T.: Gijutsu-sha no tameno Onkyo Kogaku (Engineering Acoustic for Engineers). Maruzen Co.Ltd. (1986) (in Japanese)
[44] Rossing, T.D.: The Physics of Kettledrums, Science (January 1983)
[45] Hirata, Y.: Study on Diffusional Sound Fields Consisting from Three Types of Waves. Inst. Elec. Info. Comm. Eng (IEICE), Technical Report of IEICE, EA 70-12 (1970-9) (in Japanese)
[46] Tohyama, M., Koike, T.: Fundamentals of Acoustic Signal Processing. Academic Press, London (1998)
[47] Hirata, Y.: Geometrical Acoustics for Rectangular Rooms. Acustica 43(4), 247–252 (1979)
[48] Allen, J.B., Berkley, D.A.: Image Method for Efficiently Simulating Small-room Acoustics. J. Acoust. Soc. Am. 65(4), 943–950 (1979)

[49] Lamb, H.: Hydrodynamics. The Cambridge University Press, Cambridge (1924)
[50] Johnson, M.E., Elliott, S.J., Beak, K.-H., Garcia-Bonito, J.: An Equivalent Source Technique for Calculating the Sound Field Inside an Enclosure Containing Scattering Objects. J. Acoust. Soc. Am. 104(3), 1221–1231 (1998)
[51] Tohyama, M., Itow, T.: The Relation between Geometric Acoustics and Ergodicity. Acustica 31(4), 237–239 (1974)
[52] Schroeder, M.R.: New Method of Measuring Reverberation Time. J. Acoust. Soc. Am 37(3), 409–412 (1965)
[53] Batchelder, L.: Reciprocal of the Mean Free Path. J. Acoust. Soc. Am. 36(3), 551–555 (1964)
[54] Kosten, C.W.: The Mean Free Path in Room Acoustics. Acustica 10, 245–250 (1960)
[55] Baker, J.: 50 Physiscs Ideas You Really Need to Know, Books Sales (2009)
[56] Tohyama, M.: Equivalent Sound Absorption Area in a Rectangular Reverberation Room (Sabine's Sound Absorption Factor). J. Sound and Vib. 108(2), 339–343 (1986)
[57] Thiele, R.: Richtungsverteilung und Zeitfolge der Schallrueckwuerfe in Raeumen. Acustica 3, 291–302 (1953)
[58] Schultz, T.J.: Acoustics of the Concert Hall, pp. 56–67. IEEE Spectrum (June 1965)
[59] Hirata, Y.: Reverberation Time of Listening Room and the Definition of Reproduced Sound 41(3), 222–224 (1978)
[60] Takahashi, Y., Tohyama, M., Yamasaki, Y.: Phase Response of Transfer Functions and Coherent Field in a Reverberation Room. Electronics ad Communications in Japan, Part 3 90(4), 1–8 (2007)
[61] Morse, P.M.: Vibration and Sound. McGraw-Hill Book Company, Inc., New York (1948)
[62] Tohyama, M., Suzuki, A.: Reverberation Time in an Almost-two-dimensional Diffuse Field. J. Sound and Vib. 111(3), 391–398 (1986)
[63] Tohyama, M., Yoshikawa, S.: Approximate Formula of the Averaged Sound Energy Decay Curve in a Rectangular Reverberant Room. J. Acoust. Soc. Am 70(6), 1674–1678 (1981)
[64] Nomura, H., Miyata, H.: Speech Intelligibility and Modulation Transfer Function in Non-exponential Decay Fields. Acustica 69(4), 151–155 (1989)
[65] Yegnanarayana, B., Ramakrishna, B.: Intelligibility of Speech under Nonexponential Decay Conditions. J. Acoust. Soc. Am. 58(4), 853–857 (1975)
[66] Tohyama, M., Suzuki, A.: Active Power Minimization of a Sound Source in a Closed Space. J. Sound and Vib. 119(3), 562–564 (1987)
[67] Hirata, Y.: Dependence of the Curvature of Sound Decay Curves and Absorption Distribution on Room Shapes. J. Sound and Vib. 84, 509–517 (1982)
[68] Tohyama, M., Fujisaka, Y.: Room Acoustic Chaos. J. Acoust. Soc. J. 53(2), 154–159 (1997) (in Japanese)
[69] Takahashi, Y.: Private communication (2007-2011)
[70] Tohyama, M.: Room Transfer Function. Handbook of Signal Processing in Acoustics; Part 3 Engineering Acoustics 75, 1381–1402 (2008)
[71] Berry, M.: Quantum chaology, Dynamical Chaos, Edited by I. Percival and N. Weis, pp. 183–198. Princeton University Press, Princeton (1987)
[72] Nelson, R.: Probability, Stochastic Processes, and Queueing Theory. Springer, Heidelberg (1995)
[73] Schroeder, M.R.: Statistical Parameters of the Frequency Response Curves in Large Rooms. J. Audio Eng. Soc. 35(5), 299–306 (1987)
[74] Lyon, R.H.: Statistical Analysis of Power Injection and Response in Structures and Rooms. J. Acoust. Soc. Am. 45(3), 545–565 (1969)

[75] Weaver, R.L.: Spectral Statistics in Elastodynamics. J. Acous. Soc. Am. 85(3), 1005–1013 (1989)
[76] Fujisaka, Y., Tohyama, M.: Eigenfrequency Spacing Analysis and Eigenmode Breakdown for Semi-stadium-type 2-D Fileds. J. Sound and Vib. 267, 867–878 (2003)
[77] Fujisaka, Y., Tohyama, M.: Eigenfrequency Distributions and Chaotic Properties of Acoustic Rays in Acoustic Fields Having Higher Order Curvature. J. Inst. Elect. Inf. Comm. Eng. Jpn. A J86-A(12), 1435–1441 (2003)
[78] Morse, P.M., Bolt, R.H.: Sound Waves in Rooms. Rev. Mod. Phys. 16, 69–150 (1944)
[79] Papoulis, A.: Probability, Random Variables, and Stochastic Processes. McGraw-Hill Book Company, New York (1965)
[80] Ebeling, K.J.: Properties of Random Wave Fields, Physical Acoustics, XVII, pp. 233–310. Academic Press, London (1984)
[81] Wigner, E.: Statistical Properties of Real Symmetric Matrices with Many Dimensions. In: Porter, C. (ed.) Statistical Theory of Spectra: Fluctuations, pp. 188–198. Academic Press, London (1965)
[82] Cramer, H.: Mathematical Methods of Statistics. Princeton University Press, Princeton (1945)
[83] Ando, Y.: Concert Hall Acoustics. Springer, Heidelberg (1985)
[84] Cook, R.K., Waterhouse, R.V., Berendt, R.D., Edelman, S., Thompson Jr., M.C.: Measurement of Correlation Coefficients in Reverberant Sound Fields. J. Acoust. Soc. Am. 27(6), 1072–1077 (1955)
[85] Tohyama, M., Suzuki, A.: Interaural Cross-correlation Coefficients in Stereo-reproduced Sound Fields. J. Acoust. Soc. Am. 85(2), 780–786 (1989)
[86] Baxter, S.M., Morfey, C.L.: Angular Distribution Analysis in Acoustics. Springer, Heidelberg (1986)
[87] Tohyama, M., Suzuki, A., Yoshikawa, S.: Correlation Coefficients in a Rectangular Reverberant Room. Acustica 39(1), 51–53 (1977)
[88] Tohyama, M., Suzuki, A., Yoshikawa, S.: Correlation Coefficients in a Rectangular Reverberant Room-Experimental Results. Acustica 42(3), 184–186 (1979)
[89] Sommerfeld, A.: Partial Differential Equations in Physics (translated by Ernst G. Straus). Academic Press, London (1949)
[90] Suzuki, A., Tohyama, M.: Variable-Reverberation Room Using Policylinders (in Japanese with English abstract). NTT R&D 36(8), 1081–1087 (1987)
[91] Yanagawa, H., Yamasaki, Y., Itow, T.: Effect of Transient Signal Length on Cross-correlation Functions in a Room. J. Acoust. Soc. Am. 84(5), 1728–1733 (1988)
[92] Yanagawa, H.: Transient Interaural Cross Correlation Function in Rooms. Music and Concert Hall Acoustics, 251–259 (1997)
[93] Takahashi, Y., Ando, A.: Audio Content Down-Mixing Using Sound Field Reproduction Method Based on Spatial Covariance. Applied Acoustics 71, 1177–1184 (2010)
[94] Tohyama, M., Imai, A., Tachibana, H.: The Relative Variance in Sound Power Measurements using Reverberation Rooms. J. Sound and Vib. 128(1), 57–69 (1989)
[95] Maling Jr., G.C.: Guidelines for Determination of the Average Sound Power Radiated by Discrete-Frequency Sources in a Reverberation Room. J. Acoust. Soc. Am. 53(4), 1064–1069 (1973)
[96] Waterhouse, R.V.: Interference Patterns in Reverberant Sound Fields. J. Acoust. Soc. Am. 27(2), 247–258 (1955)
[97] Suzuki, A., Tohyama, M.: Sound Power Measurements of Narrow Band Sources by Rectangular Room Corner Method. Autumn Meeting of Acoust. Soc. Jpn., 1-4-19 (1983) (in Japanese)

[98] Chu, W.T.: Eigenmode Analysis of the Interference Patterns in Reverberant Sound Fields. J. Acoust. Soc. Am. 68(1), 184–190 (1980)
[99] Tohyama, M., Suzuki, A.: Space Variances in the Mean-square Pressure at the Boundaries of a Rectangular Reverberation Room. J. Acoust. Soc. Am. 80(3), 828–832 (1986)
[100] Bartel, T.W., Yaniv, S.L., Flynn, D.R.: Use of Corner Microphones for Sound Power Measurements in a Reverberant Chamber. J. Acoust. Soc. Am. 74(6), 1794–1800 (1983)
[101] Waterhouse, R.V.: Estimation of Monopole Power Radiated in a Reverberation Chamber. J. Acoust. Soc. Am. 64(5), 1443–1446 (1978)
[102] Waterhouse, R.V., Cook, R.K.: Diffuse Sound Fields: Eigenmode and Free-wave Models. J. Acoust. Soc. Am. 59(3), 576–581 (1976)
[103] Nelson, P.A., Elliott, S.J.: Active Noise Control of Sound. Academic Press, London (1992)
[104] Tohyama, M., Suzuki, A., Sugiyama, K.: Active Power Minimization of a Sound Source in a Reverberant Closed Space. IEEE Trans. on Signal Processing 39(1), 246–248 (1991)
[105] Elliott, S.J.: The Influence of Modal Overlap in the Active Control of Sound and Vibration, The University of Southampton, ISVR Memorandum, 695 (December 1989)
[106] Tohyama, M.: Response Statistics of Rooms. In: Crocker, M.J. (ed.) Encyclopedia of Acoustics 2(77), 913–923. John Wiley & Sons, Inc., Chichester (1997)
[107] Tohyama, M., Lyon, R.H.: Zeros of a Transfer Function in a Multi-degree-of-freedom Vibrating System. J. Acoust. Soc. Am. 86(5), 1854–1863 (1989)
[108] Manabe, T., Kanda, Y., Koike, T., Tohyama, M., Lyon, R.H.: Reverberation Phase and Zeros in a 1-D and 2-D Sound Fields (in Japanese with English abstract), Inst. Elec. Info. Comm. Eng (IEICE), Technical Report of IEICE, EA 95-68 (1995-12)
[109] Tohyama., M., Lyon, H., Koike, T.: Reverberant Phase in a Room and Zeros in the Complex Frequency Plane. J. Acoust. Soc. Am. 89(4), 1701–1707 (1991)
[110] Manabe, T., Koike, T., Tohyama, M.: Residue Sign Model of a Room Transfer Function and its Zeros (in Japanese with English abstract), Inst. Elec. Info. Comm. Eng (IEICE), Technical Report of IEICE, EA 96-102 (1997-03)
[111] Schroeder, M.R.: Frequency Correlation Functions of Frequency Responses in Rooms. J. Acoust. Soc. Am. 34, 1819–1823 (1962)
[112] Tohyama, M., Lyon, R.: Phase Variabilities and Zeros in a Reverberant Transfer Function. J. Acoust. Soc. Am. 95(1), 286–323 (1994)
[113] Takahashi, Y., Tohyama, M.: Artificial Reverberation Control Using Cepstrum Decomposition while Maintaining Locations of Peaks and Dips on the Frequency Responses. J. Audio Eng. Soc. 53(12), 1142–1151
[114] Liu, L., He, J., Palm, G.: Effects of Phase on the Perception of Intervocalic Stop Consonants. Speech Communication 22, 403–417 (1997)
[115] Schroeder, M.R.: Computer Speech. Springer, Heidelberg (1999)
[116] Oppenheim, A.V., Lim, J.S.: The Importance of Phase in Signals. Proc. IEEE 69(5), 529–541 (1981)
[117] Gotoh, S.: Private communication (2010)
[118] Terada, T., Nakajima, H., Tohyama, M., Hirata, Y.: Nonstationary Waveform Analysis and Synthesis Using Generalized Harmonic Analysis. In: IEEE-SP Int. Symp. Time-Frequency and Time-Scale Analysis, pp. 429–432 (1994)
[119] Ohnishi, T., Kazama, M., Tohyama, M.: Acoustic Signal Processing Using Multi-Windowed STFT and Harmonics Sieving. In: Fifth Int. Cong. Sound and Vibration, pp. 2167–2174 (1997)

[120] Kazama, M., Yoshida, K., Tohyama, M.: Signal Representation Including Waveform Envelope by Clustered Line-Spectrum Modeling. J. Audio Eng. Soc. 51(3), 123–137 (2003)
[121] Kazama, M., Tohyama, M.: Estimation of Speech Components by ACF Analysis in a Noisy Environment. J. Sound and Vib. 241(1), 41–52 (2001)
[122] Meddis, R., Hewitt, M.J.: Virtual Pitch and Phase Sensitivity of a Computer Model of the Auditory Periphery. II: Phase Sensitivity. J. Acoust. Soc. Am. 89(6), 2883–2894 (1991)
[123] Hara, Y., Matsumoto, M., Miyoshi, K.: Method for Estimating Pitch Independently from Power Spectrum Envelope for Speech and Musical Signal. J. Temporal Design in Archtecture and the Environment 9(1), 121–124 (2009)
[124] Flanagan, J.L., Golden, R.M.: Phase Vocoder, Bell Syst. Bell Syst. Tech. J. 45(11), 149–162 (1966)
[125] Yoshida, K., Kazama, M., Tohyama, M.: Pitch and Speech-Rate Conversion using Envelope Modulation Modeling. In: Int. Conf. Acoustics, Speech, and Signal Processing, I, SP-P04.04, pp. 425–428 (2003)
[126] Terada, K., Tohyama, M., Houtgast, T.: The Effect of Envelope or Carrier Delays on the Precedence Effect. Acustica 91(6), 1016–1019 (2005)
[127] Yasojima, O., Takahashi, Y., Tohyama, M.: Resonant Bandwidth Estimation of Vowels Using Clustered-Line Spectrum Modeling For Pressure Speech Waveforms. In: Int. Symp. Signal Processing and Information Technology, pp. 589–593 (2006)
[128] Hasegawa, T.: Private communication (2010-2011)
[129] Kumaresari, R., Rao, A.: On Minimum/Maximum/All-pass Decomposition in Time and Frequency Domains. IEEE Trans. SP 48, 2973–2976 (2000)
[130] Nakajima, H., Tanaka, M., Tohyama, M.: Signal Representation and Inverse Filtering Using Recursive Vector Projection (in Japanese with English abstract). J. IEICE Jpn. J83-A(4), 353–360 (2000)
[131] Hirata, Y., Tohyama, M., Matsumoto, M., Gotoh, S.: Health Monitoring for a Structure Using Its Nonstationary Vibration. Advances in Acoustics and Vibration, Article ID 696512 (2010)
[132] Tohyama, M., Lyon, R., Koike, T.: Pulse Waveform Recovery in a Reverberant Condition. J. Acoust. Soc. Am. 91(5), 2805–2812 (1992)
[133] Tohyama, M., Lyon, R.H., Koike, T.: Statistics on Reverbernant Transfer Functions. In: Second Int. Cong. on Recent Developments in Air- and Structure-borne Sound and Vib., pp. 869–876 (1992)
[134] Shaw, E.A.G., Vaillancourt, M.M.: Transformation of Sound Pressure Level from the Free Field to the Eardrum Presented in Numerical Form. J. Acoust. Soc. Am. 78(3), 1120–1123 (1985)
[135] Tohyama, M.: Room Transfer Functions and Sound Field Control. Active 95, 15–20 (1995)
[136] Ushiyama, S., Hirai, T., Tohyama, M., Shimizu, Y.: Howling Suppression by Smoothing the Open-Loop Transfer Function (in Japanese with English abstract). Technical Report of IEICE Jpn. EA94-4 (1994-04)
[137] Takahashi, Y., Tohyama, M., Yamasaki, Y.: Cumulative Spectral Analysis for Transient Decaying Signals in a Transmission System Including a Feedback Loop. J. Audio. Eng. Soc. 54(7/8), 620–629 (2006)
[138] Berman, J.M., Fincham, L.R.: The Application of Digital Techniques to the Measurement of Loudspeakers. J. Audio. Eng. Soc. 25(6), 370–384 (1977)
[139] Nomura, H., Tohyama, M., Houtgast, T.: Loudspeaker Arrays for Improving Speech Intelligibility in a Reverberant Space. J. Audio Eng. Soc. 39(5), 338–343 (1991)

Index

GPSR Compliance

The European Union's (EU) General Product Safety Regulation (GPSR) is a set of rules that requires consumer products to be safe and our obligations to ensure this.

If you have any concerns about our products, you can contact us on ProductSafety@springernature.com

In case Publisher is established outside the EU, the EU authorized representative is:

Springer Nature Customer Service Center GmbH
Europaplatz 3
69115 Heidelberg, Germany

Batch number: 10682207

Printed by Printforce, the Netherlands